Sources and Studies
in the History of Mathematics and
Physical Sciences



Springer-Verlag London Ltd.

## Sources and Studies in the History of Mathematics and Physical Sciences

Bos H.J.M.
**Redefining Geometrical Exactness: Descartes' Transformation of the Early Modern Concept of Construction**

Dale A.I.
**A History of Inverse Probability: From Thomas Bayes to Karl Pearson**

Dale A.I.
**Pierre-Simon Laplace: Philosophical Essay on Probabilities**

Federico P.J.
**Descartes on Polyhedra: A Study of the *De Solidorum Elementis***

Graßhof G.
**The History of Ptolemy's Star Catalogue0**

Grootendorst A.W.
**Jan De Witt's *Elementa Curvarum***

Hawkins T.
**Emergence of the Theory of Lie Groups: An Essay in the History of Mathematics 1869-1926**

Hermann A., Meyenn K. von, Weisskopf V.F. (Eds)
**Wolfgang Pauli: Scientific Correspondence I: 1919-1929**

Hoyrup J.
**Lengths, Widths, Surfaces: A Portait of Old Babylonion Algebra and Its Kin**

Jones A. (Ed.)
**Pappus Of Alexandria - Book VII of the Collection**

Kheirandish E.
**The Arabic Version of Euclid's Optics**

Lützen J.
**Joseph Liouville 1809-1882**

Ian Tweddle

# James Stirling's *Methodus Differentialis*

## An Annotated Translation of Stirling's Text.

With 12 Figures

Springer

Ian Tweddle
Department of Mathematics
University of Strathclyde
Livingstone Tower
26 Richmond Street
Glasgow G1 1XH
UK

*Sources Editor:*
Gerald J. Toomer
2800 South Ocean Boulevard, 21F
Boca Raton, FL 33432
USA

ISBN 978-1-4471-1127-6

British Library Cataloguing in Publication Data
Stirling, James, 1692 - 1770
James Stirling's Methodus differentialis : an annotated translation of Stirling's text. – (Sources and studies in the history of mathematics and the physical sciences)
1. Stirling, James, 1692 – 1770. Methodus differentialis
2. Calculus – Early works to 1800
I. Title II. Tweddle, Ian
515
ISBN 978-1-4471-1127-6 ISBN 978-1-4471-0021-8 (eBook)
DOI 10.1007/978-1-4471-0021-8

Library of Congress Cataloging-in-Publication Data
A catalog record for this book is available from the Library of Congress


Originally published by Springer-Verlag London Limited in 2003
Softcover reprint of the hardcover 1st edition 2003



Typesetting: Camera-ready by author

12/3830-543210 Printed on acid-free paper SPIN 10867721

# Preface

In 1988 I published a little book [70] on the later, largely unpublished work of the Scottish mathematician James Stirling (1692–1770). Much of the background work for the present volume was also done about that time and my original intention was to produce an edition of Stirling's *Methodus Differentialis* (1730) soon after the publication of my earlier book. However, the demands of other mathematical projects led to my neglecting Stirling and the *Methodus Differentialis* throughout much of the 1990s (but see [71]). I returned to the task with some trepidation in 2000 but I soon found great pleasure in the work and I believe that the long fallow period has resulted in a much better account of Stirling's important work than anything I could have produced 10 years ago. I hope that mathematical historians, analysts, numerical analysts and others will find something of interest and even present-day relevance in what I have produced.

I am indebted to many people and institutions for assistance, advice and encouragement. I would like to record my thanks to the following in particular:

Col. James Stirling of Garden who allowed me to consult Stirling's own annotated copy of the *Methodus Differentialis* and some of his notebooks and to quote material from these sources;

Professor G.J. Toomer for his helpful comments and especially for his guidance on some of the finer points of translation;

an anonymous reviewer for some valuable suggestions on presentation;

the Department of Special Collections, Glasgow University Library, for providing me with much of my material;

the Special Collections staff of the Andersonian Library, University of Strathclyde, for help with material and references;

the Royal Society for access to its archives and permission to quote from its Journal Book;

my wife, Grace, and my son, Edward, for their general support;

the University of Strathclyde for providing me with suitable technical facilities and other support for my work.

Finally, it is a great pleasure to acknowledge my indebtedness to all the editorial and production staff of Springer-Verlag London with whom I have worked.

IT
Glasgow
January 2003

# Contents

# Introduction

## Background

The *Methodus Differentialis: sive Tractatus de Summatione et Interpolatione Serierum Infinitarum* (London, 1730) of James Stirling (1692–1770) is surely one of the early classics of numerical analysis. It contains not only the results and ideas for which Stirling is chiefly remembered today (the Stirling numbers, Stirling's interpolation formula, Stirling's formula for $\ln n!$), but also a wealth of material on transformations of series and limiting processes. Inverse factorial series, especially hypergeometric series, are much in evidence and asymptotic series (including that for $\ln n!$) also appear.[1] Interpolation and quadrature are discussed and there is an impressive collection of calculations throughout to illustrate the efficacy of the methods presented.

Stirling's book was well received by his contemporaries. For example, Euler expressed his admiration for the work (see the note on Proposition 14), De Moivre employed some of Stirling's results to improve earlier work of his own and to develop his own series for $\ln(n-1)!$ (see the notes on Proposition 23 and Proposition 28, and the Appendix), and MacLaurin discussed or referred to several of Stirling's results in his *Treatise of Fluxions* (1742) (see [39, Articles 357, 361, 838, 842, 844]). Two enthusiastic reviews ([1], [2]) appeared in 1732 and 1734.[2]

However, the book's influence has extended over more than two and a half centuries. A vast number of articles have been devoted to "Stirling's formula" (proofs, extensions, pedagogical aspects) and the modern development of combinatorial theory has ensured a central place for the Stirling numbers. In the 1860s K. Schellbach [53] developed the process for obtaining limits which Stirling presents in his Proposition 30 and this in turn was discussed as the "Stirling–Schellbach algorithm" by I.J. Schoenberg in 1981 [56]. Nielsen's *Handbuch der Theorie der Gammafunktion* (1906) [51] has many references to Stirling's work and includes a chapter entitled "Methoden von Stirling".

[1] G.A. Gibson remarks in his important article on the history of mathematics in Scotland [18]: "next to Newton I would place Stirling as the man whose work is specially valuable where series are in question."

[2] Although these reviews were published anonymously, we know from correspondence that the author of the first was Louis Castel (see [74, p. 151]).

More recently (1976) R.W. Gosper asserted in his article "A calculus of series rearrangements" [20]: "We will be taking up almost exactly where James Stirling left off ...".

Stirling's work is founded on Newton's "Methodus Differentialis", which was published in 1711 under the editorship of William Jones as the penultimate item (pp. 93–101) in [47].[3] In this short treatise Newton presents six propositions and a scholium in which he discusses differences, interpolation formulae and quadrature. Roger Cotes (1682–1716), starting from results in Newton's *Principia*, had already produced related material in 1708 and was stimulated by the publication of Newton's treatise to pursue the matter further. Unfortunately, Cotes's papers were only published posthumously in 1722 [13], as a result of which his priority in certain results and the importance of his contributions have not been generally recognised.[4] Another early proponent of Newton's ideas was Brook Taylor (1685–1731), who developed "Taylor's theorem" as a limiting case of an interpolating series [66].

Stirling's first contribution to this area was his paper "Methodus Differentialis Newtoniana Illustrata" [61], which was communicated to the Royal Society[5] at its meeting on 18 June 1719. The first part is, as the title suggests, essentially an explanation of Newton's treatise. In the second part[6] we have an early attempt at a problem which he pursued with conspicuous success in his later book, namely the transformation of a slowly converging series into one which converges rapidly to the same sum. He also gives there a rather obscure account of a process for finding limits, to which he returns with greater clarity in the book (Proposition 30 – the Stirling–Schellbach algorithm); this may be explained in terms of interpolation (see the note on Proposition 30). The paper concludes with the indication of a possible sequel, but nothing more seems to have appeared in print before the publication of the book in 1730.

By 1725 Stirling had settled in London, where he became a teacher and later a partner in Watt's Academy in Little Tower Street, Covent Garden. He was soon established in the scientific community and was elected a Fellow of the Royal Society in 1726.[7] Apparently he had already been working on the book by this time, but his duties at the Academy did not allow rapid

[3] Concerning the origins of Newton's work see [77, Vol. VIII, pp. 236–257].

[4] Cotes's work is discussed in [21]. See especially its pp. 112–133.

[5] The first record I have found of Stirling's involvement with the Royal Society is of his attendance at the meeting on 4 April 1717, when Taylor presented a paper on the numerical solution of equations and the construction of logarithms (Journal Book of the Royal Society). Stirling's first publication [60] appeared in the same year. He was at that time a student at Balliol College, Oxford, but was shortly to leave for the Venetian Republic, where he remained until 1721 or 1722.

[6] I have discussed the second part of Stirling's paper in [71].

[7] The minutes of the meeting of the Royal Society on 27 October 1726 record that "Mr Sterling was proposed for a Fellow by Dr Arbuthnot and recommended by Sir Alexander Cuming." He was elected on 3 November and admitted on 8 December (Journal Book of the Royal Society).

progress and the final production was rather rushed. For example, on 22 July 1729 he wrote to his brother John Stirling ([74, p. 14]):

> I designed to have spent some time this summer among you, but on second thoughts I choose to publish some papers during my Leisure time which have long lain by me.

And, in a letter to Gabriel Cramer written in September 1730 and accompanying copies of the book for Cramer and Nikolaus Bernoulli,[8] Stirling said of its contents ([74, pp. 118–119]):

> This first part has been written 8 or 9 years ago, so that if I were to write it again I should scarce change anything in it; But indeed that is more than I can say for the second part, because there was not above one half of it finished, when the beginning of it was sent to the Printer. And altho' I am not conscious of any Errors in it but typographical ones, yet I am sensible that it might have been better done.

Stirling's *Methodus Differentialis* [62] was printed by William Bowyer for G. Strahan, London. It was reissued in 1753 by Richard Manby, Ludgate Hill, London and in 1764 by J. Whiston and B. White, Fleet Street, London. Apart from the title pages and variations in the setting of pages 1–4,[9] the three issues are identical.

A translation by Francis Holliday,[10] Master of the Free-School at Haughton Park, near Retford, Nottinghamshire was published in 1749 [64]. It was

[8] One of Stirling's notebooks contains the names of intended recipients of the book, which provide an idea of Stirling's mathematical connections. One list is headed "Books given to" and contains the names Smith, Halley, Jones, Foulks, Pemberton, Moivre, Ouchterlony, Robins, Grahame, Klingenstierna, Campbell, Dr Stuart, Dr Arbuthnot, Mr Montague, Dr Taylor. The names in the other list, headed "Not yet given", are Dr Johnston, Symson, Mclaurin, Sanderson, N. Bernoulli, Cramer, Bradley as well as several also on the first list. Most of these can be identified with certainty (see [59] and the Royal Society website). Edmund Halley, William Jones, Abraham De Moivre and Brook Taylor are all mentioned in the *Methodus Differentialis*. Other British mathematicians on the lists are (probably) Robert Smith (Cotes's cousin, responsible for [13]), Benjamin Robins, George Campbell, Robert Simson, Colin MacLaurin and Nicholas Sanderson or Saunderson (the blind Lucasian Professor); Henry Pemberton and John Arbuthnot were physicians who also had mathematical training (Pemberton superintended the third edition of Newton's *Principia*) and James Bradley was a celebrated astronomer. George Graham was a famous instrument maker and Alexander Ouchterlony a London merchant (Stirling proposed him for Fellowship of the Royal Society in November 1733). Martin Folkes was at that time Vice-President of the Royal Society. Dr Stuart is probably Charles Stuart. The inclusion of the Swedish mathematician Samuel Klingenstierna is interesting: he became a Fellow of the Royal Society in April 1730 and in April 1731 a paper of his was communicated to the Royal Society by Stirling. With the exception of Robert Simson (Professor at Glasgow) all those I have identified, as well as Bernoulli and Cramer, were or became Fellows of the Royal Society.

[9] Presumably copies of the original printing of these pages were not available for the later issues.

[10] Holliday also published a translation [63] of Stirling's 1719 paper in [30].

printed for E. Cape, St John's Gate, London. The translation was made with Stirling's approval and a certain amount of cooperation: there are in existence two letters from Holliday to Stirling concerning typographical and other errors in the original – in the first of these, dated 15 April 1747, Holliday includes a list of errors he had found, in the second (undated) he acknowledges receipt of Stirling's list of corrections. Stirling's own copy of his book has been annotated with these corrections and I have incorporated them in the present work. The few additional points which I believe require correction are identified in my notes.

My translation is fairly literal except that I have occasionally recast a phrase or sentence to avoid unsatisfactory English. I thought at one time of reformulating the mathematics in modern notation but rejected this idea as being too destructive, although I have used a modern system of bracketing and have written squares in the form $x^2$ rather than $xx$, which is often used by Stirling. I have employed modern notation in the notes for explanations where this seemed desirable. Stirling's proofs are often contracted and sometimes, in my opinion, a little obscure, so I have tried to remedy these deficiencies in the notes, which also contain material on historical background. Stirling's methods succeed spectacularly with his chosen examples, for which the true values are often known. However, he rarely says anything about error bounds, one of the key concepts of numerical analysis. I have therefore endeavoured to justify the perceived accuracy of his methods by calculating effective error bounds for almost all of his calculations.

In Parts I and II Stirling presents his results in thirty-three propositions, which are illustrated by examples and supported by notes in the form of scholia.[11] Many of the propositions contain formal statements of results which are then established, while others are just statements of intent and depend on the accompanying examples for their substance. Usually I have discussed a proposition, its examples and scholion in a single note. Occasionally, however, a scholion contains substantial new material and merits its own note. The page number of the start of a note is given in the left-hand margin opposite the heading of the item to which the note refers. Each note is intended to be read as a whole. However, I have italicised the terms proposition, corollary, example, scholion, and case where this might help the reader to locate particular points of interest.

It is a humbling experience to study the work of Stirling and his contemporaries and to discover how much they were able to achieve without the understanding, developments and facilities which we have today. However, we must resist the temptation to assume that, because they knew a certain result, they must also have been aware of what we regard as its immediate consequences, or that they obtained a stated result in a way which we now regard as standard. In attempting to explain Stirling's results in a manner

[11] Stirling uses the Greek singular form *scholion*, which I have retained, although *scholium* seems to be more common in English.

acceptable to the modern reader, I am not seeking to attribute to Stirling any results or techniques other than those which are to be found explicitly in his book or other writings.

The definitive work on James Stirling is Charles Tweedie's *James Stirling: a sketch of his life and works along with his scientific correspondence* [74]. In it Tweedie gives a useful commentary on the principal results of Stirling's *Methodus Differentialis.* My own book [70] is chiefly concerned with Stirling's later unpublished work, but it does contain an updated biographical sketch and has many points of contact with the present work. In addition to works already cited we should note the following books and papers which discuss material from Stirling's text: [10], [14], [19], [28], [34], [37], [52], [54], [55], [67], [69], [72], [73]. Specific references will be given in the notes.

My aim in producing this annotated translation has been to make Stirling's work more generally accessible. It certainly has historical importance, but I believe that the reader will find in it much that is still relevant today.

## Some Mathematical Points

### 1. Hypergeometric Series

Many of Stirling's results and examples can be explained in terms of the hypergeometric series[12]

$$F(a,b;c;z) = \sum_{n=0}^{\infty} \frac{(a)_n (b)_n}{n!\,(c)_n} z^n,$$

where

$$(d)_n = \begin{cases} 1 & \text{for } n = 0, \\ d(d+1)\dots(d+n-1) & \text{for } n = 1, 2, \dots. \end{cases}$$

If none of $a$, $b$, $c$ is a negative integer or zero, the series has radius of convergence equal to 1. We will require Gauss's theorem which states that

$$F(a,b;c;1) = \frac{\Gamma(c)\,\Gamma(c-a-b)}{\Gamma(c-a)\,\Gamma(c-b)}, \tag{1}$$

provided $c \neq 0, -1, -2, \dots$ and $\mathrm{Re}(c-a-b) > 0$ (see, for example, [40, pp. 261–262]).

We will also encounter cases where $F(a,b;c;1)$ does not converge. The partial sums will then be of interest and for these we have Whipple's result [76] (see also [3, p. 94]) that when $a+b-c > -1$

[12] Stirling's contributions to hypergeometric series and some of his other work are discussed in [14].

$$\sum_{n=0}^{m-1} \frac{(a)_n(b)_n}{n!\,(c)_n} = \frac{\Gamma(1+a-c)\,\Gamma(1+b-c)}{\Gamma(1-c)\,\Gamma(1-c+a+b)} \times \left\{1 - \frac{(a)_m(b)_m}{m!\,(c-1)_m}\, {}_3F_2\left[\begin{matrix}1-a,\ 1-b,\ m\\ 2-c,\ m+1\end{matrix}\right]\right\}, \tag{2}$$

where

$${}_3F_2\left[\begin{matrix}\alpha,\ \beta,\ \gamma\\ \delta,\ \epsilon\end{matrix}\right] = \sum_{n=0}^{\infty} \frac{(\alpha)_n(\beta)_n(\gamma)_n}{n!\,(\delta)_n(\epsilon)_n}.$$

## 2. Inverse Factorial Series

Stirling may first have learned of these series and their applications from the work of the French mathematician François Nicole.[13] They are discussed in Stirling's Introduction and find application in various transformations. It will be useful to summarise some of their general properties here – for proofs and further details see [40, Chapter X].

An inverse factorial series has the form[14]

$$\sum_{n=0}^{\infty} \frac{\alpha_n}{z(z+1)(z+2)\ldots(z+n)} = \frac{\alpha_0}{z} + \frac{\alpha_1}{z(z+1)} + \frac{\alpha_2}{z(z+1)(z+2)} + \ldots .$$

Either the series diverges for all $z \in \mathbb{C}$ or there exists $\lambda$ such that $-\infty \le \lambda < \infty$ and the series

(a) diverges if $\operatorname{Re} z < \lambda$,
(b) converges if $\operatorname{Re} z > \lambda$ and $z$ is not a negative integer or zero.

The parameter $\lambda$ is called the *abscissa of convergence* of the series.

Where the series does have points of convergence there is a second parameter $\mu$, the *abscissa of absolute convergence*, such that $\lambda \le \mu \le \lambda + 1$ and the series converges

(c) absolutely if $\operatorname{Re} z > \mu$,
(d) only conditionally if $\lambda < \operatorname{Re} z < \mu$,

where in both cases $z$ is not a negative integer or zero.

Moreover, if the series converges for $z = z_0$ and $\epsilon$, $r$ are any positive quantities, the series converges uniformly on the set

$$\{z \in \mathbb{C} : \operatorname{Re} z \ge \operatorname{Re} z_0 + \epsilon\} \setminus \bigcup_{n=0}^{\infty} \{z \in \mathbb{C} : |z+n| < r\}.$$

It then follows that, if the series has abscissa of convergence $\lambda$, its sum function is analytic on the domain

[13] See the note on Stirling's Introduction.

[14] It is customary to write $\alpha_n$ as $a_n n!$ but this has no advantage for the present discussion.

$$\{z \in \mathbb{C} : \operatorname{Re} z > \lambda\} \backslash \{0, -1, -2, \ldots\}$$

and has simple poles at any of the points 0, $-1$, $-2$, ... which are greater than $\lambda$.

The more general series

$$\sum_{n=0}^{\infty} \frac{\beta_n}{az(az+k)(az+2k)\ldots(az+nk)},$$

where $a$, $k$ are nonzero constants, may be reduced to the inverse factorial case by a simple change of variable.

## 3. Stirling's Summation Procedure

Stirling's aim is to find accurate values for the sums of numerical series and to this end he develops various transformations which accelerate convergence. He has

$$\begin{aligned}\sum_{n=n_0}^{\infty} a_n &= \sum_{n=n_0}^{n_0+N-1} a_n + \sum_{n=n_0+N}^{\infty} a_n \\ &= \sum_{n=n_0}^{n_0+N-1} a_n + \sum_{n=n_0+N}^{\infty} b_n,\end{aligned}$$

where the $b_n$ are obtained from the $a_n$ by an appropriate transformation, and he takes a suitable number of terms of $\sum b_n$ to give the required accuracy. There are therefore two questions to be answered:

(i) How big should $N$ be?
(ii) How many terms of $\sum b_n$ should be taken?

It is unfortunate, but perhaps not surprising, that Stirling has little to say about these questions. For instance, in Example 5 of Proposition 2 he remarks, "Therefore I substitute for $z$ its fourteenth value $13\frac{1}{2}$, so that $z$ is sufficiently large to make the series converge rapidly"; here $z = n + \frac{1}{2}$, so he is proposing to add the first thirteen terms directly and to transform from the fourteenth term, but there is no justification of why the chosen value is sufficiently large – presumably trial and error is the approach. The unstated stopping criterion for the second question appears to be that individual terms are to be neglected when they have only zeros in the required decimal places. This succeeds in practice because the transformed series converge very rapidly in the chosen examples.

## 4. Error Bounds

As noted above I have tried to provide effective error bounds for Stirling's calculations, which in most cases amounts to finding bounds for the sum of the terms of the transformed series which are neglected. For this purpose we note some simple inequalities.

For a decreasing, non-negative function we have[15]

$$\int_m^\infty f(x)\,dx \le \sum_{k=m}^\infty f(k) \le \int_{m-1}^\infty f(x)\,dx. \tag{3}$$

The midpoint rule asserts that

$$\int_{x_0}^{x_0+h} g(x)\,dx = hg(x_0 + \tfrac{1}{2}h) + \frac{h^3}{24}g''(\xi)$$

for some $\xi$ between $x_0$ and $x_0 + h$. Thus if $g''$ is non-negative we have

$$\int_{x_0}^{x_0+h} g(x)\,dx \ge hg(x_0 + \tfrac{1}{2}h). \tag{4}$$

Usually our decreasing, non-negative function has non-negative second derivative, being concave-up, in which case we can replace the upper bound in (3) by

$$\int_{m-\frac{1}{2}}^\infty f(x)\,dx, \tag{5}$$

since we have from (4) with $x_0 = k - \frac{1}{2}$, $h = 1$ that

$$f(k) \le \int_{k-\frac{1}{2}}^{k+\frac{1}{2}} f(x)\,dx.$$

Occasionally the terms of the transformed series alternate in sign, in which case we may be able to use the following simple bounds which apply when $f$ is non-negative and decreases to zero (cf. Leibniz's Test):

$$f(m) - f(m-1) \le \sum_{k=m}^\infty (-1)^{k-m} f(k) \le f(m). \tag{6}$$

---

[15] These inequalities provide the proof of the Integral Test for convergence of a series.

## 5. Notation and Terminology

Stirling employs *series* for the modern idea of sequence, whether or not he wishes to add up the terms, and he sometimes writes series in what he refers to as the *Newtonian style* in which a single letter is used to denote the entire preceding term. For example (Stirling's Introduction, p. 30),

$$\frac{1}{z(z+1)}+\frac{1-n}{z+2}A+\frac{2-n}{z+3}B+\frac{3-n}{z+4}C+\ldots$$

stands for

$$\frac{1}{z(z+1)}+\frac{1-n}{z(z+1)(z+2)}+\frac{(1-n)(2-n)}{z(z+1)(z+2)(z+3)}$$
$$+\frac{(1-n)(2-n)(3-n)}{z(z+1)(z+2)(z+3)(z+4)}+\ldots.$$

It is generally clear from the context whether $A$, $B$, $C$, ... are being used in this way or are simply constants.

As was common practice at the time, Stirling often includes an asterisk in a formula to indicate the absence of an expected term. For example (Example 2 of Proposition 26, p. 146), in

$$T=A\sqrt[3]{z}\times\left(1-\frac{1}{9z}\quad *+\frac{10}{2187z^3}+\frac{11}{19683z^4}-\frac{77}{59049z^5}+\&c.\right)$$

the asterisk indicates that there is no term in $z^{-2}$.

Stirling sometimes uses a system of dots to divide decimal digits into groups, for example .43429.44819.03252 (Proposition 28). The initial dot is of course the decimal point, but this and leading zeros may be omitted in tabulated calculations as in the examples of Proposition 11.

I have retained the following terms as used by Stirling: affected, assignable, dimension, fluent, fluxion. An *affected equation* is one involving three or more terms (see [68, II, p. 42 (7)]) and the *dimension* of an equation or a quantity is its degree. A quantity is *assignable* if there is a formula from which it can be calculated. *Fluxions* were conceived in terms of flow: the ordinate $y$ and the abscissa $x$ both have fluxions (parametric derivatives), but usually $x$ is assumed to *flow uniformly* ($\dot{x}$ is constant) or, more particularly, $\dot{x}$ is taken as 1, in which case we have an ordinary derivative. In Stirling's text we may interpret fluxion (fluxional) as derivative (differential). A *fluent* is just a varying quantity, but the term generally refers to the solution of a fluxional equation; in particular, fluents arise as antiderivatives.

The term *parabola* or *parabolic curve* is applied to any curve with equation $y=p(x)$, where $p(x)$ is a polynomial – here the $p(x)$ arise as interpolating polynomials; the corresponding interpolating series are also allowed. Stirling uses *hyperbola* and *hyperbolic* in ways that may not be familiar. Such usages are explained in the text or in my notes.

## 6. Checking Stirling's Calculations

I have made use of the computer algebra system Maple as well as programs written in Basic (double precision) and Fortran (quadruple precision). Values obtained by such means are identified in the notes where appropriate. I have used the abbreviation DP to denote places after the decimal point.

# Summary of the Contents of Stirling's Text

**Introduction.** This has three sections: (i) *On the Relation of Terms*, (ii) *On Difference Equations Which Define Series*, and (iii) *On the Form and Reduction of Series.* The first two are largely concerned with notation and simple illustrations. The third section is much more substantial. It deals first with the representation of polynomials, in particular $z$, $z^2$, $z^3$, ... , in terms of the factorial polynomials $z$, $z(z-1)$, $z(z-1)(z-2)$, ... and then the representation of $z^{-2}$, $z^{-3}$, $z^{-4}$, ... by means of inverse factorial series. The Stirling numbers arise in connection with these operations and short tables are given along with rules for the calculation of these numbers. The section ends with a rule for transforming an inverse factorial series into a series of reciprocal powers.

**Part I (Propositions 1–15).** The initial sections, (i) *On Simpler Series*, (ii) *On Series Which Converge More Rapidly*, and (iii) *On Successive Sums*, establish notation and basic ideas. Next, Propositions 1–3 deal with the summation of series where the terms are given by certain factorial expressions. Propositions 4 and 5 are concerned with the determination of series when there is a known relationship between the successive sums. In Proposition 6 the situation where the ratio of successive sums is a rational function is considered. Propositions 7 and 8 find the sum of series where the terms satisfy certain two-term recurrence relations. The problem of determining the terms of one series from those of another given a relation between the successive sums of the two series and a relation between the terms of the other series is considered in Proposition 9. Propositions 10–12 and their scholia give rules for the transformation of certain types of series to improve convergence. The MacLaurin series of a rational function which is analytic at the origin and the relation between the terms of such a series are the subject of Proposition 13. The important idea of the *ultimate relation of the terms*, a limiting case of the relation determining the terms, is used in Proposition 14 to split a suitable series into two parts. Proposition 15 discusses techniques for finding equations, algebraic or fluxional (differential), which are satisfied by a given series. Part I ends with some observations on fluxional equations.

**Part II (Propositions 16–33).** Propositions 16–18 and the section which follows them, *On the Differences of Quantities*, lay down some basic priciples to be used in interpolation. Next we have the section *On the Description of Curves Through Given Points*, which refers to Newton's contributions and presents in Proposition 19 Newton's forward difference formula and in Proposition 20 what have become known as Stirling's and Bessel's interpolation formulae. Propositions 21 and 22 are concerned with practical aspects of interpolation. In Proposition 23 we have four series from which the quantity $\binom{2n}{n}/2^{2n}$ may be determined; here the methods are analytic in contrast to the interpolation methods by which this problem was treated in the first example of the preceding proposition. Propositions 24 and 25 are concerned with properties of what is in effect the Beta function: in the former Stirling discusses the sequence $A_n = B(r+n, p-r)$ and in the latter the sequence $B_m = B(p-m+1, r+m)$. In Proposition 26 he finds an asymptotic series for a function which may be expressed as $\Gamma(z+r-p)\,\Gamma(p)/(\Gamma(z)\,\Gamma(r))$. Proposition 27 returns to inverse factorial series; here Stirling uses them to solve a two-term quadratic difference equation. Proposition 28 and its second example contain perhaps the most famous of all Stirling's results, although it is generally De Moivre's form which is quoted, with Stirling's name erroneously attached: "To find the sum of any number of logarithms, whose arguments are in arithmetic progression." The general Newton–Lagrange interpolation formula is given in Proposition 29 and Proposition 30 applies what may be interpreted as interpolation techniques to find limits. Proposition 31 deals with quadrature and presents in its scholion several Newton–Cotes formulae with corrections. In Proposition 32 Stirling shows how to approximate to any one of $n+1$ equidistant ordinates given the other $n$ of them by equating to zero the $n$th difference. Finally in Proposition 33 Stirling deals with "interpolation to halves".

## Stirling's Principal Calculations

**Part I** (All by transformation of the stated series.)

**Proposition 2.**

*Example 5.* $\ln 2 = \sum_{n=1}^{\infty} \frac{1}{2n(2n-1)}$ (9DP) (Brouncker).

*Example 6.* $\frac{\pi^2}{6} = \sum_{n=1}^{\infty} \frac{1}{n^2}$ (9DP).

**Proposition 3.**

*Example 1.* $\frac{\pi}{4} = \sum_{r=0}^{\infty} \frac{(-1)^r}{2r+1}$ (10DP) (Leibniz).

**Proposition 7.**

*Example 1.* $\frac{\pi}{4} = \frac{1}{2}\,F(1,1\,;\frac{3}{2}\,;\frac{1}{2}) = \sin^{-1}\frac{1}{\sqrt{2}}$ (11DP).

*Example 2.* $\frac{1}{\sqrt{2}} = F(1,\frac{1}{2}\,;1\,;-1) = (1+1)^{-1/2}$ (10DP).

**Proposition 8.**

*Example 1.* $\frac{\pi}{2} = \sin^{-1}1 = 1 + \sum_{k=1}^{\infty} \frac{\prod_{r=1}^{k}(2r-1)^2}{(2k+1)!}$ (9DP).

*Example 2.* $\ln 2 = \sum_{n=1}^{\infty} \frac{1}{2n(2n-1)}$ (no numerical work) (Brouncker).

**Proposition 11.**

*Example 1.* $\frac{\pi^2}{6} = \sum_{n=1}^{\infty} \frac{1}{n^2}$ (17DP).

*Example 2.* $\frac{\pi}{2} = F(\frac{1}{2},\frac{1}{2}\,;\frac{3}{2}\,;1)$ (17DP).

*Example 3.* $\frac{\sqrt{2}\pi^{3/2}}{\left(\Gamma(\frac{1}{4})\right)^2} = \frac{1}{3}\,F(\frac{1}{2},\frac{3}{4}\,;\frac{7}{4}\,;1)$ (17DP)

(Elastica/lemniscate constants).

*Example 4.* $\frac{1}{4\sqrt{2\pi}}\left(\Gamma(\frac{1}{4})\right)^2 = F(\frac{1}{2},\frac{1}{4}\,;\frac{5}{4}\,;1)$ (17DP)

*Scholion, Example.* $\frac{2}{\pi} = \frac{1}{2}\,F(\frac{1}{2},\frac{1}{2}\,;2\,;1)$ (13DP).

**Proposition 12.**

*Example 1.* $\pi = 6\tan^{-1}\frac{1}{\sqrt{3}} = 2\sqrt{3}\sum_{i=0}^{\infty} \frac{(-1)^i}{(2i+1)3^i}$ (21DP) (Halley).

*Example 2.* $\frac{\pi^2}{12} = \sum_{i=0}^{\infty} \frac{(-1)^i}{(1+i)^2}$ (17DP).

*Example 3.* $\frac{\pi}{4} = \sum_{i=0}^{\infty} \frac{(-1)^i}{2i+1}$ (17DP) (Leibniz).

**Part II** (Interpolations, partial sums, asymptotic series, quadrature.)

**Proposition 21.** (From Bessel's interpolation formula.)

*Example 2.* $\Gamma(\frac{3}{2}) = \dfrac{\sqrt{\pi}}{2}$ (10DP), $\Gamma(\frac{1}{2}) = \sqrt{\pi}$ (10DP).

**Proposition 22.**

*Example 1.* $2^{200}/\binom{200}{100}$ (5DP).

**Proposition 23.**

*First Solution.* $2^{100}/\binom{100}{50} = \sqrt{50\pi F(\frac{1}{2}, \frac{1}{2}; 51; 1)}$ (10DP).

*Second Solution.* $\binom{200}{100}/2^{100} = \sqrt{\dfrac{2}{101\pi} F(\frac{1}{2}, \frac{1}{2}; 51.5; 1)}$ (13DP).

**Proposition 26.** (Evaluation of leading coefficient in asymptotic series.)

*Example 1.* $\dfrac{1}{\sqrt{\pi}}$ (12DP).

*Example 2.* $\dfrac{\left(\Gamma(\frac{1}{3})\right)^2 \sqrt{3}}{2\pi}$ (6DP).

**Proposition 27.** (Evaluation of leading coefficient in asymptotic series.)

*Example.* $\dfrac{\pi}{2}$ (8DP).

**Proposition 28.** (From Stirling's formula.)

*Example 1.* $\sum_{k=0}^{9} \log_{10}(101 + 2k)$ (11DP).

*Example 2.* $\sum_{k=11}^{1000} \log_{10} k$ (10DP), $\log_{10}(1000!)$ (10DP).

*Example 3.* $\binom{1000}{499}$ (10DP).

**Proposition 30.** (From Stirling–Schellbach algorithm.)

*Example.* $\pi = \lim_{z \to -\infty} \dfrac{\sin\left(\pi 2^{z-1}\right)}{2^{z-1}}$ (14DP).

**Proposition 31.** (From quadrature formulae.)

*Example.* $\ln 2 = \int_0^1 \dfrac{1}{1+x}\, dx$ (8DP).

*Methodus Differentialis:*

SIVE

# TRACTATUS

DE

# SUMMATIONE

ET

# INTERPOLATIONE

## SERIERUM INFINITARUM

---

AUCTORE *JACOBO STIRLING*, R.S.S.

---

**The Method of Differences:**
**or a Treatise on Summation and Interpolation**
**of Infinite Series**

*by*

**James Stirling, F.R.S.**

(p.169)

# Preface

*Since it happens very often that series converge to the true value so slowly that they are no more use for the intended purpose than if they were in fact divergent, I have presented certain theorems in the first part of this treatise, by means of which one may arrive promptly at the values of those series which approximate most slowly of all: and indeed with that intention that Problems which depend upon quadratures can be considered for solution with the same authority as those which are reduced to affected equations. For here I do not examine which series are summable (as some may perhaps infer from the title) but by what methods we may come up with the values of those which are not able to be summed.*

James Gregory *published a treatise of this type in the Appendix to* Vera Circuli et Hyperbolae Quadratura, *where he taught a very easy method of approximating to the areas of these curves from a very few given polygons of a small number of sides. In this way he has rendered the method of exhaustion of* Archimedes *so convenient that, if the same method could be extended to other curves with equal success, further effort on the determination of areas would be expended to no purpose. And we present here almost the same as* Gregory *did there for series of polygons, but in a more general way for any other series which enjoys a simple relation of the terms.*

*That* Newton *had previously thought about this matter is confirmed by his first letter to* Oldenburg *which is found printed in the* Commercium Epistolicum *of* Collins. *Certainly, after he had produced the area of the circle to sixteen decimals by a computation, he says,* If I had used other methods, I could have attained many more places of figures, perhaps twenty-five or more, using the same number of terms of the series: but the intention here was to show what could be achieved through simple computation of a series. *Nevertheless, among his writings (at least those published up to this time) there is not even a trace from which we may make a conjecture about these methods, although he had a conspicuous opportunity for presenting them in this very letter. Also in the first letter to* Oldenburg, *published in the same place, he relates that he had thought out certain things about the reduction of infinite series to finite series where the nature of the thing permits: If these survive among his posthumous works, without doubt they will shed no small light on this topic: For general theorems which provide the values of series accurately, when it can be done, will necessarily approximate in other cases, provided they are applied correctly.*

*The principle which is commonly applied for this purpose is the taking of the difference of two successive values of some quantity, so that terms may then be formed whose sum was known before; in fact this is the same principle as* Newton *used previously for obtaining the ordinate of a curve from its given area. Although this is universal in quadratures, it is only particular in summations; in fact it is only applicable to those series whose terms can be assigned: however the assignment of sums and terms is equally easy in those series which usually arise for the most part in quadratures.*

Newton's Method of Differences *has provided a much more general foundation: in particular, he describes the parabolic curve through the extremities of any number of ordinates or terms, and in that way he assigns the value of any intermediate one by means of an infinite series; but this will not approach the true value if that term is far removed from the beginning. Therefore, in order that I may obtain very remote terms of series, I have described the hyperbolic figure through the extremities of the terms; and the matter has turned out successfully, the value of a term as far removed as you wish resulting from a convergent series. But with this problem having been solved in general, its easiest case, namely the determination of the term at an infinite distance from the beginning, did not escape notice; in fact this is equally effective for the summation of series. However, the description of any geometrical curve through given points suffices for only one type of series: and yet there are countless others which cannot be dealt with at all on this basis. For the value of the term found by means of the parabola or the hyperbola does not approximate except where the differences taken according to the* Newtonian *rules form a sufficiently rapidly decreasing progression.*

*When these things had been examined, I devoted myself finally to the consideration of the relation of the terms, a very significant and simple property of series, which is commonly applied for their continuation: For I was not unaware that* De Moivre *had introduced this property of the terms into algebra with the greatest success, as the basis for solving very difficult problems concerning recurrent series: And so I decided to find out whether it could also be extended to others, which of course I doubted since there is so great a difference between recurrent and other series. But, the practical test having been made, the matter has succeeded beyond hope, for I have found out that this discovery of* De Moivre *contains very general and also very simple principles not only for recurrent series but also for any others in which the relation of the terms varies according to some regular law. For even if the relation of the terms is variable, it is, however, easily assignable: And then summations and interpolations and other more difficult problems of that type are reduced to a certain class of equations, which apart from the root to be extracted involve other unknown quantities which cannot be eliminated; in spite of this, the resolution of these equations is sometimes carried out with the greatest ease, but sometimes it does not succeed unless the discoveries of* De Moivre *concerning the assignation of terms in recurrent series are applied. And this little book deals with almost everything about this topic.*

*The problem about finding the middle coefficient in a very large power of the binomial had been solved by* De Moivre *some years before I considered it: And it is probable that to this very day I would not have thought about it, unless that most esteemed man, Mr* Alex. Cuming, *had not stated that he very much doubted that it could be solved by* Newton's Method of Differences.

# A TREATISE
## ON
## SUMMATION & INTERPOLATION
## *OF INFINITE SERIES*

---

(p.169)

## *INTRODUCTION*

Just as curves are not determined by some given ordinates no matter how many, but by the general relation between the abscissae and the ordinates, so series are not determined by some given terms no matter how many, but by the relation between successive terms. For any quantities which are finite in number can form terms in different series: in fact the series is unique which has the same initial terms and the same law for forming the remaining terms up to infinity. Therefore in the first place the relations of the terms have to be investigated; then when these have been found they are to be specified by difference equations just as *Des Cartes* has defined curves by algebraic equations: when these things have been obtained, problems about summation and interpolation and other matters of that type concerning series will be solved by an analysis no less exact than common algebra is.

### *On the Relation of Terms*

The terms of a series taken two at a time, three at a time, or in greater number will maintain to a large extent a certain relation among themselves which is simple and obvious and by which the series is determined and continued indefinitely. Thus if one is divided by $1 - x$, the geometric progression will come forth in which any subsequent term is to that immediately preceding it as $x$ is to one. For by this property the series

$$1 + x + x^2 + x^3 + x^4 + x^5 + \&c.$$

is distinguished from any other and is continued to infinity.

Suppose that the fraction $\dfrac{1}{r + sx + tx^2}$ is to be resolved into a series: to that end set $y = \dfrac{1}{r + sx + tx^2}$; and by multiplying both parts by the denominator

$$y \times (r + sx + tx^2) = 1, \quad \text{or} \quad y \times (r + sx + tx^2) - 1 = 0,$$

will be obtained. For $y$ substitute a series of the required form

$$A + Bx + Cx^2 + Dx^3 + Ex^4 + Fx^5 + \&c.$$

and

$$\left.\begin{array}{ll} rA & +rB \\ -1 & +sA \end{array}\right\} x \left.\begin{array}{l} +rC \\ +sB \\ +tA \end{array}\right\} x^2 \left.\begin{array}{l} +rD \\ +sC \\ +tB \end{array}\right\} x^3 \left.\begin{array}{l} +rE \\ +sD \\ +tC \end{array}\right\} x^4 \left.\begin{array}{l} +rF \\ +sE \\ +tD \end{array}\right\} x^5 + \&c. = 0.$$

will result.

Here, by setting like members equal to zero for the determination of the assumed coefficients, there will be $rA - 1 = 0$, $rB + sA = 0$; then $rC + sB + tA = 0$, $rD + sC + tB = 0$, $rE + sD + tC = 0$, etc. and so on to infinity. From these it is obvious that the same relation holds everywhere between any three successive terms. Similarly, by using for $y$ a series of this form $Ax^{-2} + Bx^{-3} + Cx^{-4} + Dx^{-5} + \&c.$ and substituting it into the equation,

$$\left.\begin{array}{ll} tA & +tB \\ -1 & +sA \end{array}\right\} x^{-1} \left.\begin{array}{l} +tC \\ +sB \\ +rA \end{array}\right\} x^{-2} \left.\begin{array}{l} +tD \\ +sC \\ +rB \end{array}\right\} x^{-3} \left.\begin{array}{l} +tE \\ +sD \\ +rC \end{array}\right\} x^{-4} + \&c. = 0.$$

will result.

And now for determining the coefficients we have the equations $tA - 1 = 0$, $tB + sA = 0$; then $rA + sB + tC = 0$, $rB + sC + tD = 0$, $rC + sD + tE = 0$, and so on with the rest. It is clear that here the same relations come forth as in the previous calculation, except that the coefficients have been taken in the reverse order. And the quantities $r$, $s$, $t$, which show the relation of the terms, are the same as those in the denominator of the fraction. Now, however obvious it may be, it was Mr *De Moivre* who first of all brought this property of these series into use in the solution of problems concerning infinite series, which otherwise would have been very involved.

But in very many series the relation of the terms is not constant as in those which result from division; but it very often varies according to some well known law which is obvious at a glance; examples of this matter are series which are commonly produced by quadrature, and countless others. Thus in this series

$$1 + \tfrac{2}{3}x + \tfrac{8}{15}x^2 + \tfrac{16}{35}x^3 + \tfrac{128}{315}x^4 + \&c.$$

the terms are continued to infinity by repeated multiplication of these fractions $\frac{2}{3}$, $\frac{4}{5}$, $\frac{6}{7}$, $\frac{8}{9}$, etc. And in this

$$1 + \tfrac{1}{6}x + \tfrac{3}{40}x^2 + \tfrac{5}{112}x^3 + \tfrac{35}{1152}x^4 + \&c.$$

it is by multiplication of

$$\tfrac{1\times1}{2\times3},\quad \tfrac{3\times3}{4\times5},\quad \tfrac{5\times5}{6\times7},\quad \tfrac{7\times7}{8\times9},\quad \&c.$$

And these fractions vary according to a law which is obvious to anyone; and so there will be no difficulty in assigning them.

## *On Difference Equations Which Define Series*

The equation defining the series is that which assigns the relation of the terms generally from their distances given from the beginning. Now the terms are to be imagined as standing above a straight line which is given in position, just like so many ordinates, whose common distance is one. And for the sake of simplicity I use one everywhere in what follows for the common interval: let it suffice to have advised of this once.

I denote the initial terms of the series by the initial letters of the alphabet $A$, $B$, $C$, $D$, etc. $A$ is the first, $B$ is the second, $C$ is the third, and so on. And I denote an arbitrary term generally by the letter $T$ and the remaining terms following it in order by the same letter $T$ with the *Roman* numerals I, II, III, IV, V, VI, VII, etc. attached to distinguish them. Thus if $T$ is the tenth term, then $T'$ will be the eleventh, $T''$ will be the twelfth, $T'''$ will be the thirteenth, and so on. And in general, whatever term is defined by $T$, the succeeding ones will be defined universally by $T'$, $T''$, $T'''$, $T^{\text{iv}}$, etc.

I denote by the indeterminate quantity $z$ the distance of the term $T$ from any given term, or from any given intermediate point between any two terms: in this way the distances of the terms $T'$, $T''$, $T'''$, etc. from the specified term or point will be $z+1$, $z+2$, $z+3$, etc. For the increment of the abscissa $z$ is equal to the common interval of the terms standing above the abscissa: and the quantities $z$, $z+1$, $z+2$, $z+3$, etc. follow each other as the subsequent terms follow the preceding ones.

These things having been stated, let us consider the series

$$1, \quad \tfrac{1}{2}x, \quad \tfrac{3}{8}x^2, \quad \tfrac{5}{16}x^3, \quad \tfrac{35}{128}x^4, \quad \tfrac{63}{256}x^5, \quad \text{\&c.},$$

where the relations of the terms are

$$B = \tfrac{1}{2}Ax, \quad C = \tfrac{3}{4}Bx, \quad D = \tfrac{5}{6}Cx, \quad E = \tfrac{7}{8}Dx, \quad \text{\&c.};$$

in general the relation will be defined by the equation $T' = \dfrac{z+\frac{1}{2}}{z+1}Tx$, where $z$ denotes the distance of $T$ from the first term of the series. For by writing 0, 1, 2, 3, 4, etc. successively for $z$, the relations of the terms in the proposed series will come out. Likewise, if $z$ denotes the distance of $T$ from the second term of the series, the equation will be $T' = \dfrac{z+\frac{3}{2}}{z+2}Tx$; as will be established by writing the numbers $-1$, 0, 1, 2, 3, etc. successively for $z$. Or if the indeterminate $z$ denotes the position of the term $T$ in the series, its successive values will be 1, 2, 3, 4, etc. and the equation will be $T' = \dfrac{z-\frac{1}{2}}{z}Tx$, as will be clear to anyone who tries it.

Therefore countless different difference equations can define the same series, according as the beginning of the abscissa $z$ is taken at this or that point. And on the other hand the same equation defines countless different series as a result of using different successive values for $z$. For in the equation

$T' = \frac{z-\frac{1}{2}}{z} Tx$, which defines the series which we have now been discussing, when 1, 2, 3, 4, etc. are the values of the abscissa in increasing order, write successively $1\frac{1}{2}$, $2\frac{1}{2}$, $3\frac{1}{2}$, $4\frac{1}{2}$, etc. for $z$ and the relations of the terms will come forth as

$$B = \tfrac{2}{3}Ax, \quad C = \tfrac{4}{5}Bx, \quad D = \tfrac{6}{7}Cx, \quad \&c.$$

Whence the series is

$$A, \quad \tfrac{2}{3}Ax, \quad \tfrac{8}{15}Ax^2, \quad \tfrac{16}{35}Ax^3, \quad \tfrac{128}{315}Ax^4, \quad \&c.$$

which is different from the first. But the equation always determines the series from the given values of the abscissa and at the same time from the first term, where the equation involves only two terms of the series. As in the last series all terms are given once the first is given. However, where the equation involves three terms, it is necessary for the determination of the series that two be given, and three where it involves four, and so on.

Now let us consider the series

$$x, \quad \tfrac{1}{6}x^3, \quad \tfrac{3}{40}x^5, \quad \tfrac{5}{112}x^7, \quad \tfrac{35}{1152}x^9, \quad \&c.,$$

where the relations of the terms are

$$B = \tfrac{1\times 1}{2\times 3}Ax^2, \quad C = \tfrac{3\times 3}{4\times 5}Bx^2, \quad D = \tfrac{5\times 5}{6\times 7}Cx^2, \quad \&c.$$

The equation for the same series will be

$$T' = \frac{(2z-1)\times(2z-1)}{2z\times(2z+1)} Tx^2, \quad \text{or} \quad T' = \frac{4z^2-4z+1}{4z^2+2z} Tx^2,$$

where the successive values of the indeterminate $z$ are 1, 2, 3, 4, etc. Therefore in the equation defining the series the abscissa $z$ can be of one, two, or more dimensions.

Series whose terms are assignable can be defined by the equations which assign the terms. Thus the series

$$1 - \tfrac{1}{2}x + \tfrac{1}{3}x^2 - \tfrac{1}{4}x^3 + \tfrac{1}{5}x^4 - \&c.$$

is defined by the equation $T = \frac{(-x)^z}{z+1}$, as will be confirmed by substituting 0, 1, 2, 3, etc. for $z$. And in the same way the series

$$x + \tfrac{1}{4}x^2 + \tfrac{1}{9}x^3 + \tfrac{1}{16}x^4 + \tfrac{1}{25}x^5 + \&c.$$

is specified by this expression $T = \frac{x^z}{z^2}$. Such equations can always be reduced to those of the other type: for where the terms are assignable, their relations will also be assignable. And the difference between the latter and the former will rarely be so great that one may not safely proceed to either as seems appropriate.

It follows from what has been said up to this point that the relations of the following terms are derived from those of the preceding terms by writing for $z$ its successive value $z+1$ in the difference equation. Consider the equation $T' = \frac{z+n}{z} T$; write $z+1$ for $z$, $T'$ for $T$, and $T''$ for $T'$; and $T'' = \frac{z+n+1}{z+1} T'$ will arise, which is the relation between the terms $T'$ and $T''$. In this last expression write the subsequent values of the variables $z+1$, $T''$, $T'''$ for the preceding values $z$, $T'$, $T''$, and you will obtain $T''' = \frac{z+n+2}{z+2} T''$, which is indeed the relation between $T''$ and $T'''$.

But also by the reverse operation one may go back to the relation of the preceding terms given that of the subsequent ones. Let the equation be

$$T'' = \frac{z^2 - 1}{z^3 + 3z^2 + 3z + 2} T',$$

and in this write $T$ for $T'$, $T'$ for $T''$, and $z - 1$ for $z$; and you will have $T' = \frac{z^2 - 2z}{z^3 + 1} T$. By going backwards and forwards in this way series can be continued on this side or that side to infinity where their nature allows: and even if it is not known which terms are denoted by $T$, $T'$, $T''$, etc. one can undertake the calculation upon them, just as if they had been completely known.

The equations which we have been discussing up to this point involve only two terms of the series; but they can involve more, and the terms just as the indeterminate $z$ can be of several dimensions. But I only deal with the simpler cases in this account.

## *On the Form and Reduction of Series*

After we have converted series to difference equations, it has to be shown how these can be resolved in numbers. For it is the task of the analyst to extract quantities, however they have been prescribed, exactly or very closely. Now the roots of difference equations are resolved most appropriately into series of the following types

$$A + Bz + Cz(z-1) + Dz(z-1)(z-2) + Ez(z-1)(z-2)(z-3) + \&\text{c.}$$

$$A + \frac{B}{z} + \frac{C}{z(z+1)} + \frac{D}{z(z+1)(z+2)} + \frac{E}{z(z+1)(z+2)(z+3)} + \&\text{c.}$$

Indeed, where $z$ is a small quantity, it will be appropriate to use the first form, and the latter where $z$ is large. And these series, which are built up from factors in arithmetic progression, are much more suitable for this task than the ordinary series which are made up from increasing or decreasing powers of an indeterminate quantity. Besides, the latter form has this convenient property that as a rule $z$ can be as large as you please in it, which makes the series very rapidly convergent.

But if the forms of these series are changed by any operations whatsoever, they have to be referred back to the same form as they had previously, in order that the terms may be expressed in the same way and they can be collected together as the matter requires. Suppose therefore that the following equation is being used:

$$T = A + Bz + Cz(z-1) + Dz(z-1)(z-2) + Ez(z-1)(z-2)(z-3) + \&c.$$

When this has been multiplied by $z$ it loses its previous form, and assumes a new one, namely

$$Tz = Az + Bz^2 + Cz^2(z-1) + Dz^2(z-1)(z-2) + Ez^2(z-1)(z-2)(z-3) + \&c.$$

It is clear that from this its terms cannot be compared with the corresponding terms in the former series. And so, in order that the required form may be restored, I make use of

$$\begin{array}{lllll}
Az & = Az, & & & \\
Bz^2 & = Bz + Bz(z-1), & & & \\
Cz^2(z-1) & = \ * & 2Cz(z-1) + Cz(z-1)(z-2), & & \\
Dz^2(z-1)(z-2) & = \ * & * & 3Dz(z-1)(z-2) & \\
 & & & + Dz(z-1)(z-2)(z-3), & \\
Ez^2(z-1)(z-2)(z-3) = & * & * & * \ \ 4Ez(z-1)(z-2)(z-3) & \\
 & & & + \&c. &
\end{array}$$

And so by collecting like terms into one, the series will be brought back to the original form

$$Tz = \left.\begin{array}{r}+A\\+B\end{array}\right\} z \ \left.\begin{array}{r}+\ B\\+2C\end{array}\right\} z(z-1) \ \left.\begin{array}{r}+\ C\\+3D\end{array}\right\} z(z-1)(z-2)$$

$$\left.\begin{array}{r}+\ D\\+4E\end{array}\right\} z(z-1)(z-2)(z-3) + \&c.$$

Since without doubt the identity of the terms depends in no way on the coefficients $A$, $B$, $C$, $D$, etc. but entirely on the indeterminate $z$, the first term in this series $\left.\begin{array}{r}+A\\+B\end{array}\right\} z$ can be compared with the second term $Bz$ in the other series, likewise the second term in this series can be compared with the third term in that series, and so on with the remaining terms.

Similarly, if in the first equation

$$T = A + Bz + Cz(z-1) + Dz(z-1)(z-2) + \&c.$$

the succeeding values of the variables are written for the preceding values, that is $T'$ for $T$ and $z+1$ for $z$, then

$$T' = A + B(z+1) + C(z+1)z + D(z+1)z(z-1) + E(z+1)z(z-1)(z-2) + \&c.$$

arises.

Now there is

$$\begin{array}{llll}
A & = A, & & \\
B(z+1) & = B + Bz, & & \\
C(z+1)z & = * \quad 2Cz + Cz(z-1), & & \\
D(z+1)z(z-1) & = * \quad * & 3Dz(z-1) & \\
 & & & + Dz(z-1)(z-2), \\
E(z+1)z(z-1)(z-2) & = * \quad * & * & 4Ez(z-1)(z-2) + \&c.
\end{array}$$

And thence

$$T' = \left.\begin{array}{l}+A+B\\+B+2C\end{array}\right\}z \quad \left.\begin{array}{l}+C\\+3D\end{array}\right\}z(z-1) \quad \left.\begin{array}{l}+D\\+4E\end{array}\right\}z(z-1)(z-2)$$

$$\left.\begin{array}{l}+E\\+5F\end{array}\right\}z(z-1)(z-2)(z-3) \ + \&c.$$

This is the desired form.

Now here is the basis of these operations. The quantity to be reduced is reduced by multiplication to powers of the indeterminate $z$: then operate in the manner of the following example. Let $(z-3)z(z+1)(z+4)$ be the quantity to be reduced; form

$$(z-3)z(z+1)(z+4) = az(z-1)(z-2)(z-3) + bz(z-1)(z-2) + cz(z-1) + dz.$$

Here the greatest number of factors in the resolved quantity is equal to the number of the same in the quantity to be resolved. Let both quantities be reduced to powers of the indeterminate, the multiplication having been carried out, and

$$z^4 + 2z^3 - 11z^2 - 12z = az^4 \ \left.\begin{array}{r}-6a\\+b\end{array}\right\}z^3 \ \left.\begin{array}{r}+11a\\-3b\\+c\end{array}\right\}z^2 \ \left.\begin{array}{r}-6a\\+2b\\-c\\+d\end{array}\right\}z$$

will be obtained.

And by comparing like terms we will obtain

$$a = 1, \quad b - 6a = 2, \quad c - 3b + 11a = -11, \quad d - c + 2b - 6a = -12,$$

from which is obtained $a = 1$, $b = 8$, $c = 2$, $d = -20$; hence the quantity set forth is

$$z^4 + 2z^3 - 11z^2 - 12z = z(z-1)(z-2)(z-3) + 8z(z-1)(z-2) + 2z(z-1) - 20z.$$

And one may proceed in exactly the same manner in other cases. But for the sake of brevity take the following rule. Divide one repeatedly by the terms of this progression $n-1$, $n-2$, $n-3$, $n-4$, etc., that is, divide one by $n-1$, and the resulting quotient by $n-2$, and this quotient by $n-3$, and so on. Then display all the quotients produced in this way regularly in a table as you see, the powers of $n$ having been omitted and only the coefficients retained, these alone being useful for this task; and you will have

*First Table*

| | | | | | | | | | |
|---|---|---|---|---|---|---|---|---|---|
| 1 | 1 | 1 | 1 | 1 | 1 | 1 | 1 | 1 | &c. |
| | 1 | 3 | 7 | 15 | 31 | 63 | 127 | 255 | &c. |
| | | 1 | 6 | 25 | 90 | 301 | 966 | 3025 | &c. |
| | | | 1 | 10 | 65 | 350 | 1701 | 7770 | &c. |
| | | | | 1 | 15 | 140 | 1050 | 6951 | &c. |
| | | | | | 1 | 21 | 266 | 2646 | &c. |
| | | | | | | 1 | 28 | 462 | &c. |
| | | | | | | | 1 | 36 | &c. |
| | | | | | | | | 1 | &c. |
| | | | | | | | | | &c. |

Now take for the coefficients the numbers in the descending columns, and you will have the following values of the powers:

$$
\begin{aligned}
z &= z, \\
z^2 &= z + z(z-1), \\
z^3 &= z + 3z(z-1) + z(z-1)(z-2), \\
z^4 &= z + 7z(z-1) + 6z(z-1)(z-2) + z(z-1)(z-2)(z-3), \\
z^5 &= z + 15z(z-1) + 25z(z-1)(z-2) + 10z(z-1)(z-2)(z-3) \\
&\quad + z(z-1)(z-2)(z-3)(z-4),
\end{aligned}
$$

&c.

And so, once this table has been obtained, any quantity is reduced to the form sought without tedious computation. Let the expression previously reduced $z^4 + 2z^3 - 11z^2 - 12z$ be set forth. Extract the values of the powers from the table, and when these have been multiplied by their respective coefficients $-12$, $-11$, $+2$, and 1, you will obtain

$$\begin{aligned}
-12z &= -12z,\\
-11z^2 &= -11z - 11z(z-1),\\
+2z^3 &= +2z + 6z(z-1) + 2z(z-1)(z-2),\\
+z^4 &= +z + 7z(z-1) + 6z(z-1)(z-2)\\
&\quad + z(z-1)(z-2)(z-3),
\end{aligned}$$

$$\left.\begin{matrix} z^4+2z^3 \\ -11z^2-12z \end{matrix}\right\} = -20z + 2z(z-1) + 8z(z-1)(z-2) + z(z-1)(z-2)(z-3).$$

And the values of the members collected into one sum give the value of the whole expression as it has now come forth. It is to be noted that an infinite series which is made up from ascending powers of the indeterminate cannot in general be reduced to another of the stated form: for each coefficient would be an infinite series. However, in finite series, the matter succeeds as has been shown above.

Series of the other form are also reduced similarly. For let us suppose some quantity sought to be

$$T = \frac{A}{z} + \frac{B}{z(z+1)} + \frac{C}{z(z+1)(z+2)} + \frac{D}{z(z+1)(z+2)(z+3)} + \&c.$$

Then if the need arises to investigate the successive value of $T$, write $z+1$ for $z$, and the successive value comes out as

$$T' = \frac{A}{z+1} + \frac{B}{(z+1)(z+2)} + \frac{C}{(z+1)(z+2)(z+3)} + \frac{D}{(z+1)(z+2)(z+3)(z+4)} + \&c.$$

And in order that this series may be brought back to the first form, I make use of the following:

$$\begin{aligned}
\frac{A}{z+1} &= \frac{A}{z} - \frac{A}{z(z+1)},\\
\frac{B}{(z+1)(z+2)} &= * \quad \frac{B}{z(z+1)} - \frac{2B}{z(z+1)(z+2)},\\
\frac{C}{(z+1)(z+2)(z+3)} &= * \quad * \quad \frac{C}{z(z+1)(z+2)} - \frac{3C}{z(z+1)(z+2)(z+3)},\\
\frac{D}{(z+1)(z+2)(z+3)(z+4)} &= * \quad * \quad * \quad \frac{D}{z(z+1)(z+2)(z+3)} +\\
&\&c.
\end{aligned}$$

And I have

$$T' = \frac{A}{z} + \frac{B-A}{z(z+1)} + \frac{C-2B}{z(z+1)(z+2)} + \frac{D-3C}{z(z+1)(z+2)(z+3)} + \&c.,$$

where the denominators are now the same as in the value of $T$; and for this reason, it is possible to undertake the comparison of terms just as the occasion demands. Now operations of this type are demonstrated as follows. Put $\dfrac{1}{z+1} = \dfrac{1}{z} - \dfrac{a}{z(z+1)}$, where $a$ is a quantity which has to be found straightaway; then by multiplying by the denominator $z(z+1)$, there will come forth $z = z+1-a$, or by eliminating $z$ in both places, $0 = 1-a$ and $a = 1$; hence by substituting one for $a$, $\dfrac{1}{z+1} = \dfrac{1}{z} - \dfrac{1}{z(z+1)}$ will be obtained. Likewise I form

$$\frac{1}{(z+1)(z+2)} = \frac{1}{z(z+1)} - \frac{a}{z(z+1)(z+2)},$$

and by multiplying by the denominator there will be $z = z+2-a$, or $a = 2$, and from this

$$\frac{1}{(z+1)(z+2)} = \frac{1}{z(z+1)} - \frac{2}{z(z+1)(z+2)}.$$

Now let us consider

$$\frac{A}{z+2} + \frac{B}{(z+2)(z+3)} + \frac{C}{(z+2)(z+3)(z+4)} + \frac{D}{(z+2)(z+3)(z+4)(z+5)} + \&c.,$$

which it is required to reduce to another of the required form. The operation is undertaken as just shown, and you will find

$$\begin{array}{llllll}
\frac{A}{z+2} & = \frac{A}{z} & - \frac{2A}{z(z+1)} & + \frac{2A}{z(z+1)(z+2)}, & & \\
\frac{B}{(z+2)(z+3)} & = * & \frac{B}{z(z+1)} & - \frac{4B}{z(z+1)(z+2)} & + \frac{6B}{z(z+1)(z+2)(z+3)}, & \\
\frac{C}{(z+2)(z+3)(z+4)} & = * & * & \frac{C}{z(z+1)(z+2)} & - \frac{6C}{z(z+1)(z+2)(z+3)} & + \&c. \\
\frac{D}{(z+2)(z+3)(z+4)(z+5)} & = * & * & * & \frac{D}{z(z+1)(z+2)(z+3)} & - \&c.
\end{array}$$

And in the required form the proposed series will be found to be

$$\frac{A}{z} + \frac{B-2A}{z(z+1)} + \frac{C-4B+2A}{z(z+1)(z+2)} + \frac{D-6C+6B}{z(z+1)(z+2)(z+3)} + \&c.$$

And the reduction in other cases is undertaken in this way. If the fraction to be reduced is $\dfrac{1}{z+1}$, there will be two members in its value, as in the first example above. If it is $\dfrac{1}{z+2}$, there will be three, as in the latter example. And in general in the value of $\dfrac{1}{z+n}$ after reduction to the required form the number of members will exceed the number $n$ by one. However, I suppose here that $n$ is a positive integer; for if it is fractional or negative, the value of the fraction $\frac{1}{z+n}$ will go off to infinity.

But the general rule for transformations of this type is that which follows. Multiply the terms of this progression $n$, $1+n$, $2+n$, $3+n$, etc. repeatedly by themselves, and let the results be arranged in the following table in order of the powers of the number $n$, only the coefficients having been retained, and there will result:

*Second Table*

| | | | | | | | | | |
|---|---|---|---|---|---|---|---|---|---|
| 1 | | | | | | | | | |
| 1 | 1 | | | | | | | | |
| 2 | 3 | 1 | | | | | | | |
| 6 | 11 | 6 | 1 | | | | | | |
| 24 | 50 | 35 | 10 | 1 | | | | | |
| 120 | 274 | 225 | 85 | 15 | 1 | | | | |
| 720 | 1764 | 1624 | 735 | 175 | 21 | 1 | | | |
| 5040 | 13068 | 13132 | 6769 | 1960 | 322 | 28 | 1 | | |
| 40320 | 109584 | 118124 | 67284 | 22449 | 4536 | 546 | 36 | 1 | |
| &c. | &c. | &c. | &c. | &c. | &c. | &c. | &c. | &c. | &c. |

Then by taking the coefficients from the descending columns, you will obtain the values of the powers,

$$\frac{1}{z^2} = \frac{1}{z(z+1)} + \frac{1}{z(z+1)(z+2)} + \frac{2}{z(z+1)(z+2)(z+3)} + \frac{6}{z(z+1)(z+2)(z+3)(z+4)} + \&c.$$

$$\frac{1}{z^3} = \frac{1}{z(z+1)(z+2)} + \frac{3}{z(z+1)(z+2)(z+3)} + \frac{11}{z(z+1)(z+2)(z+3)(z+4)} + \&c.$$

$$\frac{1}{z^4} = \frac{1}{z(z+1)(z+2)(z+3)} \times \left(1 + \frac{6}{z+4} + \frac{35}{(z+4)(z+5)} + \frac{225}{(z+4)(z+5)(z+6)} + \&c.\right)$$

And continue in this way for the remaining powers. And so, if a given series is made up of powers, it can always be reduced to another of the desired form by use of this table.

Or, if the series $\frac{A}{z^2}+\frac{B}{z^3}+\frac{C}{z^4}+\frac{D}{z^5}+$ &c. is to be considered, take the coefficients from the transverse columns and put

$$\begin{aligned}
a &= A,\\
b &= A+B,\\
c &= 2A+3B+C,\\
d &= 6A+11B+6C+D,\\
e &= 24A+50B+35C+10D+E,\\
f &= 120A+274B+225C+85D+15E+F,\\
&\text{\&c.}
\end{aligned}$$

And the series made up of powers will be transformed into the following series of the required form

$$\frac{a}{z(z+1)}+\frac{b}{z(z+1)(z+2)}+\frac{c}{z(z+1)(z+2)(z+3)}+\frac{d}{z(z+1)(z+2)(z+3)(z+4)}+\text{\&c.}$$

Now let us consider the fraction $\frac{1}{z^2+nz}$; I first resolve it using division into the ordinary series

$$\frac{1}{z^2}-\frac{n}{z^3}+\frac{n^2}{z^4}-\frac{n^3}{z^5}+\frac{n^4}{z^6}-\text{\&c.}$$

whence

$$A=1,\quad B=-n,\quad C=+n^2,\quad D=-n^3,\quad E=+n^4,\quad \text{\&c.}$$

and, when these values have been substituted, there will come out

$$a=1,\quad b=1-n,\quad c=2-3n+n^2,\quad d=6-11n+6n^2-n^3,\quad \text{\&c.}$$

and so

$$\frac{1}{z^2+nz}=\frac{1}{z(z+1)}+\frac{1-n}{z(z+1)(z+2)}+\frac{2-3n+n^2}{z(z+1)(z+2)(z+3)}+\frac{6-11n+6n^2-n^3}{z(z+1)(z+2)(z+3)(z+4)}+\text{\&c.}$$

That is

$$\frac{1}{z^2+nz}=\frac{1}{z(z+1)}+\frac{1-n}{z+2}A+\frac{2-n}{z+3}B+\frac{3-n}{z+4}C+\frac{4-n}{z+5}D+\text{\&c.},$$

where the quantites $A$, $B$, $C$, $D$, etc. now denote terms of this series in the style of *Newton*. And it is clear that the series terminates whenever $n$ is a positive integer. In other examples let $z$ also denote the least factor in the denominator, and the series will always terminate in this method where its nature allows. Thus if the fraction is $\frac{1}{x(x-3)(x+2)}$, I put $z = x - 3$, the least of the three factors; then there will be $x = z + 3$ and $x + 2 = z + 5$. And the fraction will become $\frac{1}{z(z+3)(z+5)}$, or when the multiplication has been carried out, $\frac{1}{z^3+8z^2+15z}$, which by division is

$$\frac{1}{z^3} - \frac{8}{z^4} + \frac{49}{z^5} - \frac{272}{z^6} + \frac{1441}{z^7} - \frac{7448}{z^8} + \frac{37969}{z^9} - \&c.$$

Whence

$$A = 0,\ B = 1,\ C = -8,\ D = +49,\ E = -272,\ F = +1441,\ G = -7448,\ \&c.$$

And from these come forth

$$a = 0,\ b = 1,\ c = -5,\ d = 12,\ e = -12;$$

but $f$ and the remaining cofficients are zero: and in consequence the series terminates, being exactly

$$\frac{1}{z(z+3)(z+5)} = \frac{1}{z(z+1)(z+2)} \times \left(1 - \frac{5}{z+3} + \frac{12}{(z+3)(z+4)} - \frac{12}{(z+3)(z+4)(z+5)}\right).$$

In an arbitrary fraction

$$\frac{1}{z(z+a)(z+b)(z+c)\,\&c.}$$

let $z$ be the least of the factors, and, provided that $a$, $b$, $c$, etc. are positive and also integral, the series will be terminated, otherwise it will go off to infinity. Now where the series terminates, it can be found in very many ways and that more elegantly than by the above general rule: certainly it is far removed from elegance to reduce first a finite fraction to an infinite series so that its value may then be obtained in a finite number of terms: we have done this here in order to illustrate the general rule, not to teach the best method for the case where the series terminates.

If in the first table the numbers from the ascending columns are taken out and we put

$$a = A,$$
$$b = B - A,$$
$$c = C - 3B + A,$$
$$d = D - 6C + 7B - A,$$
$$e = E - 10D + 25C - 15B + A,$$
$$f = F - 15E + 65D - 90C + 31B - A,$$
&c.

then a series of this form

$$\frac{A}{z(z+1)} + \frac{B}{z(z+1)(z+2)} + \frac{C}{z(z+1)(z+2)(z+3)} + \&c.$$

will transform into the following

$$\frac{a}{z^2} + \frac{b}{z^3} + \frac{c}{z^4} + \frac{d}{z^5} + \&c.,$$

which is made up from powers.

In these transformations we have had no discussion of the term $\frac{A}{z}$, since without any transformation it belongs ambiguously both to the series of powers and to that of the factors.

# PART ONE

## *On the Summation of Series*

In this first part I have tried to shorten the calculations in the quadrature of curves, and also in more difficult problems, namely by attaining the values of infinite series more readily than by simple addition of terms as is commonly done. For rapidly converging series this indeed amply achieves the purpose, and there is no need for another method: but where they converge slowly, immeasurable work is required for the most part, and it is indeed greater according as the convergence is less; and therefore if they approximate very slowly, they become wholly intractable. For it is very well known that sometimes more than a thousand terms are required in order that the sum may be obtained exact to two or three figures. Therefore we will demonstrate in what follows a method which is easy to apply for transforming those which are slowest converging of all into others which approximate very rapidly; it is clear that from these the sums can be calculated with very little effort to very many places of figures.

Indeed the transformed series will terminate where those whose sums are to be found are summable; and in that case the transformation will result in the summation. But I am less concerned about summable series and I only touch upon them in passing as they usually appear only rarely in quadratures. For here I have devoted effort, not to producing useless series which can be summed by available theorems, but to obtaining theorems by which useful series can be readily summed to as many places of figures as any applications require.

(p.174) ## *On Simpler Series*

It is not only the convergence of a series but also its simplicity which contributes most to the carrying out of calculations. For this reason let us consider transformations first of all. It should be known that the *Newtonian* series in the Treatise on Quadrature of Curves not only terminate where the nature of the thing allows, but are also the simplest of all when they go on to infinity, and consequently they are to be preferred to those which are found by the common method, namely reducing ordinates to convergent series, so that areas may then be calculated.

Let $x^{\theta-1} \times (e + fx^{\eta})^{\lambda-1}$ be the ordinate of a curve, in which $x$ is the abscissa, $e$ and $f$ are coefficients, and $\theta - 1$, $\lambda - 1$, and $\eta$ are the indices of the powers: put $r = \dfrac{\theta + \eta}{\eta}$, $s = \dfrac{\theta + \lambda\eta}{\eta}$, and following *Newton* the area will be

$$\frac{x^{\theta}}{\theta e} \times (e + fx^{\eta})^{\lambda} - \frac{s}{r}A\frac{fx^{\eta}}{e} - \frac{s+1}{r+1}B\frac{fx^{\eta}}{e} - \frac{s+2}{r+2}C\frac{fx^{\eta}}{e} - \frac{s+3}{r+3}D\frac{fx^{\eta}}{e} - \&c.,$$

where $A$, $B$, $C$, $D$, etc. denote the terms, each in its order from the beginning: thus

$$A = \frac{x^\theta}{\theta e} \times (e + fx^\eta)^\lambda, \quad B = -\frac{s}{r}A\frac{fx^\eta}{e}, \quad C = -\frac{s+1}{r+1}B\frac{fx^\eta}{e}, \quad \text{and so on.}$$

Now let it be proposed to find the arc from the given right sine $x$, or what is equivalent, the quadrature of the curve whose ordinate is $\dfrac{1}{\sqrt{1-x^2}}$ : expressed in the required form, this is $x^0 \times (1-x^2)^{-1/2}$, which on comparison with the general ordinate gives

$$e = 1, \quad f = -1, \quad \eta = 2, \quad \theta - 1 = 0, \quad \lambda - 1 = -\tfrac{1}{2};$$

and so $\theta = 1$, $\lambda = \frac{1}{2}$: and hence $r = \frac{3}{2}$, $s = 1$; when these values have been substituted into the theorem, the series

$$x\sqrt{1-x^2} + \tfrac{2}{3}Ax^2 + \tfrac{4}{5}Bx^2 + \tfrac{6}{7}Cx^2 + \tfrac{8}{9}Dx^2 + \tfrac{10}{11}Ex^2 + \&c.$$

arises for the arc. But if the proposed ordinate is first resolved into a series by *Newton's* Theorem for developing the binomial and then the fluent of each term is taken, the series

$$x + \tfrac{1\times1}{2\times3}Ax^2 + \tfrac{3\times3}{4\times5}Bx^2 + \tfrac{5\times5}{6\times7}Cx^2 + \tfrac{7\times7}{8\times9}Dx^2 + \tfrac{9\times9}{10\times11}Ex^2 + \&c.$$

will result for the same arc. Whence it is clear that the first series is much simpler and consequently it can more easily be continued to infinity. For example, if the arc required is an eighth part of the whole circumference, its sine $x$ will be equal to $\sqrt{\frac{1}{2}}$ and when this has been substituted the series becomes

$$\text{First series:} \quad \tfrac{1}{2} + \tfrac{1}{3}A + \tfrac{2}{5}B + \tfrac{3}{7}C + \tfrac{4}{9}D + \tfrac{5}{11}E + \&c.$$

$$\text{Second series:} \quad \sqrt{\tfrac{1}{2}} + \tfrac{1}{12}A + \tfrac{9}{40}B + \tfrac{25}{84}C + \tfrac{49}{144}D + \tfrac{81}{220}E + \&c.$$

In this case the former is to be preferred for two reasons, first since it is produced by simpler factors, and secondly because it is free from the surd which is found in the latter. Nevertheless, where $x$ is a rational quantity and at the same time $x\sqrt{1-x^2}$ is an irrational quantity, the latter series is to be chosen, but only if $x$ is of a sufficiently negligible magnitude which produces a very rapidly converging series; for in this way the extraction of the square root is avoided. Besides, if $x = 1$, from necessity we have to come back to the second, since in that case the quantity $\sqrt{1-x^2}$, by which the first is multiplied, vanishes.

## (p.175) *On Series Which Converge More Rapidly*

Where an indeterminate quantity quickly becomes very large as the quantity sought increases, and eventually becomes infinitely large, the terms of the series made up from it will be alternately negative and positive, and they will approximate more slowly than where the indeterminate quantity cannot increase beyond a given magnitude. Thus, if a circular area or arc is required, it is better to use the right sine, which cannot exceed the radius, than the tangent, which quickly increases to an immense length, as *Newton* previously observed. But on the other hand, tangents are to be preferred in the case of the hyperbola, in as much as they cannot exceed a given magnitude, but are contained between prescribed limits, and so are sufficiently confined. But what we have said here does not in the least mean that an area or arc of moderate or insignificant magnitude cannot be sought out just as seems appropriate to any one of them: for the difference is only noteworthy in those cases in which the quantities sought are large. And series whose terms are alternately negative and positive are more tractable than others as far as summation is concerned. Moreover, the things that have been said here about binomial curves will also apply to those involving a larger number of terms.

It is indeed true that rapidly converging series can be encountered in many places, when *Newton's* Method of Differences has been used. But the more they converge, the more complicated they usually are: consequently I prefer simpler series even if they are more slowly convergent.

(p.176)

## *On Successive Sums*

By successive sum I understand the quantity which follows the sum of all the terms, when the subsequent terms come down in place of the preceding ones. Thus, if the sum is $T+T'+T''+T'''+T^{\text{iv}}+T^{\text{v}}+$&c., write the latter terms for the former, and you will have the successive sum $T'+T''+T'''+T^{\text{iv}}+T^{\text{v}}+$&c. and if once more the following terms are substituted for the preceding ones in this, the sum $T''+T'''+T^{\text{iv}}+T^{\text{v}}+T^{\text{vi}}+$&c. will result which follows the most recent one, and so on.

Hence, if $S$, $S'$, $S''$, $S'''$, etc. denote the successive sums, there will be

$$\begin{aligned} S &= T + T' + T'' + T''' + T^{\text{iv}} + \text{\&c.} \\ S' &= \ldots . T' + T'' + T''' + T^{\text{iv}} + \text{\&c.} \\ S'' &= \ldots\ldots\ldots T'' + T''' + T^{\text{iv}} + \text{\&c.} \\ S''' &= \ldots\ldots\ldots\ldots\ldots T''' + T^{\text{iv}} + \text{\&c.} \end{aligned}$$

That is to say, from any infinite series let the first term be taken away, and from what is left let the first term also be taken away, then from what remains again let the first term be taken away, and so on indefinitely; the series produced in this way by removing the first terms step by step will be the successive sums; that is,

$$S' = S - T, \quad S'' = S' - T', \quad S''' = S'' - T'', \quad \text{\&c.},$$

or,

$$\begin{aligned} S' &= S - T, \\ S'' &= S - T - T', \\ S''' &= S - T - T' - T'', \\ S^{\text{iv}} &= S - T - T' - T'' - T''', \quad \text{\&c.} \end{aligned}$$

Here we have been speaking about sums of all terms up to infinity, which begin at some given term; for whatever term $T$ may be, $S$ will be the sum of it and all subsequent terms, likewise $S'$ will be the sum of $T'$ and all the rest. In fact these things will hold where we are concerned with the sum of an infinite number of terms: But if there is only a finite number of terms, $S$ will be the sum of all terms from the beginning to some given term $T$, and $S'$ will be the sum of the same terms less $T$, and $S''$ will be the sum of the same terms less both $T$, $T'$, and so on.

Hence, if $z$ is the length of the abscissa which corresponds to the sum $S$ in the summation of the terms from some given term to infinity, then $z+1$, $z+2$, $z+3$, etc. will be respectively its lengths corresponding to the successive sums $S'$, $S''$, $S'''$, etc. But on the other hand, in the summation of the terms from some given term back to the beginning of the series, the lengths $z-1$, $z-2$, $z-3$, etc. will correspond to the sums $S'$, $S''$, $S'''$, etc., when the abscissa $z$ corresponds to $S$ itself. For in the first case the distances of the sums from the beginning increase constantly by the increment of the abscissa, and they decrease by the same decrement in the latter.

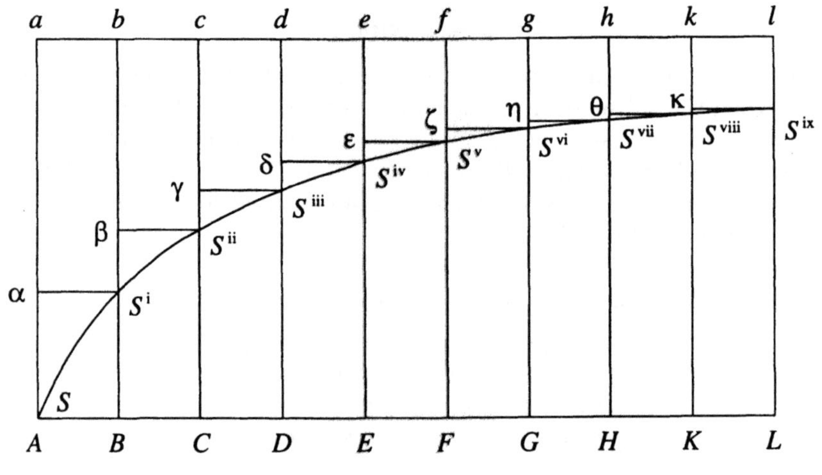


Let $SS'$ be any curve whose asymptote is $ab$ and is such that the abscissa $AB$ is parallel to it. Let the abscissa be divided into infinitely many parts all equal to each other, $AB$, $BC$, $CD$, etc. And from the points of division $A$, $B$, $C$, $D$, etc. let perpendiculars to the asymptote be erected, cutting the curve in the points $S$, $S'$, $S''$, etc. and the asymptote in $a$, $b$, $c$, etc. From the points $S'$, $S''$, $S'''$, etc. to the immediately preceding ordinates let $S'\alpha$, $S''\beta$, $S'''\gamma$, $S^{iv}\delta$, etc. be drawn parallel to the abscissa; thus $S\alpha$, $S'\beta$, $S''\gamma$, $S'''\delta$, etc. are the differences of the ordinates, both of those which extend from the curve to the asymptote and of those which extend from the curve to the abscissa. Therefore the ordinates intercepted between the curve and the asymptote will express the sums and by continuing from these to infinity the differences will represent the terms. That is to say, if $eS^{\text{iv}}$ denotes the sum, the successive ones will be $fS^{\text{v}}$, $gS^{\text{vi}}$, $hS^{\text{vii}}$, etc., and the differences of these $\epsilon S^{\text{iv}}$, $\zeta S^{\text{v}}$, $\eta S^{\text{vi}}$, etc. up to infinity are the terms whose sum is $eS^{\text{iv}}$. And similarly if $ES^{\text{iv}}$, $DS'''$, $CS''$, etc. denote successive sums, whose first is $ES^{\text{iv}}$, the preceding differences $\delta S'''$, $\gamma S''$, etc. will represent the finitely many terms continuing from the ordinate $ES^{\text{iv}}$ back to the beginning of the series. Therefore the summation of series is reduced to finding the ordinates from their given differences. But it is to be noted that the final sum must be zero in both cases; this always happens when the curve passes through the point $A$ on the abscissa, and at the same time has $ab$ as asymptote. This caution is to be applied so that the sums which have to be investigated by the methods presented are true, requiring no correction at all, as is very often the case in the quadrature of curves.

(p.176)

## Proposition 1

*If the terms of some series are formed by writing the numbers* 1, 2, 3, 4, 5, *etc. for* $z$ *in the quantity*

$$A + Bz + Cz(z-1) + Dz(z-1)(z-2) + Ez(z-1)(z-2)(z-3) + \&\text{c.},$$

*then the sum of the terms from the beginning whose number is* $z$ *will be*

$$Az + (z+1) \times \left(\tfrac{1}{2}Bz + \tfrac{1}{3}Cz(z-1) + \tfrac{1}{4}Dz(z-1)(z-2) + \tfrac{1}{5}Ez(z-1)(z-2)(z-3) + \&\text{c.}\right).$$

Here it is to be noted that the quantity $z+1$ is multiplied into the whole series which immediately follows it. Now the Proposition is demonstrated as follows. Form the sum

$$S = Az + (z+1)\left(\tfrac{1}{2}Bz + \tfrac{1}{3}Cz(z-1) + \tfrac{1}{4}Dz(z-1)(z-2) + \&\text{c.}\right)$$

or

$$S = Az + \tfrac{1}{2}B(z+1)z + \tfrac{1}{3}C(z+1)z(z-1) + \tfrac{1}{4}D(z+1)z(z-1)(z-2)(z-3) + \&\text{c.}$$

Then write the next values of the variables for the present values; that is, $S-T$ for $S$, and $z-1$ for $z$; and you will obtain

$$S - T = A(z-1) + \tfrac{1}{2}Bz(z-1) + \tfrac{1}{3}Cz(z-1)(z-2) + \tfrac{1}{4}Dz(z-1)(z-2)(z-3) + \&\text{c.}$$

Now subtract this equation from the former, and

$$T = A + Bz + Cz(z-1) + Dz(z-1)(z-2) + \&\text{c.}$$

will remain. Hence conversely, if this value of the term is given as in the Proposition, the sum will be that which has been assigned. Moreover, this sum evaluates to zero when $z$ is zero: and so the Theorem is established. Q.E.D.

### Example 1

Let the series of natural numbers 1, 2, 3, 4, 5, 6, etc. be given; these are formed by writing 1, 2, 3, etc. for $z$ in that very quantity $z$, and so by comparison with the term in the Theorem there will be $A = 0$, $B = 1$, and $C$, $D$, $E$, and the subsequent coefficients will be zero; when these values have been substituted the sum comes forth as $(z+1) \times \frac{1}{2}z$, or $\dfrac{z^2+z}{2}$, for the aggregate of as many terms as there are units in $z$. Thus, if $z = 6$, then $\frac{36+6}{2} = 21$ will come out as the sum of the first six terms.

### Example 2

Now let the series of odd numbers 1, 3, 5, 7, 9, 11, etc. be given; these are formed by writing 1, 2, 3, 4, etc. in the quantity $2z - 1$, that is, $-1 + 2z$, and having been compared with the value of the general term, this gives $A = -1$, $B = 2$, and $C$, $D$, $E$, etc. all zero; when these have been written in the sum, $-z + (z+1) \times \dfrac{2z}{2}$, or $z^2$, comes out for the aggregate of as many terms as $z$ enumerates. And so indeed is the situation in the present case, for the successive sums are the squares of the natural numbers.

### Example 3

Suppose that the series of squares is to be summed, namely 1, 4, 9, 16, 25, 36, 49, etc., which are formed by the expression $z^2$. By what is explained in the Introduction the quantity $z^2$, reduced to the form of the Theorem, becomes $z + z(z-1)$; and so $A = 0$, $B = 1$, $C = 1$, and thence the sum is $(z+1) \times \left(\dfrac{z}{2} + \dfrac{z(z-1)}{3}\right)$, that is $\dfrac{z(z+1)(2z+1)}{6}$. For example, write 7 for $z$, and you will have $\frac{7.8.15}{6} = 140$, which is the aggregate of seven terms.

### Example 4

Now let us consider the squares of the odd numbers, 1, 9, 25, 49, 81, 121, 169, etc. which are formed by writing 1, 2, 3, 4, etc. successively in the expression $1 + 4z^2 - 4z$; when this is written as $1 + 4z(z-1)$, it gives $A = 1$, $B = 0$, $C = 4$, and $D$, $E$ etc. zero: and when these have been substituted, the sum comes out as

$$z + (z+1) \times \tfrac{4}{3}z(z-1), \quad \text{or} \quad \frac{4z^3 - z}{3}.$$

## Example 5

If the cubes 1, 8, 27, 64, 125, 216, etc. are given which $z^3$ assigns, let $z^3$ be reduced to the required form $z + 3z(z-1) + z(z-1)(z-2)$; and there will be $A = 0$, $B = 1$, $C = 3$, $D = 1$, and the remaining coefficients will be zero; and consequently the sum is

$$(z+1) \times \left(\tfrac{1}{2}z + z(z-1) + \tfrac{1}{4}z(z-1)(z-2)\right),$$

which after simplification becomes $\dfrac{z^2}{4} \times (z+1)^2$. And hence it follows that the sums of these cubes are the squares of the numbers 1, 3, 6, 10, 15, etc., namely of the triangular numbers.

## Scholion

Series of this type are more easily summed by the differences of the terms; for let $A$, $A_2$, $A_3$, etc. denote the series to be summed; collect together the first differences of the terms $B$, $B_2$, $B_3$, etc., the second $C$, $C_2$, $C_3$, etc., the third $D$, $D_2$, etc., and so on until the last has been reached which here is $E$:

$$\begin{array}{ccccccccc} A & & A_2 & & A_3 & & A_4 & & A_5 \\ & B & & B_2 & & B_3 & & B_4 & \\ & & C & & C_2 & & C_3 & & \\ & & & D & & D_2 & & & \\ & & & & E & & & & \end{array}$$

and the sum of the terms, $z$ in number, will be

$$A\frac{z}{1} + B\frac{z}{1} \times \frac{z-1}{2} + C\frac{z}{1} \times \frac{z-1}{2} \times \frac{z-2}{3} + D\frac{z}{1} \times \frac{z-1}{2} \times \frac{z-2}{3} \times \frac{z-3}{4} + \&c.$$

But it has to be noted that the differences have to be formed by taking the former from the latter, that is, by putting $B = A_2 - A$, $B_2 = A_3 - A_2$, etc., then $C = B_2 - B$, etc. Now the demonstration of this depends on the *Newtonian* method of differences.

Let us consider the series 1, −1, 0, 8, 27, 61, 114, 190, etc. When the differences have been collected together in the manner explained above, it will be found that $A = 1$, $B = -2$, $C = 3$, $D = 4$, while the remaining coefficients are zero:

$$\begin{array}{ccccccccccccccc} 1 & & -1 & & 0 & & 8 & & 27 & & 61 & & 114 & & 190 \\ & -2 & & 1 & & 8 & & 19 & & 34 & & 53 & & 76 & \\ & & 3 & & 7 & & 11 & & 15 & & 19 & & 23 & & \\ & & & 4 & & 4 & & 4 & & 4 & & 4 & & & \end{array}$$

and so the sum comes out as

$$\frac{z}{1} - 2 \times \frac{z}{1} \times \frac{z-1}{2} + 3 \times \frac{z}{1} \times \frac{z-1}{2} \times \frac{z-2}{3} + 4 \times \frac{z}{1} \times \frac{z-1}{2} \times \frac{z-2}{3} \times \frac{z-3}{4},$$

which in simplified form is $\frac{(z-3)(z-2)z(z+2)}{6}$. Moreover, the series is formed by writing 0, 1, 2, 3, 4, etc. in the quantity $\frac{4z^3-3z^2-13z+6}{6}$.

(p.177)

## Proposition 2

*If the terms of any series are formed by writing any numbers which differ by one in the quantity*

$$\frac{A}{z(z+1)}+\frac{B}{z(z+1)(z+2)}+\frac{C}{z(z+1)(z+2)(z+3)} + \frac{D}{z(z+1)(z+2)(z+3)(z+4)}+\&c.,$$

*the sum of all the terms beginning at any given term and continuing up to infinity will be*

$$\frac{A}{z}+\frac{B}{2z(z+1)}+\frac{C}{3z(z+1)(z+2)}+\frac{D}{4z(z+1)(z+2)(z+3)}+\&c.$$

Put the sum

$$S=\frac{A}{z}+\frac{B}{2z(z+1)}+\frac{C}{3z(z+1)(z+2)}+\frac{D}{4z(z+1)(z+2)(z+3)}+\&c.$$

Then write the values of $S$ and $z$, following for preceding, that is $S-T$ for $S$ and $z+1$ for $z$, since we are now dealing with an infinite number of terms: and

$$S-T=\frac{A}{z+1}+\frac{B}{2(z+1)(z+2)}+\frac{C}{3(z+1)(z+2)(z+3)} + \frac{D}{4(z+1)(z+2)(z+3)(z+4)}+\&c.$$

will result. When this equation has been subtracted from the previous one, there remains

$$T=\frac{A}{z(z+1)}+\frac{B}{z(z+1)(z+2)}+\frac{C}{z(z+1)(z+2)(z+3)} + \frac{D}{z(z+1)(z+2)(z+3)(z+4)}+\&c.$$

Hence conversely, if this term is given, the sum will be what is assigned in the Proposition. Q.E.D.

*Corollary* 1. If the term is $\frac{p}{z(z+1)(z+2)(z+3)(z+4)\&c.}$, throw away the last factor, then divide what is left by the number of factors which are left behind, and you will have the sum of the terms. Let the term be $\frac{A}{z(z+1)}$; throw away the last factor $z+1$, and $\frac{A}{z}$ will remain; and since there is a single factor $z$ left, $\frac{A}{z}$ will be the sum of all the terms.

Now let the term consist of three factors $\frac{B}{z(z+1)(z+2)}$; throw away the last factor $z+2$, and $\frac{B}{z(z+1)}$ will remain, which on division by 2, namely the number of factors which are left behind, will produce $\frac{B}{2z(z+1)}$ for the sum.

Similarly, if from the term $\frac{C}{z(z+1)(z+2)(z+3)}$, which is made up of four factors, the last factor $z+3$ is thrown away, and what is left is divided by 3, the sum will be obtained as $\frac{C}{3z(z+1)(z+2)}$.

If the term is $\frac{A}{z}$, throw away the factor $z$, and since nothing remains, divide $A$ by zero, and you will have for the sum an infinitely large quantity, as is known. And as far as I know it was Mr *Taylor* who first dealt with this matter in the Method of Increments. The same topic was also discussed in greater detail and most elegantly by M. *Nicol* in the Acts of the Royal Academy of Paris.

*Corollary* 2. By the things which are presented in the Introduction concerning this material, it is known that any term $\frac{A}{z(z+a)(z+b)(z+c)\&c.}$ can always be resolved into two or perhaps more summable terms, finite in number, when $a$, $b$, $c$, etc. are whole numbers; therefore in that case the series will be summable. Thus, if the term is $\frac{1}{z(z+3)}$, it is resolved into three summable terms

$$\frac{1}{z(z+1)} - \frac{2}{z(z+1)(z+2)} + \frac{2}{z(z+1)(z+2)(z+3)}.$$

Hence by the preceding Corollary, the sum will be

$$\frac{1}{z} - \frac{1}{z(z+1)} + \frac{2}{3z(z+1)(z+2)},$$

which when combined are $\frac{3z^2+6z+2}{3z^3+9z^2+6z}$. And likewise, if the term is of this form

$$\frac{A+Bz^2+Cz^3+Dz^4+\&c.}{z(z+a)(z+b)(z+c)(z+d)\&c.},$$

the series will be summable as long as $a$, $b$, $c$, $d$, etc. are whole numbers and the number of factors in the denominator exceeds the highest power of $z$ in the numerator by at least two. But I except the cases in which two or more factors in the denominator are equal to each other; in those cases the series are not summable.

### Example 1

Suppose that the series

$$\frac{1}{1.4.7}+\frac{1}{4.7.10}+\frac{1}{7.10.13}+\frac{1}{10.13.16}+\frac{1}{13.16.19}+\&c.$$

is to be summed. The terms of this series are assigned by the quantity

$$\frac{1}{3z(3z+3)(3z+6)},$$

as will be clear on writing $\frac{1}{3}$, $1\frac{1}{3}$, $2\frac{1}{3}$, $3\frac{1}{3}$, etc. successively for $z$, that is,

$$\frac{1}{27z(z+1)(z+2)};$$

hence the sum is $\dfrac{1}{54z(z+1)}$. For instance, on writing for $z$ its first value $\frac{1}{3}$ in this, $\frac{1}{24}$ will result for the sum of the whole series. If for $z$ its second value $1\frac{1}{3}$ is written, $\frac{1}{168}$ will result for the sum of the whole series less the first term. If for $z$ its third value $2\frac{1}{3}$ is written, $\frac{1}{420}$ will result for the sum of the whole series less the first two terms. And so on to infinity.

### Example 2

Let us consider the series

$$\frac{1}{1.4}+\frac{1}{2.5}+\frac{1}{3.6}+\frac{1}{4.7}+\frac{1}{5.8}+\&c.$$

The terms of this are assigned by the quantity $\dfrac{1}{z(z+3)}$, in which 1, 2, 3, 4, etc. are to be written successively for $z$. Now the quantity $\dfrac{1}{z(z+3)}$ is reduced to three summable terms, namely

$$\frac{1}{z(z+1)}-\frac{2}{z(z+1)(z+2)}+\frac{2}{z(z+1)(z+2)(z+3)}.$$

Hence the sum will be

$$\frac{1}{z}-\frac{1}{z(z+1)}+\frac{2}{3z(z+1)(z+2)} \quad \text{or} \quad \frac{3z^2+6z+2}{3z(z+1)(z+2)}.$$

Now if you wish the aggregate of every one of the terms, substitute one for $z$ in the sum, and you will obtain $\frac{3+6+2}{3.1.2.3}$, that is $\frac{11}{18}$, for the value of the series set forth.

### Example 3

Now let the series

$$\frac{1}{2.3.4.5}+\frac{4}{3.4.5.6}+\frac{9}{4.5.6.7}+\frac{16}{5.6.7.8}+\frac{25}{6.7.8.9}+\&c.$$

be given, where the numerators are the squares of the natural numbers; in general each term will be assigned by the expression $\dfrac{z^2-2z+1}{z(z+1)(z+2)(z+3)}$, the successive values of the indeterminate being 2, 3, 4, 5, etc. And that quantity is resolved into three summable terms, namely

$$\frac{1}{z(z+1)}-\frac{7}{z(z+1)(z+2)}+\frac{16}{z(z+1)(z+2)(z+3)};$$

and thence the sum is

$$\frac{1}{z}-\frac{7}{2z(z+1)}+\frac{16}{3z(z+1)(z+2)}, \quad \text{that is,} \quad \frac{6z^2-3z+2}{6z(z+1)(z+2)},$$

and if you substitute 2 for $z$ in this, you will have $\frac{5}{36}$ for the value of the series.

### Example 4

Let the value of the series

$$\frac{1}{1.2.3.4.5}+\frac{27}{2.3.4.5.6}+\frac{125}{3.4.5.6.7}+\frac{343}{4.5.6.7.8}+\&c.$$

be required, where the numerators are the cubes of the odd numbers 1, 3, 5, 7, etc. Now if 1, 2, 3, 4, 5 etc. are put for the successive values of $z$, the terms will be assigned by the expression $\dfrac{8z^3-12z^2+6z-1}{z(z+1)(z+2)(z+3)(z+4)}$, which is resolved into

$$\frac{8}{z(z+1)}-\frac{84}{z(z+1)(z+2)}+\frac{386}{z(z+1)(z+2)(z+3)} \\ -\frac{729}{z(z+1)(z+2)(z+3)(z+4)}.$$

Therefore the sum is

$$\frac{8}{z}-\frac{84}{2z(z+1)}+\frac{386}{3z(z+1)(z+2)}-\frac{729}{4z(z+1)(z+2)(z+3)},$$

that is, $\dfrac{96z^3+72z^2+80z-3}{12z(z+1)(z+2)(z+3)}$, and if you write the first value of $z$, that is one for $z$, you will find $\frac{245}{288}$ to be the value of the series.

**Example 5**

Let us consider the series

$$\frac{1}{1.2}+\frac{1}{3.4}+\frac{1}{5.6}+\frac{1}{7.8}+\frac{1}{9.10}+\&\text{c.},$$

which Viscount *Brouncker* found for the quadrature of the hyperbola; in general each term is assigned by the expression $\dfrac{1}{4z(z+\frac{1}{2})}$, where the values of $z$ are $\frac{1}{2}$, $1\frac{1}{2}$, $2\frac{1}{2}$, $3\frac{1}{2}$, etc. And having been reduced to summable form, the quantity becomes

$$\frac{1}{4z(z+1)}+\frac{1}{8z(z+1)(z+2)}+\frac{1.3}{16z(z+1)(z+2)(z+3)}$$
$$+\frac{1.3.5}{32z(z+1)(z+2)(z+3)(z+4)}+\&\text{c.}$$

Indeed it expands into an infinite series because the difference of the factors in the expression which assigns the terms is a fraction; in any case this is an indication that the series is not summable. But going back from the term to the sum,

$$\frac{1}{4z}+\frac{1}{16z(z+1)}+\frac{1.3}{48z(z+1)(z+2)}+\frac{1.3.5}{128z(z+1)(z+2)(z+3)}+\&\text{c.}$$

will be obtained. This is a series which converges more rapidly the larger $z$ is. But for an easier calculation put

$$A=\frac{1}{4z},\ B=\frac{A}{2z+2},\ C=\frac{3B}{2z+4},\ D=\frac{5C}{2z+6},$$
$$E=\frac{7D}{2z+8},\ F=\frac{9E}{2z+10},\ \&\text{c.},$$

and the sum will be $A+\frac{1}{2}B+\frac{1}{3}C+\frac{1}{4}D+\frac{1}{5}E+$ &c. If in this for $z$ its first value $\frac{1}{2}$ is substituted, the value of the whole series which has to be summed will be obtained: if for $z$ its second value is substituted, the sum of all the terms less the first will come out; if for $z$ its third value is substituted, the sum of all terms except the first two will come out, and so on. Therefore I substitute for $z$ its fourteenth value $13\frac{1}{2}$, so that $z$ is sufficiently large to make the series converge rapidly; and I have

$$A=\tfrac{1}{54},\ B=\tfrac{1}{29}A,\ C=\tfrac{3}{31}B,\ D=\tfrac{5}{33}C,\ E=\tfrac{7}{35}D,\ F=\tfrac{9}{37}E,\ \&\text{c.},$$

in which case the sum $A+\frac{1}{2}B+\frac{1}{3}C+\frac{1}{4}D+$ &c. will be equal to

$$\frac{1}{27.28} + \frac{1}{29.30} + \frac{1}{31.32} + \&c.,$$

in fact to the whole series to be summed less its first thirteen terms. Therefore I seek the sum of these by addition, and I find it to be .674285961. Then, in order to obtain the sum of the rest, I extract by calculation $A$, $B$, $C$, $D$, etc. to as many places of decimals as is desired; and when these have been found I divide them by 1, 2, 3, 4, 5, etc. respectively as you see:

| | | |
|---|---|---|
| $A =$ | .018518519 | .018518519 |
| $B =$ | 638570 | 319285 |
| $C =$ | 61797 | 20599 |
| $D =$ | 9363 | 2341 |
| $E =$ | 1873 | 375 |
| $F =$ | 455 | 76 |
| $G =$ | 128 | 18 |
| $H =$ | 41 | 5 |
| $I =$ | 14 | 1 |
| | | .018861219 |

And in that way I obtain .018861219 for the sum of all the terms after the thirteenth; finally, when this has been added to the aggregate of the initial terms found first of all, it makes up .693147180 for the value of the series which had to be summed, that is, for the hyperbolic logarithm of two.

The more terms are collected together at the beginning, the more rapidly the series which gives the sum of the remaining terms will converge on account of $z$ being so much the greater. And the superiority of this method is especially conspicuous in the fact that, by adding terms to the aggregate of the initial terms, $z$ is increased by so many units and as a rule the series transformed at will in this way will converge.

Now it will be clear from the following calculation that in practice it is impossible to come up with the sums of these series by straightforward collection of the terms; here the sum of a hundred, a thousand, ten thousand, and so on up to ten thousand hundred-thousands of terms are given:

| The sum of | | terms is | |
|---|---|---|---|
| | 100 | | .690653446 |
| | 1000 | | .692897242 |
| | 10000 | | .693122181 |
| | 100000 | | .693144680 |
| | 1000000 | | .693146930 |
| | 10000000 | | .693147155 |
| | 100000000 | | .693147178 |
| | 1000000000 | | .693147180 |

From this calculation it appears that a hundred terms give the sum accurate to two figures; and by collecting together at each stage ten times the previous number of terms, only one figure more or less is gained: thus, if someone should wish to work out the value of this series accurate to nine places of figures using no technique other than addition, about ten thousand hundred-thousands of terms would be required. And this series converges far more rapidly than very many others whose values are finite quantities.

**Example 6**

Suppose that the series $1 + \frac{1}{4} + \frac{1}{9} + \frac{1}{16} + \frac{1}{25} + \frac{1}{36} + \&c.$ is to be summed, where the denominators are the squares of the numbers 1, 2, 3, 4, etc. and the term is in general $\frac{1}{z^2}$: reduced to summable form, this becomes

$$\frac{1}{z(z+1)} + \frac{1}{z(z+1)(z+2)} + \frac{1.2}{z(z+1)(z+2)(z+3)} + \frac{1.2.3}{z(z+1)(z+2)(z+3)(z+4)} + \&c.$$

Therefore the sum is equal to

$$\frac{1}{z} + \frac{1}{2z(z+1)} + \frac{1.2}{3z(z+1)(z+2)} + \frac{1.2.3}{4z(z+1)(z+2)(z+3)} + \&c.$$

By putting

$$A = \frac{1}{z},\ B = \frac{A}{z+1},\ C = \frac{2B}{z+2},\ D = \frac{3C}{z+3},\ E = \frac{4D}{z+4},\ \&c.$$

this becomes $A + \frac{1}{2}B + \frac{1}{3}C + \frac{1}{4}D + \&c.$ Now if for $z$ its thirteenth value 13 is substituted, the sum of all terms in the series to be summed after the twelfth term will be obtained; in this case there will be

$$A = \tfrac{1}{13},\ B = \tfrac{1}{14}A,\ C = \tfrac{2}{15}B,\ D = \tfrac{3}{16}C,\ E = \tfrac{4}{17}D,\ \&c.$$

And $A + \frac{1}{2}B + \frac{1}{3}C + \frac{1}{4}D + \&c.$ will be the sum of the terms

$$\frac{1}{169} + \frac{1}{196} + \frac{1}{225} + \frac{1}{256} + \&c.$$

Now the calculation is as follows:

| | | |
|---|---|---|
| $A =$ | .076923077 | .076923077 |
| $B =$ | 5494505 | 2747252 |
| $C =$ | 732601 | 244200 |
| $D =$ | 137363 | 34341 |
| $E =$ | 32321 | 6464 |
| $F =$ | 8978 | 1496 |
| $G =$ | 2835 | 405 |
| $H =$ | 992 | 124 |
| $I =$ | 378 | 42 |
| $K =$ | 155 | 16 |
| $L =$ | 67 | 6 |
| $M =$ | 31 | 3 |
| $N =$ | 15 | 1 |
| | | .079957427 |

Hence .079957427 comes out for the sum of the terms $\frac{1}{169}+\frac{1}{196}+\frac{1}{225}+\&c.$ which when added to the aggregate of the twelve initial terms, or 1.564976638, produces 1.644934065 for the value of the whole series $1+\frac{1}{4}+\frac{1}{9}+\frac{1}{16}+\&c.$ Now this series converges less rapidly than that of *Brouncker* in the previous example.

(p.182)

## Proposition 3

*If the terms of any series are formed by writing any numbers which differ by one for $z$ in the quantity*

$$x^{z+n}\times\left(\frac{a}{z}+\frac{b}{z(z+1)}+\frac{c}{z(z+1)(z+2)}+\frac{d}{z(z+1)(z+2)(z+3)}+\&c.\right),$$

*the sum will be equal to*

$$x^{z+n}\times\left(\frac{a}{(1-x)z}+\frac{b-Ax}{(1-x)z(z+1)}+\frac{c-2Bx}{(1-x)z(z+1)(z+2)}\right.$$
$$\left.+\frac{d-3Cx}{(1-x)z(z+1)(z+2)(z+3)}+\&c.\right).$$

The quantities $A$, $B$, $C$, $D$, etc. denote the coefficients of the terms preceding those in which they are found; thus,

$$A=\frac{a}{1-x},\quad B=\frac{b-Ax}{1-x},\quad C=\frac{c-2Bx}{1-x},\quad \&c.$$

But I except the case in which $x$ is equal to one: where this happens, the series will be summable by the previous proposition. Now here is the demonstration. Form the sum

$$S=x^{z+n}\times\left(\frac{A}{z}+\frac{B}{z(z+1)}+\frac{C}{z(z+1)(z+2)}\right.$$
$$\left.+\frac{D}{z(z+1)(z+2)(z+3)}+\&c.\right),$$

then write the next values of the variables $S-T$ and $z+1$ for their predecessors $S$ and $z$ respectively, and you will have

$$S-T=x^{z+n+1}\times\left(\frac{A}{z+1}+\frac{B}{(z+1)(z+2)}+\frac{C}{(z+1)(z+2)(z+3)}\right.$$
$$\left.+\frac{D}{(z+1)(z+2)(z+3)(z+4)}+\&c.\right).$$

This is

$$S - T = x^{z+n} \times \left(\frac{Ax}{z+1} + \frac{Bx}{(z+1)(z+2)} + \frac{Cx}{(z+1)(z+2)(z+3)} + \frac{Dx}{(z+1)(z+2)(z+3)(z+4)} + \&c.\right),$$

which, when reduced to the form of $S$, becomes

$$S - T = x^{z+n} \times \left(\frac{Ax}{z} + \frac{Bx - Ax}{z(z+1)} + \frac{Cx - 2Bx}{z(z+1)(z+2)} + \frac{Dx - 3Cx}{z(z+1)(z+2)(z+3)} + \&c.\right).$$

Now subtract the value of $S - T$ from the value of $S$ and the term

$$T = x^{z+n} \times \left(\frac{A(1-x)}{z} + \frac{B(1-x) + Ax}{z(z+1)} + \frac{C(1-x) + 2Bx}{z(z+1)(z+2)} + \frac{D(1-x) + 3Cx}{z(z+1)(z+2)(z+3)} + \&c.\right)$$

will be left. Finally, if this value is compared with that in the proposition, it gives

$$A(1-x) = a, \quad B(1-x) + Ax = b, \quad C(1-x) + 2Bx = c, \quad \text{and so on.}$$

These equations show the values of the coefficients as above. Thus the value of the sum is correctly assigned. Q.E.D.

**Example 1**

Suppose that the series

$$1 + \tfrac{1}{3}t + \tfrac{1}{5}t^2 + \tfrac{1}{7}t^3 + \tfrac{1}{9}t^4 + \&c.$$

has to be summed. The equation for this is $T = t^{z-\frac{1}{2}} \times \frac{\frac{1}{2}}{z}$, for the terms of the series will come out if we write $\frac{1}{2}$, $1\frac{1}{2}$, $2\frac{1}{2}$, $3\frac{1}{2}$, etc. successively for $z$. Now by comparing this term with that in the theorem, there will be $x = t$, $n = -\frac{1}{2}$, $a = \frac{1}{2}$, while $b$, $c$, $d$, $e$, and the remaining coefficients are zero. And finally, when these values have been written in, there arises

$$S = t^{z-\frac{1}{2}} \times \left(\frac{\frac{1}{2}}{(1-t)z} + \frac{At}{(t-1)z(z+1)} + \frac{2Bt}{(t-1)z(z+1)(z+2)} + \frac{3Ct}{(t-1)z(z+1)(z+2)(z+3)} + \&c.\right).$$

Let us suppose, for example, that $t = -1$; and the series to be summed will be

$$1 - \frac{1}{3} + \frac{1}{5} - \frac{1}{7} + \frac{1}{9} - \&c.,$$

and the sum will turn out to be

$$S = \pm 1 \times \left( \frac{1}{4z} + \frac{A}{2z(z+1)} + \frac{2B}{2z(z+1)(z+2)} + \frac{3C}{2z(z+1)(z+2)(z+3)} + \&c. \right)$$

or

$$S = \pm 1 \times \left( \frac{1}{4z} + \frac{A}{2z+2} + \frac{2B}{2z+4} + \frac{3C}{2z+6} + \frac{4D}{2z+8} + \&c. \right),$$

where $A$, $B$, $C$, $D$, etc. now denote whole terms in the *Newtonian* manner, and no longer just coefficients. And one with the ambiguous sign by which the whole series is multiplied will be positive where $z - \frac{1}{2}$ is an even number, and negative where it is odd. Now collect together twelve initial terms, or equivalently six in this series

$$\frac{2}{1.3} + \frac{2}{5.7} + \frac{2}{9.11} + \&c.,$$

each pair in the former series having been combined: and you will find their sum to be .7646006915. Then write for $z$ its thirteenth value $12\frac{1}{2}$, and you will obtain

$$S = \tfrac{1}{50} + \tfrac{1}{27}A + \tfrac{2}{29}B + \tfrac{3}{31}C + \tfrac{4}{33}D + \tfrac{5}{35}E + \&c.,$$

which is a simple and rapidly converging series: for ten terms give $S = .0207974719$ as is clear from the adjacent calculation: and when this quantity has been added to the aggregate of the initial terms, .7853981634 will result for the value of the series to be summed: however it would never be possible to attain this by addition of terms. And by collecting together more initial terms, the value of $S$ will approximate far more rapidly. And so, by means of this proposition the circumference of the circle can be produced with very little effort to a great many figures from this series, albeit it a slowly converging one; some time ago *Leibniz* was very desirous of such a result.

| |
|---:|
| .0200000000 |
| 7407407 |
| 510856 |
| 49438 |
| 5992 |
| 856 |
| 139 |
| 25 |
| 5 |
| 1 |
| $S = .0207974719$ |

The circumference of the circle will also be obtained very accurately by means of the following series of *Newton*

$$1 + \frac{1}{3} - \frac{1}{5} - \frac{1}{7} + \frac{1}{9} + \frac{1}{11} - \frac{1}{13} - \frac{1}{15} + \&c.,$$

where each pair of terms are alternately negative and positive. The same is also achieved by this series

$$1+\frac{1}{2}-\frac{1}{4}-\frac{1}{5}+\frac{1}{7}+\frac{1}{8}-\frac{1}{10}-\frac{1}{11}+\&\text{c.},$$

in which the denominators form the progression of natural numbers with every third one removed. The former is equal to a fourth and the latter to a third part of the whole circumference if it is assumed that the chords of these arcs are one. However, before they can be treated by this proposition, both have to be separated into two, the former into

$$1-\frac{1}{5}+\frac{1}{9}-\frac{1}{13}+\frac{1}{17}-\frac{1}{21}+\&\text{c.},$$
$$\frac{1}{3}-\frac{1}{7}+\frac{1}{11}-\frac{1}{15}+\frac{1}{19}-\frac{1}{23}+\&\text{c.},$$

and the latter into

$$1-\frac{1}{4}+\frac{1}{7}-\frac{1}{10}+\frac{1}{13}-\frac{1}{16}+\&\text{c.},$$
$$\frac{1}{2}-\frac{1}{5}+\frac{1}{8}-\frac{1}{11}+\frac{1}{14}-\frac{1}{17}+\&\text{c.}$$

Then each of these four series is to be considered separately, and the operation is to be set up as in the above example.

## Example 2

If the series is

$$\frac{x}{1.2}+\frac{x^2}{3.4}+\frac{x^3}{5.6}+\frac{x^4}{7.8}+\frac{x^5}{9.10}+\&\text{c.},$$

the equation will be $T=x^z\times\dfrac{1}{4z(z-\frac{1}{2})}$, which, when resolved into a series, becomes

$$T=x^z\times\left(\frac{1}{4z(z+1)}+\frac{3}{8z(z+1)(z+2)}+\frac{15}{16z(z+1)(z+2)(z+3)}\right.$$
$$\left.+\frac{105}{32z(z+1)(z+2)(z+3)(z+4)}+\&\text{c.}\right).$$

Hence on comparing the members, there will be

$$n=0,\quad a=\tfrac{1}{4},\quad b=\tfrac{3}{8},\quad c=\tfrac{15}{16},\quad d=\tfrac{105}{32},\quad e=\tfrac{945}{64},\quad \&\text{c.}$$

And so

$$S = x^z \times \left( \frac{1}{4(1-x)z} + \frac{3-8Ax}{8(1-x)z(z+1)} + \frac{15-32Bx}{16(1-x)z(z+1)(z+2)} \right.$$

$$\left. + \frac{105-96Cx}{32(1-x)z(z+1)(z+2)(z+3)} + \&c. \right).$$

The progression of this series is clear to anyone. And where the value of $x$ is given in any particular case, the sum will be given as accurately as desired; namely, by first adding a sufficient number of initial terms with the purpose that $z$ may be sufficiently big to cause the value of $S$ to converge rapidly. And now that these things have been set forth about series whose terms are assignable, it is appropriate to move on to those which are determined by the relation of the terms.

(p.187)

## Proposition 4

*To find the relation between the terms when that between the successive sums has been given.*

In the equation defining the relation between the sums substitute for $S'$, $S''$, $S'''$, etc. their actual values $S-T$, $S-T-T'$, $S-T-T'-T''$, etc. and you will have an equation involving the one sum $S$; in this write the following values of the variables for the preceding ones, and you will have a new equation involving that sum $S$: finally, let $S$ be eliminated by means of these equations, and what results will show the relation of the terms. Q.E.I.

### Example 1

Let the equation for the sums be $(z-n)S = (z-1)S'$; substitute for $S'$ its value $S-T$, and the equation will become $(n-1)S = (z-1)T$; in this write the following values of the variables for the preceding ones, that is, $S-T$ for $S$, $T'$ for $T$, and $z+1$ for $z$; and $(n-1)S = (n-1)T + zT'$ will result; subtract this from the first equation $(n-1)S = (z-1)T$, and there will remain $(z-n)T = zT'$, which is the equation for the terms of the series.

### Example 2

Consider the equation for the sums $S \times (8z^2+20z+9) + 3S' \times (8z^2+4z-3) = 0$; substitute $S-T$ for $S'$, and you will find

$$S = \frac{3}{32}T \times \frac{8z^2+4z-3}{z^2+z};$$

then, in accordance with the method of differences, write $S-T$ for $S$, $T'$ for $T$, and $z+1$ for $z$, and

$$S-T = \frac{3}{32}T' \times \frac{8z^2+20z+9}{z^2+3z+2}$$

will be produced; when $S$ has been eliminated by means of these equations, $(z+2)T + 3T'z = 0$ will be obtained, which is the equation showing the relation of the terms.

And by the same method three or more successive sums can be eliminated.

(p.187)

## Proposition 5

*To find as many summable series as you wish.*

The equation for the sums will give the sum of the terms, while that for the terms will give the series; the former is taken at will and from it the latter is deduced by the previous proposition: therefore the terms and their sum are obtained. Q.E.I.

### Example 1

Let the equation for the sums be $(z-n)S = (z-1)S'$ as in the first example of the preceding proposition; you will find that for the terms to be $(z-n)T = zT'$. But on substituting $S-T$ for $S'$, the equation for the sums will give the sum $S = \frac{z-1}{n-1}T$. Now let $A$, $B$, $C$, $D$, etc. denote the terms of this series, and in the equation for them write $m$, $m+1$, $m+2$, $m+3$, etc. successively for $z$, where $m$ is any number, integer or fraction, negative or positive; and the relations of the terms will come out as

$$B = \frac{m-n}{m}A, \quad C = \frac{m-n+1}{m+1}B, \quad D = \frac{m-n+2}{m+2}C, \quad E = \frac{m-n+3}{m+3}D,$$

&c.

Then in the equation $S = \frac{z-1}{n-1}T$ write the first term of the series, that is $A$, for $T$, and the first value of $z$, that is $m$, for $z$, and you will find

$$S = \frac{m-1}{n-1}A = A + \frac{m-n}{m}A + \frac{m-n+1}{m+1}B + \frac{m-n+2}{m+2}C + \&c.,$$

where any numbers may be substituted for $m$ and $n$. For example, let $m = 5$, $n = 2$, $A = \frac{1}{12}$, and

$$\tfrac{1}{3} = \tfrac{1}{12} + \tfrac{3}{5}A + \tfrac{4}{6}B + \tfrac{5}{7}C + \tfrac{6}{8}D + \&c.$$

will result, that is

$$\frac{1}{3} = \frac{1}{3.4} + \frac{1}{4.5} + \frac{1}{5.6} + \frac{1}{6.7} + \frac{1}{7.8} + \&c..$$

Here it is clear that the terms are assignable, which will always happen when $n$ is an integer: indeed there will be as many factors in the denominators as there are units in $n$. Thus, in the present example $n = 2$, and because of that there are two factors in the denominators of the terms.

Now let $m = 2$, $n = \frac{3}{2}$, $A = 1$, and

$$S = 2 = 1 + \tfrac{1}{4}A + \tfrac{3}{6}B + \tfrac{5}{8}C + \tfrac{7}{10}D + \tfrac{9}{12}E + \&c.$$

will result, that is,

$$S = 2 = 1 + \tfrac{1}{4} + \tfrac{1}{8} + \tfrac{5}{64} + \tfrac{7}{128} + \tfrac{21}{512} + \&c.$$

But in this case it is known that the terms are not assignable, since $n$ is a fraction. It should be noted that the series terminates and has a finite number of terms whenever $m - n$ is zero or a negative integer. And if $n - 1$ is zero or a negative number, the value of the series will be infinitely great, as is shown by the value of the sum, namely $\dfrac{m-1}{n-1}A$.

### Example 2

Let the equation for the sums be $S \times (8z^2 + 20z + 9) + 3S' \times (8z^2 + 4z - 3) = 0$ as in the last example of the previous proposition, where it was found that

$$S = \frac{3}{32} T \times \frac{8z^2 + 4z - 3}{z^2 + z},$$

and the relation of the terms was $(z + 2)T + 3zT' = 0$. And if one is taken for the first term, and the first value of $z$ is also set equal to one,

$$S = \tfrac{27}{64} = 1 - \tfrac{3}{3} + \tfrac{6}{9} - \tfrac{10}{27} + \tfrac{15}{81} - \tfrac{21}{243} + \&c.$$

will be obtained. Here the denominators are the powers of three, while the numerators are the triangular numbers. In these examples I have not digressed into the deduction of series from equations which define the relations of the terms, since I suppose this known already from the Introduction.

### Scholion

The summation of series in the Method of Differences corresponds to the quadrature of curves in the Method of Fluxions, and because of this similar difficulties usually arise in both; these require to be outlined here. We have said that series can be continued to infinity on both sides: for example, the series $1 + x + x^2 + x^3 + \&c.$ continued backwards is $\dfrac{1}{x} + \dfrac{1}{x^2} + \dfrac{1}{x^3} + \&c.$ And these two joined together form one series going off to infinity on both sides, namely

$$\&c. + \frac{1}{x^4} + \frac{1}{x^3} + \frac{1}{x^2} + \frac{1}{x} + 1 + x + x^2 + x^3 + x^4 + \&c.$$

For these terms are in an unbroken geometric progression, the preceding terms being to their successors as one to $x$. In the summation of this series we will find

$$\frac{1}{1-x} = 1 + x + x^2 + x^3 + \&c.$$

which is indeed true when $x$ is less than one; however, if it is greater than one, this series will be infinitely large, and the sum $\frac{1}{1-x}$ will no longer be the sum of these terms; but after the sign has been changed, it will be equal to the series going off on the other side; that is, there will be

$$\frac{-1}{1-x}, \text{ or } \frac{1}{x-1} = \frac{1}{x} + \frac{1}{x^2} + \frac{1}{x^3} + \&\text{c.}$$

But if $x$ is one, the sum will be $\frac{1}{1-1}$, and then both parts of the series will be infinitely large, being equal to one taken infinitely often.

In the same way if the equation for the series is $(z-n)T = zT'$, and the series is continued on both sides to infinity, one part will converge and the other part will diverge except where $n = 1$; and the sum $\frac{z-1}{n-1}T$ will always be equal to the convergent part of the series.

It is no different in quadratures: if $z^{-n}$ is the ordinate of a hyperbolic curve, the fluent $\frac{1}{1-n}z^{1-n}$ will represent the part of the area on one side or the other side of the ordinate according as $n$ is less than or greater than one: however, if $n$ is one, the area on both sides of the ordinate will be infinitely large, as in the hyperbola of *Apollonius.*

Indeed series very often become infinitely large as a result of divergence, even although the quantites which are being investigated by them are of finite magnitude, but in those cases, when continued on the other side, they sometimes converge and are equal to the roots sought or they differ from them by a prescribed amount. Sometimes they also diverge when continued on both sides: also more often they cannot go off to infinity on both sides on account of terms which are impossible or infinitely small.

Moreover, just as areas of curves have to be sometimes increased, sometimes decreased, by given quantities so that they come out true, so also sums found by this proposition sometimes differ from the true values, in which case they have to be corrected by the addition or subtraction of a given quantity. For instance, where the equation for the sums is such that it causes the last of them to be a quantity of finite or infinitely large magnitude, there is always need of correction: I will therefore show in the following proposition how the equation is to be taken which always causes the last one to be zero; and the sum found in this way will be true, requiring neither to be increased nor to be decreased, as has previously been pointed out.

(p.188)

## Proposition 6

*If the equation for the sums is*

$$S \times (z^{\theta} + az^{\theta-1} + bz^{\theta-2} + \&\text{c.}) = mS' \times (z^{\theta} + cz^{\theta-1} + dz^{\theta-2} + \&\text{c.}),$$

*the last of the sums will be of finite magnitude only in the case where* $m = 1$ *and at the same time* $a = c$.

For demonstrating this proposition, it has to be known that the sum $S$ can be investigated from the equation defining the relation between it and its successive value $S'$ in almost the same way as a fluxional quantity from its equation. To that end a series of the following form is to be taken for S,

$$\frac{z^n}{p^z} \times \left(A + \frac{B}{z} + \frac{C}{z^2} + \frac{D}{z^3} + \&\text{c.}\right),$$

where $n$, $p$, $A$, $B$, $C$, $D$, etc. are constant quantities. But in the present case where the sum sought is the last of all, and so at infinite distance, $z$ will also be infinitely large, being either equal to that distance or different from it by a finite quantity: for this reason the later terms of the series are infinitely smaller than the initial terms. Therefore to shorten the calculation, I throw away all terms after the first, these being of no use in this demonstration: thus I have $S = \frac{Az^n}{p^z}$; by writing $S'$ for $S$ and $z+1$ for $z$ in this, I obtain $S' = \frac{A(z+1)^n}{p^{z+1}}$. Substitute these values for $S$ and $S'$ in the equation for the sums, or what comes back to the same, in this expression $S \times (z+a) = mS' \times (z+c)$, the remaining members having been neglected for the reasons stated above; and

$$\frac{Az^n}{p^z} \times (z+a) = \frac{mA(z+1)^n}{p^{z+1}} \times (z+c)$$

will result, or by multiplying by $p^{z+1}$ and dividing by $A$

$$pz^n \times (z+a) = m \times (z+1)^n \times (z+c).$$

But by *Newton's* Theorem for expanding the binomial $(z+1)^n = z^n + nz^{n-1}$, and it is accurate since $z$ is infinitely large. Substitute this value, and the equation will become $pz^n \times (z+a) = m \times (z^n + nz^{n-1}) \times (z+c)$, which after division by $z^{n-1}$ is $pz \times (z+a) = m \times (z+n) \times (z+c)$, or

$$pz^2 + paz = mz^2 + (n+c) \times mz + mnc;$$

by comparing like terms in this, there will be $p = m$ and $pa = (n+c) \times m$, hence $n = a - c$: and thence the last sum $S$, which previously was $\frac{Az^n}{p^z}$, now becomes $\frac{Az^{a-c}}{m^z}$, where it is to be noted that the coefficient $A$ is not determined. Now set $m = 1$ and at the same time $a = c$; and the sum will become $\frac{Az^0}{1^z}$: now $z^0 = 1$ and $1^z = 1$, even if $z$ is infinitely large; and so the last of the sums is finite, for it is equal to the quantity $A$ where $m = 1$ and at the same time $a = c$: but there is no other case where $\frac{Az^{a-c}}{m^z}$ is a finite quantity when $z$ is infinitely large. Therefore the proposition is established.

## Corollary

If $m$ is less than one, the last sum will be infinitely large; and it will be infinitely small when $m$ is greater than one. And if $m$ is one, that sum will be infinitely large or infinitely small according as $a$ is greater or less than $c$. Therefore, if in the equation for the sums $m$ is greater than one, or it is equal to one and at the same time $a$ is less than $c$, the last sum will always be zero and no correction will be required.

## Example

Let the first sum be $A = 1$, the second $B = \frac{9}{8}A$, the third $C = \frac{25}{24}B$, the fourth $D = \frac{49}{48}C$, the fifth $E = \frac{81}{80}D$, etc. And the equation for these will be $S \times (z^2 + z + \frac{1}{4}) = S' \times (z^2 + z)$; when this has been compared with the general equation it gives $m = 1$, $a = 1$, $c = 1$, and thence $a - c = 0$; hence by this proposition the last of the sums, that is the product of all the numbers

$$1 \times \frac{9}{8} \times \frac{25}{24} \times \frac{49}{48} \times \frac{81}{80} \times \frac{121}{120} \times \text{ \&c. up to infinity,}$$

is a finite quantity.

In the equation for the sums substitute $S - T$ for $S'$, and you will find $S = -4T \times (z^2 + z)$ which, since it turns out negative with respect to $T$, shows the sum of the terms not from a given term up to infinity, but from a given term back to the beginning of the series.

So that it might be clearer, consider the two equations

$$Sz^2 = S' \times (z^2 - 1) \quad \text{and} \quad Sz = S'(z + 1);$$

both of these will give the same equation for the terms, namely $Tz = T'(z+2)$. However, $S = -T \times (z^2 - 1)$ is deduced from the former, and $S = T'(z + 2)$ from the latter. In the first case $S$ is the sum of the terms from the beginning up to $T$, while in the second it is the sum of $T$ and all subsequent terms up to infinity. Let the numbers 1, 2, 3, 4, etc. be written successively for $z$ in the equation for the terms, and let $\frac{1}{2}$ be used for the first term; and the series will come out as

$$\frac{1}{1.2} + \frac{1}{2.3} + \frac{1}{3.4} + \frac{1}{4.5} + \frac{1}{5.6} + \frac{1}{6.7} + \frac{1}{7.8} + \text{\&c.}$$

Here, if the aggregate of the first four terms is sought, write 5 for $z$ in $T \times (z^2 - 1)$, the former value of the sum $S$, and the fifth term $\frac{1}{5.6}$ for $T$, and you will obtain

$$\frac{25 - 1}{5.6} = \frac{4}{5} = \frac{1}{1.2} + \frac{1}{2.3} + \frac{1}{3.4} + \frac{1}{4.5},$$

the four initial terms. But if you wish the sum of the whole collection of terms with the exception of those four, write 5 for $z$ and $\frac{1}{5.6}$ for $T$ in $T(z + 1)$, the latter value of $S$, and you will have

$$\frac{6}{5.6} = \frac{1}{5} = \frac{1}{5.6} + \frac{1}{6.7} + \frac{1}{7.8} + \frac{1}{8.9} + \&c.$$

And these two values of $S$ added together, that is, the sum of the numbers $\frac{4}{5}$ and $\frac{1}{5}$, produces $\frac{4+1}{5} = 1$ for the value of the whole collection of terms from the beginning up to infinity.

(p.191)

## Proposition 7

*If the equation for the series is* $(z-n)T + (m-1)zT' = 0$, *there will be*

$$S = \frac{m-1}{m}T + \frac{n}{z} \times \frac{A}{m} + \frac{n+1}{z+1} \times \frac{B}{m} + \frac{n+2}{z+2} \times \frac{C}{m} + \frac{n+3}{z+3} \times \frac{D}{m} + \frac{n+4}{z+4} \times \frac{E}{m} + \&c.$$

Suppose that the sum $S$ is equal to the term $T$ multiplied by the quantity $y$, that is $S = Ty$; then write the next values of the indeterminates $S-T$, $T'$, $y'$ for the previous ones $S$, $T$, and $y$ respectively, and $S - T = T'y'$ will be obtained; when this has been subtracted from the previous expression $S = Ty$, there remains $T = Ty - T'y'$, from which $T' = T \times \dfrac{y-1}{y'}$. But from the equation for the series, namely $(z-n)T + (m-1)zT' = 0$, there is $T' = -\dfrac{(z-n)T}{z(m-1)}$. And so by equating the two values of $T'$ to each other, there will be

$$T \times \frac{y-1}{y'} = -\frac{(z-n)T}{z(m-1)};$$

when this has been divided by $T$ and multiplied by $(m-1)y'$, it becomes $(m-1)y - m + 1 = -y' + \dfrac{n}{z}y'$, or

$$(m-1)y + y' - \frac{n}{z}y' - m + 1 = 0;$$

this is a difference equation whose resolution will produce the root $y$. To that end take

$$y = a + \frac{b}{z} + \frac{c}{z(z+1)} + \frac{d}{z(z+1)(z+2)} + \frac{e}{z(z+1)(z+2)(z+3)} + \&c.$$

Then write for $y$ and $z$ their next values $y'$ and $z+1$ respectively, and you will have

$$y' = a + \frac{b}{z+1} + \frac{c}{(z+1)(z+2)} + \frac{d}{(z+1)(z+2)(z+3)} + \frac{e}{(z+1)(z+2)(z+3)(z+4)} + \&c.,$$

that is,

$$y' = a + \frac{b}{z} + \frac{c-b}{z(z+1)} + \frac{d-2c}{z(z+1)(z+2)} + \frac{e-3d}{z(z+1)(z+2)(z+3)} + \&c.$$

And after multiplication by $\frac{n}{z}$ the first value of $y'$ becomes

$$\frac{n}{z}y' = \frac{na}{z} + \frac{nb}{z(z+1)} + \frac{nc}{z(z+1)(z+2)} + \frac{nd}{z(z+1)(z+2)(z+3)} + \&c.$$

Substitute into the equation these values which have now been reduced to the same form, and

$$ma - m + 1 + \frac{mb - na}{z} + \frac{mc - (n+1)b}{z(z+1)} + \frac{md - (n+2)c}{z(z+1)(z+2)} + \&c. = 0$$

will result. On setting like members in this equal to zero,

$$a = \frac{m-1}{m}, \quad b = \frac{n}{m}a, \quad c = \frac{n+1}{m}b, \quad d = \frac{n+2}{m}c, \quad e = \frac{n+3}{m}d, \quad \&c.$$

will be obtained. And these things having been given, the value of the root $y$ will be given, which, when multiplied finally by $T$, will produce for $S$ the series stated in the proposition. Q.E.D.

*Corollary.* If $n$ is a negative integer or zero, the value of $S$ will terminate, the series being summable. And where $m$ is negative, the series will be infinitely large. But here I except the case in which $m = 0$, for then the series will be summable by Example 1 of Proposition 5.

### Example 1

Let the series

$$\tfrac{1}{2} + \tfrac{1}{3}A + \tfrac{2}{5}B + \tfrac{3}{7}C + \tfrac{4}{9}D + \tfrac{5}{11}E + \&c.$$

be proposed for summation. The equation defining it is $(z - \frac{1}{2})T - 2zT' = 0$, where the successive values of $z$ are $1\frac{1}{2}$, $2\frac{1}{2}$, $3\frac{1}{2}$, $4\frac{1}{2}$, etc., and when that has been compared with the general equation it gives $n = \frac{1}{2}$, $m - 1 = -2$, or $m = -1$; when these have been substituted,

$$S = 2T - \frac{A}{2z} - \frac{3B}{2z+2} - \frac{5C}{2z+4} - \frac{7D}{2z+6} - \frac{9E}{2z+8} - \&c.$$

arises. If any term is substituted in this for $T$ along with the corresponding value for $z$, then $S$ will be the sum of $T$ and of all the following terms up to infinity. And so I collect twelve initial terms; and their aggregate comes out as .78533961813. Then, in order that I may obtain the sum of the remaining terms, I write the thirteenth term, that is .00003029411 for $T$ and for $z$ its required value $13\frac{1}{2}$; and I have

$$S = .00006058822 - \tfrac{1}{27}A - \tfrac{3}{29}B - \tfrac{5}{31}C - \tfrac{7}{33}D - \tfrac{9}{35}E - \&c.$$

Here the terms come out alternately negative and positive; the latter I put in the first column and the former in the second as you see:

| | |
|---:|---:|
| .00006058822 | .00000224401 |
| 23214 | 3744 |
| 794 | 204 |
| 61 | 20 |
| 7 | 3 |
| 1 | 1 |
| + .00006082899 | − .00000228373 |

Then, taking the sum of the negative terms .00000228373 from that of the positive terms .00006082899, I have $S = .00005854526$, which added to the sum of the initial terms produces .78539816339 for the value of the series under consideration, that is, for the area of the circle whose diameter is one.

## Example 2

Let the value of the series

$$1 - \tfrac{1}{2}A - \tfrac{3}{4}B - \tfrac{5}{6}C - \tfrac{7}{8}D - \tfrac{9}{10}E - \&\text{c.}$$

be sought. The equation defining the relation of the terms is $(z - \frac{1}{2})T + zT' = 0$, in which the values of the abscissa $z$ are 1, 2, 3, 4, etc. and on comparing this equation with that in the theorem, there will be $n = \frac{1}{2}$, $m - 1 = 1$, and $m = 2$: and thence

$$S = \frac{1}{2}T + \frac{A}{4z} + \frac{3B}{4z+4} + \frac{5C}{4z+8} + \frac{7D}{4z+12} + \frac{9E}{4z+16} + \&\text{c.}$$

Now collect together ten initial terms of the series to be transformed, and you will find their sum to be .6168670654. Then, in order that the sum of the remaining terms may be obtained, write in the value of $S$ the eleventh term, that is, .1761970520 for $T$, and 11 for $z$; and

$$S = .0880985260$$
$$+ \tfrac{1}{44}A + \tfrac{3}{48}B + \tfrac{5}{52}C + \tfrac{7}{56}D + \tfrac{9}{60}E + \&\text{c.}$$

will come out.

And by performing the calculation as at the side, it will be found that $S = .0902397156$, which when added to the aggregate of the initial terms first found produces .7071067810 for the value of the series, that is for $\sqrt{\frac{1}{2}}$. For if $\sqrt{\frac{1}{2}}$ is written as $(1 + 1)^{-1/2}$ and expanded by *Newton's* Theorem, the series which we have just been discussing comes out.

| |
|---:|
| .0880985260 |
| 20022392 |
| 1251400 |
| 120327 |
| 15041 |
| 2256 |
| 388 |
| 74 |
| 15 |
| 3 |
| $S = .0902397156$ |

### Scholion

Every series whose terms are alternately negative and positive will change into another more rapidly convergent series whose terms are of the same sign, if it is transformed according to this proposition. And on the other hand, every series whose terms are of the same sign will change into another whose terms are alternately negative and positive; however, this will not converge more rapidly than the former except where the transformation is begun from terms sufficiently far removed from the beginning. And if the series is transformed and again the series which results from this transformation is transformed, the series first considered will come out. For example, if the series

$$1 - \frac{1}{3} + \frac{1}{5} - \frac{1}{7} + \frac{1}{9} - \&c.$$

is transformed,

$$\tfrac{1}{2} + \tfrac{1}{3}A + \tfrac{2}{5}B + \tfrac{3}{7}C + \tfrac{4}{9}D + \&c.$$

will result, and then if this last series is transformed, the first

$$1 - \frac{1}{3} + \frac{1}{5} - \frac{1}{7} + \&c.$$

will come out, that is, it translates the series for the tangent into that for the sine, or the series for the sine into that for the tangent. But in these cases the operation must be begun at the first term of the series, since we are concerned with the transformation of the whole series.

## (p.194) Proposition 8

*If the equation for the series is*

$$T' = \frac{z-m}{z} \times \frac{z-n}{z-n+1} T,$$

*and we put*

$$A = (z-n)T, \quad B = \frac{n}{z}A, \quad C = \frac{n+1}{z+1}B, \quad D = \frac{n+2}{z+2}C, \quad E = \frac{n+3}{z+3}D,$$

$$\&c.,$$

*then there will be*

$$S = \frac{A}{m} + \frac{B}{m+1} + \frac{C}{m+2} + \frac{D}{m+3} + \&c.$$

For by supposing $S = T \times (z-n) \times y$ and by the Method of Differences there will be $S - T = T' \times (z-n+1) \times y'$ and the difference of these equations will give $T = T \times (z-n) \times y - T' \times (z-n+1) \times y'$. For $T'$ write its value $\dfrac{z-m}{z} \times \dfrac{z-n}{z-n+1} T$, and

$$T = T \times (z-n) \times y - T \times \frac{z-m}{z} \times (z-n) \times y'$$

will be obtained which, when divided by $T \times (z-n)$, becomes

$$\frac{1}{z-n} = y - \frac{z-m}{z}y', \quad \text{or} \quad y - y' + \frac{m}{z}y' - \frac{1}{z-n} = 0.$$

Now the root $y$ will be extracted from this equation in the following manner. Suppose

$$y = \frac{a}{m} + \frac{b}{(m+1)z} + \frac{c}{(m+2)z(z+1)} + \frac{d}{(m+3)z(z+1)(z+2)} + \&c.$$

There will be

$$y' = \frac{a}{m} + \frac{b}{(m+1)(z+1)} + \frac{c}{(m+2)(z+1)(z+2)} + \frac{d}{(m+3)(z+1)(z+2)(z+3)} + \&c.$$

And

$$y - y' = \frac{b}{(m+1)z(z+1)} + \frac{2c}{(m+2)z(z+1)(z+2)} + \frac{3d}{(m+3)z(z+1)(z+2)(z+3)} + \&c.$$

And

$$\frac{m}{z}y' = \frac{am}{mz} + \frac{bm}{(m+1)z(z+1)} + \frac{cm}{(m+2)z(z+1)(z+2)} + \frac{dm}{(m+3)z(z+1)(z+2)(z+3)} + \&c.$$

And so

$$y - y' + \frac{m}{z}y' = \frac{a}{z} + \frac{b}{z(z+1)} + \frac{c}{z(z+1)(z+2)} + \frac{d}{z(z+1)(z+2)(z+3)} + \&c.$$

Now

$$\frac{1}{z-n} = \frac{1}{z} + \frac{n}{z(z+1)} + \frac{n(n+1)}{z(z+1)(z+2)} + \frac{n(n+1)(n+2)}{z(z+1)(z+2)(z+3)} + \&c.$$

Therefore

$$y - y' + \frac{m}{z}y - \frac{1}{z-n} = \frac{a-1}{z} + \frac{b-n}{z(z+1)} + \frac{c-n(n+1)}{z(z+1)(z+2)} + \frac{d-n(n+1)(n+2)}{z(z+1)(z+2)(z+3)} + \&c. = 0.$$

Now put the numerators equal to zero, so that the terms vanish; and for the determination of the coefficients you will have the following equations

$$a = 1, \quad b = n, \quad c = n(n+1), \quad d = n(n+1)(n+2), \quad \&c.$$

Substitute these values in place of $a$, $b$, $c$, $d$, etc. in the series which has been taken for $y$, and the value of $y$ which results when multiplied by $(z-n)T$ will provide for $S$ the series given in the theorem. Q.E.D.

*Corollary.* If $n$ is a negative integer or zero, the series will be exactly summable by this theorem. And if $m$ is zero or a negative integer, the series will be infinitely large. Moreover, this proposition and the previous one are of use for the quadrature of binomial curves. This one comes into use where in the ordinate $x^\theta(e + fx^\eta)^\lambda$ the term $e + fx^\eta = 0$, while the previous one is to be used where the contrary happens.

### Example 1

Let the value of the series

$$1 + \tfrac{1.1}{2.3}A + \tfrac{3.3}{4.5}B + \tfrac{5.5}{6.7}C + \tfrac{7.7}{8.9}D + \tfrac{9.9}{10.11}E + \&c.$$

be sought. The equation defining it is

$$T' = \frac{z - \frac{1}{2}}{z} \times \frac{z - \frac{1}{2}}{z + \frac{1}{2}} T,$$

as will be clear on writing the values 1, 2, 3, 4, etc. successively for $z$. Now when the equation in the theorem has been compared with this, it gives $m = \frac{1}{2}$, $n = \frac{1}{2}$; hence

$$A = (z - \tfrac{1}{2})T, \quad B = \frac{A}{2z}, \quad C = \frac{3B}{2z+2}, \quad D = \frac{5C}{2z+4}, \quad E = \frac{7D}{2z+6},$$

$$\&c.$$

and

$$S = \tfrac{2}{1}A + \tfrac{2}{3}B + \tfrac{2}{5}C + \tfrac{2}{7}D + \tfrac{2}{9}E + \&c.$$

To begin the calculation I find 1.407397508 to be the aggregate of twelve terms. Then I substitute the thirteenth term, that is, $\frac{1}{25} \times .161180258$, for $T$ and for $z$ its corresponding value 13, and it comes out thus

| | | |
|---|---|---|
| $A = \frac{1}{2}\times$ | .161180258, | .161180258 |
| $B = \frac{1}{2}\times$ | 6199241, | 2066414 |
| $C = \frac{1}{2}\times$ | 664204, | 132841 |
| $D = \frac{1}{2}\times$ | 110701, | 15814 |
| $E = \frac{1}{2}\times$ | 24216, | 2691 |
| $F = \frac{1}{2}\times$ | 6410, | 583 |
| $G = \frac{1}{2}\times$ | 1959, | 151 |
| $H = \frac{1}{2}\times$ | 670, | 45 |
| $I = \frac{1}{2}\times$ | 250, | 15 |
| $K = \frac{1}{2}\times$ | 102, | 5 |
| $L = \frac{1}{2}\times$ | 44, | 2 |
| $M = \frac{1}{2}\times$ | 20, | 1 |
| | | $S = .163398820$ |

From this calculation I have $S = .163398820$, that is the value of all the terms after the twelfth; and so when this has been added to the sum of the initial terms, 1.407397508, I have 1.570796328 for the value of the series which was to be summed, that is, for the length of the semicircular arc whose diameter is one.

When their sum is being worked out, the initial terms can be reduced very easily to decimal fractions by the following rule: put

$$a = 1, \quad b = a - \tfrac{1}{2}a, \quad c = b - \tfrac{1}{4}b, \quad d = c - \tfrac{1}{6}c, \quad e = d - \tfrac{1}{8}d, \quad \&c.$$

and the terms will be $a$, $\frac{1}{3}b$, $\frac{1}{5}c$, $\frac{1}{7}d$, $\frac{1}{9}e$, etc.

**Example 2**

Consider *Brouncker's* series

$$\frac{1}{1.2} + \frac{1}{3.4} + \frac{1}{5.6} + \frac{1}{7.8} + \&c.$$

The equation for this is

$$T' = \frac{z-1}{z} \times \frac{z-\frac{1}{2}}{z+\frac{1}{2}} T, \quad \text{or also} \quad T' = \frac{z-1}{z} \times \frac{z-\frac{3}{2}}{z-\frac{1}{2}} T,$$

namely by taking the beginning of the abscissa $z$ at different points. In the former the values of $z$ are $\frac{3}{2}$, $\frac{5}{2}$, $\frac{7}{2}$, $\frac{9}{2}$, etc. and

$$m = 1, \quad n = \tfrac{1}{2}, \quad A = (z - \tfrac{1}{2})T, \quad B = \frac{A}{2z}, \quad C = \frac{3B}{2z+2}, \quad D = \frac{5C}{2z+4}, \quad \&c.$$

In the latter the values of $z$ are 2, 3, 4, 5, etc. and

$$m=1,\quad n=\tfrac{3}{2},\quad A=(z-\tfrac{3}{2})T,\quad B=\frac{3A}{2z},\quad C=\frac{5B}{2z+2},\quad D=\frac{7C}{2z+4},$$

&c.

and in both cases there will be

$$S=A+\tfrac{1}{2}B+\tfrac{1}{3}C+\tfrac{1}{4}D+\text{\&c.}$$

Therefore $S$ will be given to more figures by taking for $T$ any one of the terms which are sufficiently far from the beginning. And the calculation may be set up similarly in other cases with two different methods where the the terms are assignable. But this series written in this manner

$$1-\frac{1}{2}+\frac{1}{3}-\frac{1}{4}+\frac{1}{5}-\text{\&c.}$$

will be more easily handled by Proposition 3 or Proposition 7.

## Scholion

Thus far we have been concerned with the summation of series which arise in the quadrature of binomial curves and like matters. But one may proceed in the same way in more difficult cases: for the sum of the series is determined by and can be extracted from the given relation of the terms, namely by resolving the difference equation as in the last two propositions. But it would be a wearisome task to investigate the sums independently of the terms, when the terms are not assignable; and so I have sought the quantity which when multiplied by the term $T$ produces the sum $S$. In this way also the areas of curves are more easily obtained by means of ordinates, for the series produced in this way are very simple. And these things having been set out I proceed to the method for resolving the roots of difference equations into series which are indeed more involved but at the same time far more convergent than the previous ones; I have shown these only because they are simple and sufficient for ordinary uses.

Hitherto we have denoted any sum by $S$, and its terms by $T$, $T'$, $T''$, etc. However, in what follows we will also denote series or sums by $S_2$, $S_3$, $S_4$, etc. and their terms by $T_2$, $T_2'$, $T_2''$, $T_2'''$, etc. $T_3$, $T_3'$, $T_3''$, etc. and so on: thus

$$S_2=T_2+T_2'+T_2''+\text{\&c.}$$
$$S_3=T_3+T_3'+T_3''+\text{\&c.}$$
$$S_4=T_4+T_4'+T_4''+\text{\&c.}$$

&c.

And just as in the series $S$, the successive sums are denoted by $S'$, $S''$, etc., so in the series $S_2$ they will be denoted by $S_2'$, $S_2''$, etc. and in the series $S_3$ by $S_3'$, $S_3''$, etc., and so on in the rest. Now I have been compelled to introduce this notation because of the fact that sums and terms of different series come into consideration together in the same equation.

(p.200)

# Proposition 9

*Given the relation between two sums in different series and the equation for the terms in one of them, to find the equation for the terms in the other.*

The problem is solved by going from the present relation of the variables to the following one, in order that the sums may hence be eliminated; this will be clear from examples.

## Example 1

Let $S$ and $S_2$ be two sums in different series, and let their relation be

$$S = \frac{(m+1)z - n}{m(m+1)} T + S_2;$$

and let

$$T' = \frac{(z-m)}{z} \times \frac{z-n}{z-n+1} T$$

be the relation of the terms of the series $S$: and from these given equations it is required to determine the equation for the terms of the series $S_2$. In the equation

$$S = \frac{(m+1)z - n}{m(m+1)} T + S_2,$$

which shows the relation of the sums, substitute the following values of the variables for the preceding ones, that is, $S'$ or $S - T$ for $S$, $S_2'$ or $S_2 - T_2$ for $S_2$, $T'$ for $T$ and $z+1$ for $z$; and you will have

$$S - T = \frac{(m+1)(z+1) - n}{m(m+1)} T' + S_2 - T_2,$$

and when this has been subtracted from the first equation it leaves an equation free from sums, namely

$$T = \frac{(m+1)z - n}{m(m+1)} T - \frac{(m+1)(z+1) - n}{m(m+1)} T' + T_2,$$

from which there is

$$T_2 = \frac{(m+1)(z+1) - n}{m(m+1)} T' - \frac{(m+1)(z-m) - n}{m(m+1)} T;$$

substitute in this for $T'$ its own value $\frac{(z-m)}{z} \times \frac{z-n}{z-n+1} T$, and you will obtain

$$T_2 = \frac{n(m-n+1)}{(m+1)z(z-n+1)} T.$$

Moreover, write in this value the next values of the indeterminates $T_2'$, $T'$, and $z+1$ for the present ones $T_2$, $T$, and $z$, and

$$T_2' = \frac{n(m-n+1)}{(m+1)(z+1)(z-n+2)} T'$$

will arise; here on substituting again for $T'$ its value,

$$T_2' = \frac{n(m-n+1)}{(m+1)(z+1)(z-n+2)} T \times \frac{(z-m)}{z} \times \frac{z-n}{z-n+1}$$

comes out, from which

$$T = \frac{(m+1)z(z+1)(z-n+1)(z-n+2)}{n(m-n+1)(z-m)(z-n)} T_2'$$

results. But the equation found previously

$$T_2 = \frac{n(m-n+1)}{(m+1)z(z-n+1)} T$$

gives

$$T = \frac{(m+1)z(z-n+1)}{n(m-n+1)} T_2.$$

Now let the two values of the term $T$ be equated to each other, and you will come upon the equation

$$T_2' = \frac{z-m}{z+1} \times \frac{z-n}{z-n+2} T_2,$$

which expresses the relation of the terms of the series $S_2$. Q.E.I.

**Example 2**

Now let the relation between the sums and the equation for the series $S$ be

$$S = \frac{3}{4} \times \frac{4z+3}{4z+2} T + S_2 \quad \text{and} \quad zT + 3T' \times (z+1) = 0.$$

Write the next values of the variables in the equation for the sums, and

$$S - T = \frac{3}{4} \times \frac{4z+7}{4z+6} T' + S_2 - T_2$$

will be obtained: taking this away from the former you will have

$$T = \frac{3}{4} \times \frac{4z+3}{4z+2} T - \frac{3}{4} \times \frac{4z+7}{4z+6} T' + T_2;$$

from this you will obtain

$$T_2 = \frac{1}{4} \times \frac{4z-1}{4z+2} T + \frac{3}{4} \times \frac{4z+7}{4z+6} T'.$$

But by the equation for the series $S$ there is $3T' = -\frac{z}{z+1} T$, and when this has been written in,

$$T_2 = +\frac{1}{4} \times \frac{4z-1}{4z+2} T + \frac{1}{4} \times \frac{4z+7}{4z+6} \times \frac{-z}{z+1} T,$$

or

$$T_2 = -\frac{3}{4} T \times \frac{1}{2z+1} \times \frac{1}{2z+2} \times \frac{1}{2z+3},$$

comes out. And again by proceeding to the next values of the indeterminates,

$$T_2' = -\frac{3}{4} T' \times \frac{1}{2z+3} \times \frac{1}{2z+4} \times \frac{1}{2z+5}$$

will be obtained, or on substituting $-\dfrac{z}{z+1} T$ for $3T'$,

$$T_2' = \frac{z}{2} \times \frac{1}{2z+2} \times \frac{1}{2z+3} \times \frac{1}{2z+4} \times \frac{1}{2z+5} T,$$

from which $T = \dfrac{2}{z} \times (2z+2) \times (2z+3) \times (2z+4) \times (2z+5) T_2'$; and from the value of $T_2$ this same $T = -\frac{4}{3} \times (2z+1) \times (2z+2) \times (2z+3) T_2$; when these two values have been equated to each other,

$$(z^2 + \tfrac{1}{2}z) \times T_2 + (z^2 + \tfrac{9}{2}z + 5) \times 3T_2' = 0$$

will be obtained. This is in fact the equation for the terms of the series $S_2$. And the matter is to be dealt with similarly in other cases.

**Scholion**

Infinite series can be compared with each other by means of this proposition. For the equation which shows the relation between the sums $S$ and $S_2$ will give one when the other is given: and the terms of $S$ and $S_2$ will be given by their equations, of which one is assumed while the other is extracted from the one which has been assumed, as in the examples of this proposition. Thus if the relation of the sums is $S = \dfrac{2z-1}{2} T + S_2$ and the equation for the series $S$ is $T' = \dfrac{z^2 - 2z + 1}{z^2} T$, you will find $T_2 = \dfrac{T}{2z^2}$ and $T_2' = \dfrac{z^2 - 2z + 1}{z^2 + 2z + 1} T_2$ for the equation for the series $S_2$ . Now let 2, 3, 4, 5, etc. be the successive values of the indeterminate $z$; and by taking one for $T$ you will find

$$S = 1 + \frac{1}{4} + \frac{1}{9} + \frac{1}{16} + \frac{1}{25} + \frac{1}{36} + \&c.,$$

$$S_2 = \frac{1}{8} \times \left(1 + \frac{1}{9} + \frac{1}{36} + \frac{1}{100} + \frac{1}{225} + \&c.\right).$$

The denominators in the first are the squares of the natural numbers and in the second they are the squares of the triangular numbers. Moreover, on writing 2 for $z$ and one for $T$, the relation between the series, $S = \dfrac{2z-1}{2} T + S_2$ , will give $S_2 = S - \frac{3}{2}$, and on multiplying by 8 there will be

$$8S - 12 = 8S_2 = 1 + \frac{1}{9} + \frac{1}{36} + \frac{1}{100} + \frac{1}{225} + \&c.$$

And likewise series where the denominators are the squares of the pyramidal numbers, or of the tetrahedral numbers, etc. can be compared with a series in which the denominators are the squares of the natural numbers. And in general all series can be compared which are defined by the following equations

$$T' = \frac{z^2 + a}{(z+b)(z+c)}T, \quad T' = \frac{z^2 + a}{(z+b+1)(z+c+1)}T,$$

$$T' = \frac{z^2 + a}{(z+b+2)(z+c+2)}T, \quad \&c.$$

Here the numerator remains the same, but the denominators are formed by writing repeatedly $z+1$ in place of $z$.

(p.202)

## Proposition 10

*To find approximately the value of a series which is defined by an equation of this form*

$$T \times (z^\theta + az^{\theta-1} + bz^{\theta-2} + \&c.) = rT' \times (z^\theta + cz^{\theta-1} + dz^{\theta-2} + \&c.).$$

### Case 1

First let $r = 1$, and let us suppose that the equation for the series is

$$T' = \frac{z-m}{z} \times \frac{z-n}{z-n+1}T.$$

Let us put

$$S = \frac{z+p}{q}T \text{ approximately, or } S = \frac{z+p}{q}T + S_2 \text{ exactly;}$$

and by going to the next values of the indeterminates, there will be

$$S - T = \frac{z+p+1}{q}T' + S_2 - T_2;$$

the difference of these equations will give

$$T = \frac{z+p}{q}T - \frac{z+p+1}{q}T' + T_2;$$

hence

$$T_2 = \frac{z+p+1}{q}T' - \frac{z+p-q}{q}T.$$

For $T'$ substitute its value $\dfrac{z-m}{z} \times \dfrac{z-n}{z-n+1} T$, and

$$T_2 = \frac{z+p+1}{q} \times \frac{z-m}{z} \times \frac{z-n}{z-n+1} T - \frac{z+p-q}{q} T$$

will come out: when this has been reduced to a common denominator it gives

$$T_2 = \frac{(q-m)z^2 + ((m-q)(n-1) - n - p(m+1))\, z + mn(p+1)}{qz(z-n+1)} T.$$

Now the smaller the sum $S_2$ is, the closer the quantity $\dfrac{z+p}{q} T$ approaches the true value of $S$; and the smaller its first term $T_2$, the smaller $S_2$ will be: now the term will turn out very small where the variable $z$ is of the smallest dimensions in the numerator of its value; for here $z$ is assumed to be large: therefore let the coefficients of the powers $z^2$ and $z$ be set equal to zero, and two equations will be obtained,

$$q - m = 0\,, \quad \text{and} \quad (m-q)(n-1) - n - p(m+1) = 0,$$

for the determination of the two quantities assumed $p$, $q$. The first gives $q = m$ and from this and the second $p = -\dfrac{n}{m+1}$ is extracted. Hence there will be

$$T_2 = \frac{n(m-n+1)}{(m+1)z(z-n+1)} T, \quad \text{and} \quad S = \frac{(m+1)z - n}{m(m+1)} T \quad \text{approximately.}$$

Q.E.I.

And the approximation will be found in exactly the same manner where the equation for the series is $T' = \dfrac{z^2 + az + b}{z^2 + cz + d} T$, or is more involved; but the coefficients after $b$ and $d$ do not enter into this calculation.

### Case 2

Now let the equation be $T \times (z^2 + az + b) + rT' \times (z^2 + cz + d) = 0$, where $r$ is any number apart from $-1$. Suppose that $S = p \times \dfrac{z+m}{z+n} T + S_2$; then write the next values of the variables for the first, and

$$S - T = p \times \frac{z+m+1}{z+n+1} T' + S_2 - T_2,$$

will come out, and when this has been subtracted from the previous equation, there remains

$$T = p \times \frac{z+m}{z+n} T - p \times \frac{z+m+1}{z+n+1} T' + T_2;$$

this equation will give

$$T_2 = \frac{z - pz - mp + n}{z + n} T + p \times \frac{z + m + 1}{z + n + 1} T';$$

but by the equation for the series $S$, there is $T' = -\frac{T}{r} \times \frac{z^2 + az + b}{z^2 + cz + d}$; when this has been written for $T'$, it produces

$$T_2 = \frac{z - pz - mp + n}{z + n} T - \frac{pT}{r} \times \frac{z + m + 1}{z + n + 1} \times \frac{z^2 + az + b}{z^2 + cz + d}.$$

If the members of this value are reduced to a common denominator and the coefficients of the three highest powers of $z$ are set equal to zero, the first equation will give $rp + p = r$, the second $(m - n)(r + 1) = c - a$, and finally the third $(c - a)(2n + 1) + d - b = (m - n)(rc + rn + r + n) + ma$: and these three equations give

$$p = \frac{r}{r + 1}, \quad m = c - \frac{b - d}{a - c} - \frac{1}{r + 1}, \quad \text{and } n = m + \frac{a - c}{r + 1}.$$

And so the assumed quantities $p$, $m$, and $n$ are given; and thence also the quantity $p \times \frac{z + m}{z + n} T$ which is approximately equal to the series $S$. Q.E.I.

## Proposition 11

(p.206)

*If* $T' = \frac{z - m}{z} \times \frac{z - n}{z - n + 1} T$ *is the equation expressing the relation of the terms of the series* $S$*, put*

$$T_2 = \frac{1}{m} \times \frac{m}{m + 1} \times \frac{n}{z} \times \frac{m - n + 1}{z - n + 1} T,$$

$$T_3 = \frac{2}{m + 2} \times \frac{m + 1}{m + 3} \times \frac{n + 1}{z + 1} \times \frac{m - n + 2}{z - n + 2} T_2,$$

$$T_4 = \frac{3}{m + 4} \times \frac{m + 2}{m + 5} \times \frac{n + 2}{z + 2} \times \frac{m - n + 3}{z - n + 3} T_3,$$

$$T_5 = \frac{4}{m + 6} \times \frac{m + 3}{m + 7} \times \frac{n + 3}{z + 3} \times \frac{m - n + 4}{z - n + 4} T_4,$$

$$T_6 = \frac{5}{m + 8} \times \frac{m + 4}{m + 9} \times \frac{n + 4}{z + 4} \times \frac{m - n + 5}{z - n + 5} T_5,$$

&c.

*And there will be*

$$S = \frac{(m + 1)z - 1n}{m(m + 1)} T + \frac{(m + 3)(z + 2) - 2(n + 1)}{(m + 2)(m + 3)} T_2$$

$$+ \frac{(m + 5)(z + 4) - 3(n + 2)}{(m + 4)(m + 5)} T_3 + \frac{(m + 7)(z + 6) - 4(n + 3)}{(m + 6)(m + 7)} T_4$$

$$+ \frac{(m + 9)(z + 8) - 5(n + 4)}{(m + 8)(m + 9)} T_5 + \&c.$$

By the preceding proposition the quantity $\frac{(m+1)z-n}{m(m+1)}T$ is approximately equal to the series $S$. Therefore let $S = \frac{(m+1)z-n}{m(m+1)}T + S_2$ exactly, and you will find

$$T_2 = \frac{1}{m} \times \frac{m}{m+1} \times \frac{n}{z} \times \frac{m-n+1}{z-n+1}T, \quad \text{and} \quad T_2' = \frac{z-m}{z+1} \times \frac{z-n}{z-n+2}T_2$$

for the equation for the terms of the series $S_2$: this having been given, by the previous proposition the quantity $\frac{(m+3)(z+2)-2(n+1)}{(m+2)(m+3)}T_2$ turns out approximately equal to the series $S_2$. Then by taking

$$S_2 = \frac{(m+3)(z+2)-2(n+1)}{(m+2)(m+3)}T_2 + S_3\,,$$

it will be found that

$$T_3 = \frac{2}{m+2} \times \frac{m+1}{m+3} \times \frac{n+1}{z+1} \times \frac{m-n+2}{z-n+2}T_2, \quad \text{and} \quad T_3' = \frac{z-m}{z+2} \times \frac{z-n}{z-n+3}T_3$$

is the equation for the terms of the series $S_3$: hence you will find that the quantity $\frac{(m+5)(z+4)-3(n+2)}{(m+4)(m+5)}T_3$ is approximately equal to the series $S_3$. And the process continues on in this way. Therefore the first term in the value of $S$ is approximately equal to $S$; and the second term is approximately equal to $S_2$; then the third is approximately equal to $S_3$, and so on with the rest: this is to say, the first term is approximately equal to the series whose value is being sought, the second is approximately equal to the discrepancy between the first term and the true value, the third is approximately equal to the discrepancy of the first two from the true value, the fourth is approximately equal to the discrepancy of the first three from the true value, and so on. Therefore the value of the sum $S$ is true and converges very rapidly. Q.E.D.

*Corollary.* The value of the series $S$ shown here will terminate, where $n$ or $m-n+1$ is zero or a negative integer: and in other cases it will go off to infinity approaching very rapidly the true value, except where, on account of $m$ being zero or negative, the value of the series is infinitely large.

### Example 1

Let the value of the series

$$1 + \frac{1}{4} + \frac{1}{9} + \frac{1}{16} + \frac{1}{25} + \frac{1}{36} + \&\text{c.}$$

be sought. The equation defining the relation of the terms is $T' = \frac{z-1}{z} \times \frac{z-1}{z}T$, where 2, 3, 4, 5, etc. are the values of the indeterminate $z$. Now when this has been compared with the equation in the theorem, it gives $m = 1$, $n = 1$; when these have been written in,

$$T_2 = \frac{T}{2z^2}, \quad T_3 = \frac{8T_2}{6(z+1)^2}, \quad T_4 = \frac{27T_3}{10(z+2)^2}, \quad T_5 = \frac{64T_4}{14(z+3)^2}, \quad \&c.$$

and

$$S = \frac{2z-1}{2}T + \frac{2z+2}{6}T_2 + \frac{2z+5}{10}T_3 + \frac{2z+8}{14}T_4 + \frac{2z+11}{18}T_5 + \&c.$$

come out. Now collect together the ten initial terms, and you will find their aggregate to be 1.5497.6773.1166.5406.9. Then for $z$ write its eleventh value, or 12, and the eleventh term $\frac{1}{121}$ for $T$, and you will find by calculation

| | | |
|---|---:|---:|
| $T$ = | .0082.6446.2809.9173.55 | .0950.4132.2314.0495.8 |
| $T_2$ = | 2869.6051.4233.24 | 1.2434.9556.1677.4 |
| $T_3$ = | 22.6398.8277.97 | 65.6556.6006.1 |
| $T_4$ = | 3118.7593.62 | 7128.5928.3 |
| $T_5$ = | 63.3652.70 | 123.2102.5 |
| $T_6$ = | 1.7188.93 | 2.9690.0 |
| $T_7$ = | 583.96 | 920.9 |
| $T_8$ = | 23.78 | 34.9 |
| $T_9$ = | 1.12 | 1.5 |
| | | $S$ = .0951.6633.5681.6857.4 |

From this calculation we have $S$ = .0951.6633.5681.6857.4, which, when added to the aggregate of the initial terms, will produce for the value of the proposed series 1.6449.3406.6848.2264.3.

### Example 2

Let the series to be summed be

$$1 + \tfrac{1.1}{2.3}A + \tfrac{3.3}{4.5}B + \tfrac{5.5}{6.7}C + \tfrac{7.7}{8.9}D + \tfrac{9.9}{10.11}E + \&c.$$

The equation specifying this is $T' = \dfrac{z-\frac{1}{2}}{z} \times \dfrac{z-\frac{1}{2}}{z+\frac{1}{2}}T$, as will be clear on writing the values 1, 2, 3, 4, etc. successively for $z$. Now when the equation in the theorem has been compared with this, it gives $m = \frac{1}{2}$, $n = \frac{1}{2}$, and so

$$T_2 = \frac{2.1.1}{1.3z(2z+1)}T, \quad T_3 = \frac{2.4.9}{5.7(z+1)(2z+3)}T_2,$$

$$T_4 = \frac{2.9.25}{9.11(z+2)(2z+5)}T_3, \quad \&c.$$

and

$$S = \frac{6z - 1.2}{1.3}T + \frac{14(z+2) - 3.4}{5.7}T_2 + \frac{22(z+4) - 5.6}{9.11}T_3 + \&\text{c.}$$

By addition you will find 1.3916.9464.5943.2880.5 to be the sum of the ten initial terms; then, in order that the sum of the rest may be obtained, write 11 for $z$ and the eleventh term for $T$, and you will have

| | | |
|---|---|---|
| $T$ = | .0083.9003.5809.6168.15 | .1789.9383.0605.1587.3 |
| $T_2$ = | 2210.8921.7644.71 | 1.0738.6191.4274.3 |
| $T_3$ = | 15.1604.0349.56 | 45.9406.1665.3 |
| $T_4$ = | 1963.2742.16 | 4570.9051.0 |
| $T_5$ = | 38.8835.92 | 76.0818.3 |
| $T_6$ = | 1.0484.94 | 1.8104.4 |
| $T_7$ = | 358.18 | 562.5 |
| $T_8$ = | 14.77 | 21.5 |
| $T_9$ = | 71 | 1.0 |
| | | $S$ = .1791.0168.0851.6085.6 |

Now let $S$ be added to the aggregate of the initial terms, and for the value of the series, that is, for the semicircumference of the circle whose diameter is one, 1.5707.9632.6794.8966.1 will come out.

## Example 3

Now consider the series

$$\tfrac{1}{3} + \tfrac{1.3}{2.7}A + \tfrac{3.7}{4.11}B + \tfrac{5.11}{6.15}C + \tfrac{7.15}{8.19}D + \&\text{c.},$$

which is defined by the equation $T' = \dfrac{z - \frac{1}{2}}{z} \times \dfrac{z - \frac{1}{4}}{z + \frac{3}{4}}T$, in which the values of the indeterminate $z$ are 1, 2, 3, 4, etc. And there will be $m = \frac{1}{2}$, $n = \frac{1}{4}$; and consequently

$$T_2 = \frac{5.1.1}{6z(4z+3)}T, \quad T_3 = \frac{9.2.3}{14(z+1)(4z+7)}T_2,$$

$$T_4 = \frac{13.3.5}{22(z+2)(4z+11)}T_3, \quad \&\text{c.}$$

and

$$S = \frac{6z - 1.1}{1.3}T + \frac{14(z+2) - 2.5}{5.7}T_2 + \frac{22(z+4) - 3.9}{9.11}T_3 + \&\text{c.}$$

I now collect together the nine initial terms, the sum of which comes out as .5055.0041.4718.3195.8, and on writing 10 for $z$ and the tenth term for $T$, I have

| | | |
|---|---|---|
| $T =$ | .0047.5565.5924.4791.67 | .0935.2789.9848.0902.9 |
| $T_2 =$ | 921.6387.4505.41 | 4160.5406.2053.0 |
| $T_3 =$ | 6.8760.0058.48 | 19.5167.2893.3 |
| $T_4 =$ | 995.8557.00 | 2185.7755.9 |
| $T_5 =$ | 22.0991.75 | 40.9826.8 |
| $T_6 =$ | 6653.41 | 1.0937.5 |
| $T_7 =$ | 252.54 | 379.0 |
| $T_8 =$ | 11.51 | 16.1 |
| $T_9 =$ | 61 | 8 |
| | | $S =$ .0935.6970.2649.4765.3 |

Finally, when the sum of the initial terms has been added to $S$, it makes up .5990.7011.7367.7961.1 for the value of the series, that is, for the ordinate of the elastic curve. And *Jakob Bernoulli* found correctly that this number is contained between the limits .5983 and .6004.

### Example 4

Consider the series

$$1 + \tfrac{1.1}{2.5}A + \tfrac{3.5}{4.9}B + \tfrac{5.9}{6.13}C + \tfrac{7.13}{8.17}D + \tfrac{9.17}{10.21}E + \&c.,$$

which is defined by the equation $T' = \dfrac{z-\frac{1}{2}}{z} \times \dfrac{z-\frac{3}{4}}{z+\frac{1}{4}}T$, where 1, 2, 3, 4, etc. are the successive values of the indeterminate. Now $m = \frac{1}{2}$, $n = \frac{3}{4}$ and consequently

$$T_2 = \frac{1.1.3}{2z(4z+1)}T, \quad T_3 = \frac{3.2.7}{10(z+1)(4z+5)}T_2,$$

$$T_4 = \frac{5.3.11}{18(z+2)(4z+9)}T_3, \quad \&c.$$

and

$$S = \frac{2z-1}{1}T + \frac{2z+2}{5}T_2 + \frac{2z+5}{9}T_3 + \frac{2z+8}{13}T_4 + \frac{2z+11}{17}T_5 + \&c.$$

The sum of the nine initial terms is 1.2157.0599.7306.1360.6. And, in order that the sum of the rest may be obtained, put 10 for $z$ and the tenth term for $T$, and by calculation you will obtain

| | | |
|---|---|---|
| $T =$ | .0050.1271.8406.8834.46 | .0952.4164.9730.7854.7 |
| $T_2 =$ | 1833.9213.6837.20 | 8069.2540.2083.7 |
| $T_3 =$ | 15.5605.4494.38 | 43.2237.3595.5 |
| $T_4 =$ | 2425.8219.16 | 5224.8472.0 |
| $T_5 =$ | 56.8742.44 | 103.7118.6 |
| $T_6 =$ | 1.7922.51 | 2.9017.4 |
| $T_7 =$ | 707.95 | 1047.8 |
| $T_8 =$ | 33.45 | 46.1 |
| $T_9 =$ | 1.83 | 2.3 |
| | | $S =$ .0953.2277.9839.9238.1 |

Now let $S$ be added to the aggregate of the initial terms, and you will obtain for the value of the series 1.3110.2877.7146.0598.7, that is, for the length of the elastic curve if it were straightened out into a straight line. Now *Bernoulli* showed that this number lies between the limits 1.308 and 1.315. And therefore if its ordinate is added to the length of the elastic curve, the number 1.9100.9889.4513.8559.8 will be obtained, which is the semiperimeter of the ellipse having 1 and $\sqrt{2}$ as axes. And these examples are sufficient; for I do not dwell upon series which can be summed exactly by this proposition.

(p.213)

## Scholion

With almost no effort this theorem produces the areas of binomial curves whose ordinates are comprised under the form $x^\theta \times (e + fx^\eta)^\lambda$, but in the special case where $e + fx^\eta = 0$, or $x^\eta = -\frac{e}{f}$, that is, in that case in which the series for the area converges very slowly. But where the areas do not have to be produced beyond eight or nine figures, it suffices to investigate the sum of four initial terms, for $S$ will give that of the rest with little effort; even if no initial terms are collected together, but the transformation is begun at the first term, the value of $S$ will approximate to the value of the whole series sufficiently rapidly. Now the series in the theorem is expressed more generally, and is extended to cases which do not relate to quadratures, as follows. Let the equation for the series be $T' = \dfrac{z^2 + m}{z^2 + nz + r} T$ and set

$$a = n - \tfrac{r-m}{n}, \qquad T_2 = \tfrac{T}{n-1} \times \tfrac{ma+(n-a)r}{z^2+nz+r},$$

$$b = n + 2 - \tfrac{r-m+n+1}{n+2}, \qquad T_3 = \tfrac{T_2}{n+1} \times \tfrac{mb+(n-b+2)(r+n+1)}{z^2+(n+2)z+r+n+1},$$

$$c = n + 4 - \tfrac{r-m+2n+4}{n+4}, \qquad T_4 = \tfrac{T_3}{n+3} \times \tfrac{mc+(n-c+4)(r+2n+4)}{z^2+(n+4)z+r+2n+4},$$

$$d = n + 6 - \tfrac{r-m+3n+9}{n+6}, \qquad T_5 = \tfrac{T_4}{n+5} \times \tfrac{md+(n-d+6)(r+3n+9)}{z^2+(n+6)z+r+3n+9},$$

$$e = n + 8 - \tfrac{r-m+4n+16}{n+8}, \qquad T_6 = \tfrac{T_5}{n+7} \times \tfrac{mc+(n-e+8)(r+4n+16)}{z^2+(n+8)z+r+4n+16},$$

&c. &c.

And there will be

$$S = \frac{z+a-1}{n-1}T + \frac{z+b-1}{n+1}T_2 + \frac{z+c-1}{n+3}T_3 + \frac{z+d-1}{n+5}T_4 + \frac{z+e-1}{n+7}T_5 + \&c.$$

## Example

Consider the series

$$\tfrac{1}{2} + \tfrac{1.1}{2.4}A + \tfrac{3.3}{4.6}B + \tfrac{5.5}{6.8}C + \tfrac{7.7}{8.10}D + \tfrac{9.9}{10.12}E + \&c.,$$

which is defined by the equation $T' = \frac{z}{z+\frac{1}{2}} \times \frac{z}{z+\frac{3}{2}}T$, where the successive values of the abscissa $z$ are $\frac{1}{2}$, $\frac{3}{2}$, $\frac{5}{2}$, $\frac{7}{2}$, etc. This equation cannot be compared with $T' = \frac{z-m}{z} \times \frac{z-n}{z-n+1}T$, namely the equation in the proposition; in this equation there are two factors $z-n$ and $z-n+1$ differing by one, of which one is in the numerator, the other in the denominator: however in the equation defining the proposed series the differences of the factors in the numerator and the denominator are $\frac{1}{2}$, and $\frac{3}{2}$. Therefore I multiply the factors together and $T' = \frac{z^2}{z^2+2z+\frac{3}{4}}T$ results, and I proceed to the equation in the scholion; comparing this with the other, I have $m=0$, $n=2$, and $r=\frac{3}{4}$; and when these things have been written in there arise

$$a = \frac{13}{8}, \qquad T_2 = \frac{1}{8} \times \frac{9}{4z^2+8z+3}T,$$

$$b = \frac{49}{16}, \qquad T_3 = \frac{3}{16} \times \frac{25}{4z^2+16z+15}T_2,$$

$$c = \frac{109}{24}, \qquad T_4 = \frac{5}{24} \times \frac{49}{4z^2+24z+35}T_3,$$

$$d = \frac{193}{32}, \qquad T_5 = \frac{7}{32} \times \frac{81}{4z^2+32z+63}T_4,$$

$$e = \frac{301}{40}, \qquad T_6 = \frac{9}{40} \times \frac{121}{4z^2+40z+99}T_5,$$

$$\&c. \qquad \&c.$$

And

$$S = \frac{8z+5}{1.8}T + \frac{16z+33}{3.16}T_2 + \frac{24z+85}{5.24}T_3 + \frac{32z+161}{7.32}T_4 + \&c.$$

Now find the aggregate of the six initial terms, and $.6106,6818,2373,0$ will come out. Then substitute the seventh term for $T$ and the seventh value $\frac{13}{2}$ for $z$; and you will find by calculation

| | | |
|---|---|---|
| $T =$ | .0036.3492.9656.98 | .0258.9887.3806.0 |
| $T_2 =$ | 1825.5785.11 | 5210.5053.3 |
| $T_3 =$ | 29.7131.92 | 59.6739.9 |
| $T_4 =$ | 8425.62 | 1.3879.7 |
| $T_5 =$ | 339.30 | 491.0 |
| $T_6 =$ | 17.50 | 23.1 |
| $T_7 =$ | 1.09 | 1.3 |
| | | $S =$ .0259.5158.9994.3 |

After $S$ has been calculated in this way, it is added to the sum of the initial terms and .6366.1977.2367.3 will come out for the value of the proposed series. Now if one is divided by this number, the quotient will give the area of the circle. And series which are defined by an equation of this type, $T' = \frac{z^3 + az^2 + bz + c}{z^3 + dz^2 + ez + f} T$, or of even more general type, are transformed by the same method. And it is to be noted that the equation in the scholion, $T' = \frac{z^2 + m}{z^2 + nz + r} T$, extends to no fewer cases because the term in which $z$ is of dimension one is lacking in the numerator, for that term can always be removed by changing the beginning of the abscissa, and in that way the theorem is expressed more simply.

## Proposition 12

(p.218)

*Let it be required to transform a series defined by an equation of this form*

$$T \times (z^\theta + az^{\theta-1} + bz^{\theta-2} + \&c.) + rT' \times (z^\theta + cz^{\theta-1} + dz^{\theta-2} + \&c.) = 0$$

*into another which converges very rapidly.*

Take

$$m = c - \frac{b-d}{a-c} - \frac{1}{r+1}, \quad n = m + \frac{a-c}{r+1};$$

and by Proposition 10 the quantity $\frac{r}{r+1} \times \frac{z+m}{z+n} T$ will be approximately equal to the series. And so let that quantity be the first term of the transformed series: and for finding the second, put

$$S = \frac{r}{r+1} \times \frac{z+m}{z+n} T + S_2;$$

then the equation for the terms of the series $S_2$ is sought by means of Proposition 9; and from that is determined a quantity approximately equal to the series $S_2$ by exactly the same method as the approximation to the series $S$ was first found; and that quantity will be the second term of the transformed series: and by proceeding along these steps, as many of the subsequent terms as you wish may be found. Q.E.I.

### Example 1

Let the series to be transformed be

$$\sqrt{12} \times \left(1 - \frac{1}{3.3} + \frac{1}{5.9} - \frac{1}{7.27} + \frac{1}{9.81} - \frac{1}{11.243} + \&\text{c.}\right),$$

which *Halley* used in the quadrature of the circle. The relation of the terms is defined by the equation $zT + 3T'(z+1) = 0$, in which the values of $z$ are $\frac{1}{2}$, $\frac{3}{2}$, $\frac{5}{2}$, $\frac{7}{2}$, etc. Now on comparing this with the equation in the problem, there will be $r = 3$, $a = 0$, $b = 0$, $c = 1$, $d = 0$; and thence $m = \frac{3}{4}$, $n = \frac{1}{2}$; when these have been written in, $\frac{3}{4} \times \frac{4z+3}{4z+2} T$ will be obtained for the first term of the transformed series.

Suppose

$$S = \frac{3}{4} \times \frac{4z+3}{4z+2} T + S_2.$$

Then by writing the following values of the variables for the previous ones,

$$S - T = \frac{3}{4} \times \frac{4z+7}{4z+6} T' + S_2 - T_2$$

will come out; the difference of these equations will give

$$T = \frac{3}{4} \times \frac{4z+3}{4z+2} T - \frac{3}{4} \times \frac{4z+7}{4z+6} T' + T_2,$$

and hence

$$T_2 = \frac{1}{4} \times \frac{4z-1}{4z+2} T + \frac{3}{4} \times \frac{4z+7}{4z+6} T'.$$

But by the equation for the series $T' = -\frac{1}{3} T \times \frac{z}{z+1}$; when this has been substituted, you will obtain

$$T_2 = -\frac{1}{8} \times \frac{1}{2z+1} \times \frac{2}{2z+2} \times \frac{3}{2z+3} T.$$

On writing the next values of the variables for the former in this value,

$$T_2' = -\frac{1}{8} \times \frac{1}{2z+3} \times \frac{2}{2z+4} \times \frac{3}{2z+5} T'$$

will come out, or on substituting $-\frac{1}{3} \times T \times \frac{z}{z+1}$ for $T'$ it will become

$$T_2' = +\frac{1}{8} \times \frac{1}{2z+3} \times \frac{2}{2z+4} \times \frac{1}{2z+5} \times \frac{z}{z+1} T;$$

from this value of $T_2'$,

$$T = \frac{2}{z} \times (2z+2) \times (2z+3) \times (2z+4) \times (2z+5) T_2'.$$

And from the value of $T_2$ found before,

$$T = -\tfrac{4}{3} \times (2z+1) \times (2z+2) \times (2z+3)T_2.$$

Finally, by equating these two values of the term $T$,

$$\left(z^2 + \tfrac{1}{2}z\right) \times T_2 + 3T_2' \times \left(z^2 + \tfrac{9}{2}z + 5\right) = 0$$

will result, which is the equation for the terms of the series $S_2$; when this has been compared with the equation in the proposition, it gives $r = 3$, $a = \frac{1}{2}$, $b = 0$, $c = \frac{9}{2}$, $d = 5$, and hence $m = 3$, $n = 2$; and consequently $\dfrac{3}{4} \times \dfrac{z+3}{z+2} T_2$ will be approximately equal to the series $S_2$; and so the second term of the transformed series is obtained.

For finding the third term let

$$S_2 = \frac{3}{4} \times \frac{z+3}{z+2} T_2 + S_3$$

be taken and then by the Method of Differences there will be

$$S_2 - T_2 = \frac{3}{4} \times \frac{z+4}{z+3} T_2' + S_3 - T_3,$$

and, when this has been subtracted from the former, there remains

$$T_2 = \frac{3}{4} \times \frac{z+3}{z+4} T_2 - \frac{3}{4} \times \frac{z+4}{z+3} T_2' + T_3,$$

from which arises

$$T_3 = \frac{1}{4} \times \frac{z-1}{z+2} T_2 + \frac{3}{4} \times \frac{z+4}{z+3} T_2'.$$

But by the equation for the series $S_2$,

$$T_2' = -\frac{1}{3} \times \frac{z}{z+2} \times \frac{2z+1}{2z+5} T_2;$$

when this has been substituted in the value of $T_3$,

$$T_3 = -\frac{1}{8} \times \frac{4}{2z+4} \times \frac{5}{2z+5} \times \frac{6}{2z+6} T_2$$

comes out: on writing the next values of the variables for the preceding ones in this,

$$T_3' = -\frac{1}{8} \times \frac{4}{2z+6} \times \frac{5}{2z+7} \times \frac{6}{2z+8} T_2'$$

will arise, or, on substituting for $T_2'$ its value, there will be

$$T_3' = \frac{1}{8} \times \frac{4}{2z+6} \times \frac{5}{2z+7} \times \frac{6}{2z+8} \times \frac{1}{3} \times \frac{z}{z+2} \times \frac{2z+1}{2z+5} T_2.$$

If $T_2$ is eliminated by means of the values of the terms $T_3$ and $T_3'$,

$$\left(z^2 + \tfrac{1}{2}z\right) T_3 + 3T_3' \times \left(z^2 + \tfrac{15}{2}z + 14\right) = 0$$

will result, which is the equation for the terms of the series $S_3$; and, when it has been compared with the equation in the theorem, it produces $a = \frac{1}{2}$, $b = 0$, $c = \frac{15}{2}$, $d = 14$; and so $m = \frac{21}{4}$, $n = \frac{14}{4}$, and consequently $\frac{3}{4} \times \frac{4z+21}{4z+14} T_3$ will be the approximation to the series $S_3$, or the third term of the transformed series.

And by a similar process, on putting

$$S_3 = \frac{3}{4} \times \frac{4z+21}{4z+14} T_3 + S_4,$$

you will find the fourth term to be $\frac{3}{4} \times \frac{4z+30}{4z+20} T_4$, since

$$T_4 = -\frac{1}{8} \times \frac{7}{2z+7} \times \frac{8}{2z+8} \times \frac{9}{2z+9} T_3.$$

And in exactly the same way the transformed series is continued as far as required. But now the progression of the terms is clear, being

$$\begin{aligned} T_2 &= -\frac{1}{8} \times \frac{1}{2z+1} \times \frac{2}{2z+2} \times \frac{3}{2z+3} T, \\ T_3 &= -\frac{1}{8} \times \frac{4}{2z+4} \times \frac{5}{2z+5} \times \frac{6}{2z+6} T_2, \\ T_4 &= -\frac{1}{8} \times \frac{7}{2z+7} \times \frac{8}{2z+8} \times \frac{9}{2z+9} T_3, \\ T_5 &= -\frac{1}{8} \times \frac{10}{2z+10} \times \frac{11}{2z+11} \times \frac{12}{2z+12} T_4, \\ &\text{\&c.} \end{aligned}$$

And

$$S = \frac{3}{4} \times \left( \frac{4z+3}{4z+2} T + \frac{4z+12}{4z+8} T_2 + \frac{4z+21}{4z+14} T_3 + \frac{4z+30}{4z+20} T_4 + \text{\&c.} \right).$$

The sum of ten initial terms is 3.1415.9051.0938.0800.9964.2; and when the eleventh term has been substituted for $T$ and $\frac{21}{2}$ for $z$, they give

$$\begin{aligned} +T &= .0000.0279.3565.0014.1347.8 \\ -T_2 &= 172.5274.8279.5 \\ +T_3 &= 1474.5938.7 \\ -T_4 &= 3.8136.0 \\ +T_5 &= 192.2 \\ -T_6 &= 1.5 \end{aligned}$$

| | |
|---|---|
| .0000.0214.2791.3363.1147.5 | .0000.0000.0139.7472.6106.4 |
| 1244.1885.8 | 3.3215.2 |
| 171.7 | 1.4 |
| +.0000.0214.2791.4607.3205.0 | −.0000.0000.0139.7475.9323.0 |

Now subtract the sum of the negatives from that of the positives, and there will result .0000.0214.2651.7131.3882.0, which, when added to the aggregate of the initial terms, will produce 3.1415.9265.3589.7932.3846.2 for the value of the proposed series. And these six terms of the transformed series are just as effective as thirty-two terms of the simple series. But the use of these theorems is more widely apparent in series which converge very slowly, where it is not possible to attain their values by mere addition of terms.

**Example 2**

Let the series to be transformed be

$$1-\frac{1}{4}+\frac{1}{9}-\frac{1}{16}+\frac{1}{25}-\frac{1}{36}+\&\text{c.},$$

whose equation is $z^2T+(z^2+2z+1)T'=0$, where 1, 2, 3, 4, 5, etc. are the successive values of $z$ in order. Here there will be $a=0$, $b=0$, $c=2$, $d=1$, $r=1$; hence $m=1$, $n=0$, and consequently $\frac{1}{2}\times\frac{z+1}{z}T$ is the first term of the transformed series.

In order that the second may be extracted, suppose $S=\frac{1}{2}\times\frac{z+1}{z}T+S_2$; and by the Method of Differences

$$S-T=\frac{1}{2}\times\frac{z+2}{z+1}T'+S_2-T_2$$

will come out; when this has been subtracted from the former, it leaves

$$T=\frac{1}{2}\times\frac{z+1}{z}T-\frac{1}{2}\times\frac{z+2}{z+1}T'+T_2,$$

from which

$$T_2=\frac{1}{2}\times\frac{z-1}{z}T+\frac{1}{2}\times\frac{z+2}{z+1}T'$$

is found: but the equation for the series $S$ gives $T'=\dfrac{-z^2T}{z^2+2z+1}$, and, when this has been substituted, $T_2=-\dfrac{2z+1}{2z}\times\dfrac{T}{(z+1)^3}$ will come out. Now write the next values of the variables for the preceding ones, and $T_2'=-\dfrac{2z+3}{2z+2}\times\dfrac{T'}{(z+2)^3}$ will arise, or again on putting for $T'$ its value,

$$T_2'=\frac{1}{2}\times\frac{2z+3}{(z+1)^3}\times\frac{z^2}{(z+2)^3}T.$$

By using the values of the terms $T_2$ and $T_2'$ eliminate $T$, and you will have

$$\left(z^4 + \tfrac{3}{2}z^3\right) \times T_2 + \left(z^4 + \tfrac{13}{2}z^3 + 15z^2 + 14z + 4\right) \times T_2' = 0,$$

which is the equation for the terms of series $S_2$: and, when it has been compared with that in the proposition, it gives $a = \frac{3}{2}$, $b = 0$, $c = \frac{13}{2}$, $d = 15$, $r = 1$, and consequently $m = 3$, $n = \frac{1}{2}$; therefore $\dfrac{z+3}{2z+1}T_2$ is approximately the value of the series $S_2$, or the second term of the transformed series. And by proceeding along these steps as in the above example, you will find

$$T_2 = -\frac{2z+1}{2z} \times \frac{T}{(z+1)^3}, \quad T_3 = -\frac{2z+2}{2z+1} \times \frac{8T_2}{(z+2)^3},$$

$$T_4 = -\frac{2z+3}{2z+2} \times \frac{27T_3}{(z+3)^3}, \quad \&c.$$

and

$$S = \frac{z+1}{2z}T + \frac{z+3}{2z+1}T_2 + \frac{z+5}{2z+2}T_3 + \frac{z+7}{2z+3}T_4 + \&c.$$

Or for the sake of an easier calculation, put

$$A = T,\ B = \frac{A}{(z+1)^3},\ C = \frac{8B}{(z+2)^3},\ D = \frac{27C}{(z+3)^3},\ E = \frac{64D}{(z+4)^3},\ \&c.,$$

and there will be

$$S = \frac{1}{2z} \times \Big((z+1)A - (z+3)B + (z+5)C - (z+7)D + (z+9)E - \&c.\Big).$$

The sum of ten terms of the series to be summed taken with their own signs is .8179,6217,5610,9851,3. Then in order that the sum of the rest may be obtained, substitute the eleventh term, that is, $\frac{1}{121}$, for $T$ and for $z$ its corresponding value, namely 11; and you will have

$$A = \tfrac{1}{121},\ B = \tfrac{1.1.1}{12.12.12}A,\ C = \tfrac{2.2.2}{13.13.13}B,\ D = \tfrac{3.3.3}{14.14.14}C,\ \&c.$$

and

$$S = \tfrac{1}{11} \times (6A - 7B + 8C - 9D + 10E - \&c.).$$

Now the calculation is as follows.

| | |
|---|---|
| $A =$ | .0082.6446.2809.9173.5 |
| $B =$ | 478.2675.2372.2 |
| $C =$ | 1.7415.2944.5 |
| $D =$ | 171.3604.0 |
| $E =$ | 3.2495.0 |
| $F =$ | 991.7 |
| $G =$ | 43.6 |
| $H =$ | 2.6 |
| $I =$ | 2 |

Then

| | |
|---:|---:|
| .0495.8677.6859.5041.0 | .0000.3347.8726.6605.4 |
| 13.9322.3556.0 | 1542.2436.0 |
| 32.4950.0 | 1.0908.7 |
| 523.2 | 33.8 |
| 2.8 | |
| +.0495.8691.6214.4073.0 | −.0000.3348.0269.9983.9 |

The difference of these sums divided by 11 gives $S = .0045.0485.7813.1280.8$, which, when added to the aggregate of the ten initial terms, produces for the value of the series .8224.6703.3424.1132.1. And since this number is half of that found in Example 1 of Proposition 11, it is to be concluded that both calculations have been set up correctly. For the series

$$1 - \frac{1}{4} + \frac{1}{9} - \frac{1}{16} + \&c.,$$

in which the terms are alternately negative and positive, is half of this series

$$1 + \frac{1}{4} + \frac{1}{9} + \frac{1}{16} + \&c.,$$

in which all the terms are of the same sign.

### Example 3

Consider the series

$$1 - \frac{1}{3} + \frac{1}{5} - \frac{1}{7} + \frac{1}{9} - \frac{1}{11} + \&c.$$

which is defined by the equation $zT + (z+1)T' = 0$, in which the values of the abscissa are $\frac{1}{2}$, $\frac{3}{2}$, $\frac{5}{2}$, $\frac{7}{2}$, etc. And by beginning the calculation according to this proposition, it will be reduced to the following rule. Put $A = T$, and then

$$B = \frac{1}{2}A \times \frac{1}{2z+1} \times \frac{1}{(z+1)^2},$$
$$C = \frac{1}{2}B \times \frac{2}{2z+3} \times \frac{9}{(z+2)^2},$$
$$D = \frac{1}{2}C \times \frac{3}{2z+5} \times \frac{25}{(z+3)^2},$$
$$E = \frac{1}{2}D \times \frac{4}{2z+7} \times \frac{49}{(z+4)^2},$$
$$\&c.$$

And there will be

$$S = \frac{1}{4z} \times \Big((2z+1)A - (2z+5)B + (2z+9)C - (2z+13)D + (2z+17)E - \&c.\Big).$$

The sum of twelve terms in the proposed series is .7646.0069.1481.8329.5, and the thirteenth term is $T = \frac{1}{25}$ with $z = \frac{25}{2}$; when these have been written in, we will have

$$A = \tfrac{1}{25},\ B = \tfrac{1.1.1}{13.27.27}A,\ C = \tfrac{2.3.3}{14.29.29}B,\ D = \tfrac{3.5.5}{15.31.31}C,\ E = \tfrac{4.7.7}{16.33.33}D,\ \&c.$$

and

$$S = \tfrac{1}{25} \times (13A - 15B + 17C - 19D + 21E - \&c.)\,.$$

| | |
|---|---|
| $A =$ | .0400.0000.0000.0000.0 |
| $B =$ | 422.0744.9614.9 |
| $C =$ | 6452.6422.0 |
| $D =$ | 33.5725.4 |
| $E =$ | 3776.5 |
| $F =$ | 73.4 |
| $G =$ | 2.2 |
| $H =$ | 1 |

| | |
|---|---|
| .5200.0000.0000.0000.0 | .0000.6331.1174.4223.5 |
| 10.9694.9174.0 | 637.8782.6 |
| 7.9306.5 | 1688.2 |
| 55.0 | 2.7 |
| +.5200.0010.9702.8535.5 | −.0000.6331.1812.4697.0 |

Then by dividing the difference of the sums by 25, there results $S =$ .0207.9747.1915.6153.5, which along with the aggregate of the initial terms produces for the value of the series to be summed .7853.9816.3397.4483.0.

## Scholion

Just as one equation defines infinitely many series, so one transformation applies to infinitely many series: and each individual example is to be considered as a theorem; thus the transformation in the last example applies to the general series

$$\frac{1}{m} - \frac{1}{m+n} + \frac{1}{m+2n} - \frac{1}{m+3n} + \frac{1}{m+4n} - \&c.$$

It is to be noted that the law for continuing series transformed by this proposition does not always present itself as in the examples which I have chosen here: but this in no way inconveniences the task. For after about six terms of the series to be summed have been collected together, three or four terms of the transformed series will give what is sought with sufficient accuracy for any purposes; for in practice there is rarely need to continue the calculation beyond nine or ten figures. And the matter comes back to the same, whether the terms of the series to be transformed are of the same or opposite signs, or whether they are assignable or not. For the work will always be light, except where the quantity $r$ in the equation for the series is negative and at the same time approximately equal to one; thus if $r = -\frac{9}{10}$ or $r = -\frac{99}{100}$, the calculation will be laborious. But the experienced analyst will easily avoid these cases, for which there is therefore no point in presenting a remedy.

Let me also add a few things about series of this type

$$x + \tfrac{1}{2}x^2 + \tfrac{1}{6}x^3 + \tfrac{1}{24}x^4 + \tfrac{1}{120}x^5 + \&c.$$

$$x - \tfrac{1}{6}x^3 + \tfrac{1}{120}x^5 - \tfrac{1}{5040}x^7 + \&c.$$

Here the terms when extended to infinity do not have a given ratio to each other, as is the case in the series about which we have been deliberating up to this point, but the preceding terms are infinitely greater than the subsequent ones. These series express the number from a given logarithm, or the sine from a given arc, and are the most simple of its type. They can be transformed by the principles presented above; but the matter is resolved more easily without transformations. Thus, in the series

$$x + \tfrac{1}{2}x^2 + \tfrac{1}{6}x^3 + \tfrac{1}{24}x^4 + \tfrac{1}{120}x^5 + \&c.$$

which expresses the number when its logarithm $x$ is given, if there were $x = 12.3785$, I would throw away the characteristic 12 and seek the number of the logarithm .3785 which would come out in a rapidly converging series on account of the logarithm now being less than one; when this has been given, the number of the logarithm 12.3785 would not be concealed. And it comes to the same thing whether the logarithm is tabular or hyperbolic.

And likewise in seeking the sine from a given arc, if this is greater than the quadrant, let it be subtracted from the semicircle, and an arc which is smaller than the quadrant will be left which has the same sine as the former, being its supplement to make up the semicircle. Now the arc which is smaller than the quadrant will give its sine in a rapidly converging series.

Series which are defined by equations which involve three or more terms of the series can be summed exactly or approximately from the analysis presented above: but it is enough to have laid the foundation of calculations of this type and to have shown the way to others who have the leisure and inclination to make further investigations concerning this material. But lest it may seem to be entirely neglected, we shall give a general theorem from the principles of *De Moivre*, which extends both to the summation and the transformation of this type of series.

(p.226)

## Proposition 13

*In series which arise from division there is the same relation between the terms as between the successive sums.*

Let the fraction be $\dfrac{1}{1-3x+x^2}$, which, when resolved into a series, is

$$1 + 3x + 8x^2 + 21x^3 + 55x^4 + \&c.$$

Then the successive sums will be

$$\frac{1}{1-3x+x^2} = 1 + 3x + 8x^2 + 21x^3 + 55x^4 + 144x^5 + \&c.$$

$$\frac{3x-x^2}{1-3x+x^2} = 3x + 8x^2 + 21x^3 + 55x^4 + 144x^5 + \&c.$$

$$\frac{8x^2-3x^3}{1-3x+x^2} = 8x^2 + 21x^3 + 55x^4 + 144x^5 + \&c.$$

$$\frac{21x^3-8x^4}{1-3x+x^2} = 21x^3 + 55x^4 + 144x^5 + \&c.$$

$$\frac{55x^4-21x^5}{1-3x+x^2} = 55x^4 + 144x^5 + \&c.$$

&c.

And the meaning of the proposition is that when a certain number of these sums are taken they have everywhere the same relation as the same number of terms of the series. For example, in the present case the relation between any three successive terms will be $x^2T - 3xT' + T'' = 0$: and for that reason the relation between three successive sums will also be $x^2S - 3xS' + S'' = 0$, as will be clear to anyone who tries it. Now the proposition is demonstrated as follows.

Let $r$, $s$, $t$ be given quantities, and let the equation for the sums be taken as $rS + sS' + tS'' = 0$; then on substituting the next values of the variables for the present ones, $rS' + sS'' + tS''' = 0$ will be obtained, which, when subtracted from the former, leaves $rS - rS' + sS' - sS'' + tS'' - tS''' = 0$; substitute in this $T$ for $S - S'$, $T'$ for $S' - S''$, and $T''$ for $S'' - S'''$, and $rT + sT' + tT'' = 0$ will result. And this is the same relation as that assumed first of all for the sums. And if there are more or fewer sums, the proposition will be demonstrated in exactly the same manner.

*Corollary.* Hence we have a method for summing these series given the relation of the terms, as is made clear by the following examples.

### Example 1

Let the equation for the terms $rT + sT' = 0$ be given, and by this proposition there will also be the same relation for the sums, namely $rS + sS' = 0$; for $S'$ substitute its value $S - T$, and $rS + sS - sT = 0$ will arise; hence $S = \frac{s}{r+s}T$. Consequently the sum $S$ is given when its first term $T$ is given. Thus if the series is

$$1 + \frac{2}{x} + \frac{4}{x^2} + \frac{8}{x^3} + \frac{16}{x^4} + \frac{32}{x^5} + \&c.,$$

whose equation is $2T - xT' = 0$, there will be $r = 2$, $s = -x$; when these have been written in, $S = \frac{-x}{2-x}T$, or $S = \frac{x}{x-2}T$, will result. Now let any term be substituted for $T$, and $\frac{x}{x-2}T$ will be the sum of it and of all the subsequent terms up to infinity. Let $T$ be equal to the first term, namely one, and $\frac{x}{x-2}$ will be obtained for the value of the whole series.

### Example 2

In the same way, if the equation for three terms is $rT + sT' + tT'' = 0$, the relation of the sums will be $rS + sS' + tS'' = 0$; on writing $S - T$ for $S'$ and $S - T - T'$ for $S''$ in this, $rS + sS - sT + tS - tT - tT' = 0$ will come out, and hence $S = \frac{(s+t)T + tT'}{r+s+t}$: and so $S$ is given when two of its terms have been given. Let the series to be summed be

$$1 + 3x + 8x^2 + 21x^3 + 55x^4 + 144x^5 + \&c.,$$

in which the relation of the terms is $x^2T - 3xT' + T'' = 0$, hence $r = x^2$, $s = -3x$, $t = 1$; when these have been written in, $S = \frac{(1-3x)T + T'}{1-3x+x^2}$ will be obtained. Now substitute the first term for $T$ and the second for $T'$, and $\frac{1}{1-3x+x^2}$ will come out for the value of the series.

Similarly, if the equation for the terms is $rT + sT' + tT'' + vT''' = 0$, there will be

$$S = \frac{(s+t+v)T + (t+v)T' + vT''}{r+s+t+v}.$$

And it continues in this way when the relation is between more terms.

(p.228)

### Scholion

It is to be noted that a relation of terms which is variable becomes closer to an invariable one the further removed the terms are from the beginning, and finally at infinite distance it becomes entirely invariable as in series which have arisen from division. And this I call the *ultimate relation* of the terms, to which their relation continually approximates, but it will never become exact before the terms have been removed from the beginning by an infinite interval.

Now the difference equation defining the series will give the ultimate relation of the terms if all the powers of the abscissa except the highest are rejected and the remaining equation is divided by this power. Consider the equation $x^2T \times (z^2 + 3z) = T' \times (z^2 - 5z + 2)$; reject all the powers of the abscissa lower than the square, and $x^2Tz^2 = T'z^2$ will remain, which, when divided by $z^2$, will give $x^2T = T'$. And this is the ultimate relation of the terms.

Now since the ultimate relation is constant, it provides a method for summing approximately series in which the relation of the terms is variable. If any equation $rT \times (z^2 + az + b) + sT' \times (z^2 + cz + d) = 0$ is given, the ultimate relation of the terms will be $rT + sT' = 0$, hence $S = \dfrac{s}{r+s}T$ approximately. This equation holds exactly when the term $T$ is at infinite distance from the beginning, and approximately when the distance is sufficiently great. Similarly if the equation is $rT \times (z + a) + sT' \times (z + b) + tT'' \times (z + c) = 0$, the ultimate relation will be $rT + sT' + tT'' = 0$, and $S = \dfrac{(s+t)T + tT'}{r+s+t}$ approximately. And so by collecting together some initial terms before the calculation is begun, the sum of the remaining terms will be obtained approximately by this method. From these principles one may also correct the approximation continually as in the following proposition.

(p.229)

## Proposition 14

*Every series $A + B + C + D + E +$ &c. in which the ultimate relation of the terms is $rT + sT' + tT'' = 0$ splits into the following*

$$(s+t) \times \left(\frac{A}{n} + \frac{A_2}{n^2} + \frac{A_3}{n^3} + \frac{A_4}{n^4} + \frac{A_5}{n^5} + \text{\&c.}\right)$$

$$+ t \times \left(\frac{B}{n} + \frac{B_2}{n^2} + \frac{B_3}{n^3} + \frac{B_4}{n^4} + \frac{B_5}{n^5} + \text{\&c.}\right)$$

*where $n = r + s + t$ and*

$$\begin{array}{lll} A_2 = rA + sB + tC, & A_3 = rA_2 + sB_2 + tC_2, & A_4 = rA_3 + sB_3 + tC_3, \\ B_2 = rB + sC + tD, & B_3 = rB_2 + sC_2 + tD_2, & B_4 = rB_3 + sC_3 + tD_3, \\ C_2 = rC + sD + tE, & C_3 = rC_2 + sD_2 + tE_2, & C_4 = rC_3 + sD_3 + tE_3, \\ D_2 = rD + sE + tF, & D_3 = rD_2 + sE_2 + tF_2, & D_4 = rD_3 + sE_3 + tF_3, \\ E_2 = rE + sF + tG, & E_3 = rE_2 + sF_2 + tG_2, & E_4 = rE_3 + sF_3 + tG_3, \\ \text{\&c.} & \text{\&c.} & \text{\&c.} \end{array} \quad \text{\&c.}$$

And so by means of this proposition series in which the ultimate relation of the terms is defined by an equation involving three terms will be transformed into the two series shown here; however the former of these vanishes whenever $s+t = 0$. If the relation is between two terms only, $rT + sT' = 0$, there will be $t = 0$, and for that reason the latter series will vanish, and the series which it is proposed to transform will change into a single series; in this special case this proposition will coincide with a theorem found by Mr *Monmort*. If the relation is between more terms than in this proposition, the series to be transformed will change into more series. But in every case the matter will be clear from what has now been shown. However, I do not wish to abandon simplicity by trying to embrace a great many things in a few words.

## Example 1

Let us consider the series

$$1 + 4x + 9x^2 + 16x^3 + 25x^4 + 36x^5 + \&c.,$$

where the coefficients are the squares of the natural numbers: the difference equation is $xT(z^2 + 2z + 1) - T'z^2 = 0$, from which the ultimate relation is $xT - T' = 0$, and so $r = x$, $s = -1$, $t = 0$, $n = x - 1$; and

$$\begin{array}{llll}
A = 1, & A_2 = -3x, & A_3 = 2x^2, & A_4 = 0, \\
B = 4x, & B_2 = -5x^2, & B_3 = 2x^2, & B_4 = 0, \\
C = 9x^2, & C_2 = -7x^3, & C_3 = 2x^2, & \&c. \\
D = 16x^3, & D_2 = -9x^4, & \&c. & \\
E = 25x^4, & \&c. & & \\
\&c. & & &
\end{array}$$

Therefore all terms after $A_3$ vanish; and on substitution of these values the transformed series is

$$-1 \times \left( \frac{1}{x-1} - \frac{3x}{(x-1)^2} + \frac{2x^2}{(x-1)^3} \right).$$

And, combined into a single term, these three terms give $\frac{1+x}{(1-x)^3}$ for the value of the series. If the sign of $x$ is changed both in the series and its value, the identity

$$\frac{1-x}{(1+x)^3} = 1 - 4x + 9x^2 - 16x^3 + 25x^4 - 36x^5 + \&c.$$

will be obtained. And thence also

$$\frac{1 + 6x^2 + x^4}{(1-x^2)^3} = 1 + 9x^2 + 25x^4 + 49x^6 + 81x^8 + \&c.,$$

which can be obtained directly from the proposition without these detours.

## Example 2

Let us consider the series $1 + 8x + 27x^2 + 64x^3 + 125x^4 + 216x^5 + \&c.$, where the coefficients are the cubes of the natural numbers: and the equation for the series is $xT(z^3 + 3z^2 + 3z + 1) - T'z^3 = 0$, so that the ultimate relation of the terms is $xT - T' = 0$, and thence $r = x$, $s = -1$, $t = 0$, $n = x - 1$; and

$$\begin{array}{lllll}
A = 1, & A_2 = -7x, & A_3 = 12x^2, & A_4 = -6x^3, & A_5 = 0, \\
B = 8x, & B_2 = -19x^2, & B_3 = 18x^3, & B_4 = -6x^4, & B_5 = 0, \\
C = 27x^2, & C_2 = -37x^3, & C_3 = 24x^4, & C_4 = -6x^5, & \&c. \\
D = 64x^3, & D_2 = -61x^4, & D_3 = 30x^5, & \&c. & \\
E = 125x^4, & E_2 = -91x^5, & \&c. & & \\
F = 216x^5, & \&c. & & & \\
\&c. & & & &
\end{array}$$

Here the terms after $A_4$ vanish; and the values substituted in the theorem give

$$-1 \times \left( \frac{1}{x-1} - \frac{7x}{(x-1)^2} + \frac{12x^2}{(x-1)^3} - \frac{6x}{(x-1)^4} \right),$$

and these four terms collected into one give $\dfrac{1+4x+x^2}{(1-x)^4}$ for the value of the series.

## Example 3

Let the following series be given for summation:

$$1 - 6x + 27x^2 - 104x^3 + 366x^4 - 1212x^5 + 3842x^6 - 11784x^7 + \&c.;$$

this is defined by the equation $x^2T(z+4) - 2xT'(z+2) - T''z = 0$; now let $z$ become infinitely large, and the ultimate relation of the terms will be $x^2T - 2xT' - T'' = 0$: hence let $r = x^2$, $s = -2x$, $t = -1$, $n = x^2 - 2x - 1$; and

$$\begin{array}{llll}
A = +1, & A_2 = -14x^2, & A_3 = +29x^4, & A_4 = 0, \\
B = -6x, & B_2 = +44x^3, & B_3 = -70x^5, & B_4 = 0, \\
C = +27x^2, & C_2 = -131x^4, & C_3 = +169x^6, & \&c. \\
D = -104x^3, & D_2 = +376x^5, & D_3 = -408x^7, & \\
E = +366x^4, & E_2 = -1052x^6 & \&c. & \\
F = -1212x^5, & F_2 = +2888x^7, & & \\
G = +3842x^6, & \&c. & & \\
H = -11784x^7, & & & \\
\&c. & & &
\end{array}$$

Now substitute these values for $A$, $A_2$, $A_3$, $B$, $B_2$, $B_3$, and the series set forth will transform into the following two which have finitely many terms:

$$-(2x+1) \times \left( \frac{1}{x^2-2x-1} - \frac{14x^2}{(x^2-2x-1)^2} + \frac{29x^4}{(x^2-2x-1)^3} \right)$$

$$-1 \times \left( \frac{-6x}{x^2-2x-1} + \frac{44x^3}{(x^2-2x-1)^2} - \frac{70x^5}{(x^2-2x-1)^3} \right).$$

When these have been collected into one term they give $\dfrac{1}{(1+2x-x^2)^3}$ for the value of the series.

In the same way the theorem is applied to the transformation of series which cannot be summed. In fact the demonstration is made clear by the following. Let there be any fraction

$$\frac{a + bx + cx^2 + dx^3}{(v + tx + sx^2 + rx^3)^n},$$

whose numerator is a quantity consisting of any collection of members, the number of which is however fixed. Moreover let its denominator be any power of a quantity which is also made up from any finite collection of members. Then whatever the index $n$, if the fraction is resolved into a series, the ultimate relation of the terms will always be the same as if the denominator had been the simple power $v + tx + sx^2 + rx^3$. Therefore if the series is continuously multiplied by this, the process will eventually terminate if $n$ is a positive integer: that is, if the series can be summed by a simple equation.

Let us consider the series

$$1 - 6x + 27x^2 - 104x^3 + 366x^4 - \&c.,$$

in which the ultimate relation of the coefficients is $A - 2B - C = 0$; I take the quantity $x^2 - 2x - 1$, or, the signs having been changed, $1 + 2x - x^2$, in which the coefficients are the same as in the ultimate relation, and I conclude that the series set forth, provided it is summable, is equal to some fraction whose denominator is a certain power of the quantity $1 + 2x - x^2$. Therefore I put

$$S = 1 - 6x + 27x^2 - 104x^3 + 366x^4 - \&c.$$

and I multiply both sides by $1 + 2x - x^2$; and the result is

$$S \times (1 + 2x - x^2) = 1 - 4x + 14x^2 - 44x^3 + 131x^4 - \&c.$$

which, since it does not terminate, I multiply again by the same quantity, and I have

$$S \times (1 + 2x - x^2)^2 = 1 - 2x + 5x^2 - 12x^3 + 29x^4 - \&c.$$

and then multiplying a third time by the same quantity, I obtain $S \times (1 + 2x - x^2)^3 = 1$, all the terms after the first vanishing.

It has therefore been shown that this proposition presents nothing other than a concise method of multiplication and at the same time an arrangement of the terms in the transformed series which brings about rapid convergence when they do not terminate. Hence I leave to the reader the investigation of the demonstration, which is in fact almost the same whether the denominators are trinomials, quadrinomials or have a greater number of members.

(p.233)

## Proposition 15

*To find the equation, be it algebraic or fluxional, whose root will be any given series which is defined by an equation in which the terms of the series are only of one dimension.*

The series is given if some of its initial terms are given along with the law for forming the rest: now when these have been given, the equation having that series for its root will be found, as will be perceived from the following examples.

### Example 1

To find the equation whose root is the series $A+Bx+Cx^2+Dx^3+Ex^4+$&c. in which the relation of the coefficients is constant, namely $rA+sB+tC+vD=0$. Since the relation is invariable, I conclude that the series is equal to a rational fraction, which I extract as follows. Suppose that $y$ is equal to the given series: and by taking the indices of the relation $r$, $s$, $t$, $v$ in reverse order, I multiply $y$ and its value successively by $v$, $tx$, $sx^2$, $rx^3$, and

$$\begin{aligned}
vy &= vA + vBx + vCx^2 + vDx^3 + vEx^4 + vFx^5 + \&c.\\
txy &= \qquad tAx + tBx^2 + tCx^3 + tDx^4 + tEx^5 + \&c.\\
sx^2y &= \qquad\qquad sAx^2 + sBx^3 + sCx^4 + sDx^5 + \&c.\\
rx^3y &= \qquad\qquad\qquad rAx^3 + rBx^4 + rCx^5 + \&c.
\end{aligned}$$

come out. But by hypothesis $rA+sB+tC+vD=0$, $rB+sC+tD+vE=0$, and so on to infinity. Hence all the terms in which $x$ is of degree greater than two vanish; and so there will remain

$$vy + txy + sx^2y + rx^3y = vA \left.\begin{matrix}+vB\\+tA\end{matrix}\right\} x \left.\begin{matrix}+vC\\+tB\\+sA\end{matrix}\right\} x^2.$$

This is the equation determining the value of the series. For example, let $A=2$, $B=-3$, $C=7$, $r=3$, $s=-7$, $t=6$, $v=4$; and the equation will become

$$4y + 6xy - 7x^2y + 3x^3y = 8 - 4x^2, \quad \text{or} \quad y = \frac{8-4x^2}{4+6x-7x^2+3x^3}.$$

### Example 2

To find the equation which has for a root the series

$$1 - \tfrac{2}{3}x^2 + \tfrac{8}{15}x^4 - \tfrac{16}{35}x^6 + \tfrac{128}{315}x^8 - \tfrac{256}{693}x^{10} + \&c.,$$

in which the relations of the coefficients are

$$2A+3B=0, \quad 4B+5C=0, \quad 6C+7D=0, \quad 8D+9E=0, \quad \&c.$$

In this example the equation sought will involve first fluxions, since the numbers in the relations have the same differences. Therefore I take

$$y = A + Bx^2 + Cx^4 + Dx^6 + Ex^8 + \&c.$$

There will be

$$x\dot{y} = 2Bx^2 + 4Cx^4 + 6Dx^6 + 8Ex^8 + \&c.$$
$$y + x\dot{y} = A + 3Bx^2 + 5Cx^4 + 7Dx^6 + 9Ex^8 + \&c.$$
$$2x^2y + x^3\dot{y} = 2Ax^2 + 4Bx^4 + 6Cx^6 + 8Dx^8 + \&c.$$

Now let the last equation be added to the penultimate and

$$y + x\dot{y} + 2x^2y + x^3\dot{y} = A \left.\begin{matrix} +2A \\ +3B \end{matrix}\right\} x^2 \left.\begin{matrix} +4B \\ +5C \end{matrix}\right\} x^4 \left.\begin{matrix} +6C \\ +7D \end{matrix}\right\} x^6 \left.\begin{matrix} +8D \\ +9E \end{matrix}\right\} x^8 + \&c.$$

will come out. But by hypothesis the relations of the coefficients are $2A+3B = 0$, $4B + 5C = 0$, $6C + 7D = 0$, etc. and so all terms after the first $A$ vanish, and the finite equation

$$y + x\dot{y} + 2x^2y + x^3\dot{y} = A = 1, \quad \text{or} \quad y \times (1 + 2x^2) + x\dot{y} \times (1 + x^2) = 1$$

remains, where one, or the first term of the series, has been substituted for the coefficient $A$.

### Example 3

Suppose the equation has to be found whose root is the series

$$1 - \tfrac{1}{4}x^2 - \tfrac{3}{64}x^4 - \tfrac{5}{256}x^6 - \tfrac{175}{16384}x^8 - \&c.,$$

where the relations of the coefficients are

$$-1.1A - 2.2B = 0, \quad 1.3B - 4.4C = 0, \quad 3.5C - 6.6D = 0, \quad 5.7D - 8.8E = 0, \quad \&c.$$

or

$$-A - 4B = 0, \quad 3B - 16C = 0, \quad 15C - 36D = 0, \quad 35D - 64E = 0, \quad \&c.$$

In this case the desired equation will necessarily involve second fluxions, since the numbers which determine the relation of the coefficients are of two dimensions, or the product of numbers having the same differences taken two at a time. And so let

$$y = A + Bx^2 + Cx^4 + Dx^6 + Ex^8 + \&c.$$
$$\dot{y} = 2Bx + 4Cx^3 + 6Dx^5 + 8Ex^7 + \&c.$$
$$x\ddot{y} = 2Bx + 12Cx^3 + 30Dx^5 + 56Ex^7 + \&c.$$
$$\dot{y} + x\ddot{y} = 4Bx + 16Cx^3 + 36Dx^5 + 64Ex^7 + \&c.$$
$$x^3\ddot{y} + x^2\dot{y} - xy = -Ax + 3Bx^3 + 15Cx^5 + 35Dx^7 + \&c.$$

Finally on subtracting this last equation from the penultimate

$$x\ddot{y}-x^3\ddot{y}+\dot{y}-x^2\dot{y}+xy = \left\{\begin{matrix}+A\\+4B\end{matrix}\right\}x \begin{matrix}-3B\\+16C\end{matrix}\left.\right\}x^3 \begin{matrix}-15C\\+36D\end{matrix}\left.\right\}x^5 \begin{matrix}-35D\\+64E\end{matrix}\left.\right\}x^7 + \&c.$$

will remain, that is,

$$x\ddot{y} - x^3\ddot{y} + \dot{y} - x^2\dot{y} + xy = 0, \quad \text{or} \quad \dot{y} + x\ddot{y} + \frac{xy}{1-x^2} = 0.$$

For on account of the relation of the coefficients, all members on one side of the equation vanish.

### Example 4

Let the equation be sought whose root is

$$1 + \tfrac{1}{4}Ax^2 + \tfrac{9}{16}Bx^4 + \tfrac{25}{36}Cx^6 + \tfrac{49}{64}Dx^8 + \&c.$$

where the relations of the coefficients are

$$A - 4B = 0, \quad 9B - 16C = 0, \quad 25C - 36D = 0, \quad \&c.$$

Suppose

$$y = A + Bx^2 + Cx^4 + Dx^6 + Ex^8 + Fx^{10} + \&c.$$

There will be

$$\dot{y} = 2Bx + 4Cx^3 + 6Dx^5 + 8Ex^7 + 10Fx^9 + \&c.$$

and

$$x\ddot{y} = 2Bx + 12Cx^3 + 30Dx^5 + 56Ex^7 + 90Fx^9 + \&c.$$

Then you will find by calculation

$$xy + 3x^2\dot{y} + x^3\ddot{y} = Ax + 9Bx^3 + 25Cx^5 + 49Dx^7 + 81Ex^9 + \&c.$$

$$\dot{y} + x\ddot{y} = 4Bx + 16Cx^3 + 36Dx^5 + 64Ex^7 + 100Fx^9 + \&c.$$

On subtracting the second of these equations from the first you will have

$$xy - \dot{y} \times (1 - 3x^2) - x\ddot{y} \times (1 - x^2)$$

$$= \left\{\begin{matrix}+A\\-4B\end{matrix}\right\}x \begin{matrix}+9B\\-16C\end{matrix}\left.\right\}x^3 \begin{matrix}+25C\\-36D\end{matrix}\left.\right\}x^5 \begin{matrix}+49D\\-64E\end{matrix}\left.\right\}x^7 + \&c.$$

And thence on account of the relation of the coefficients, the equation will be

$$xy = \dot{y} \times (1 - 3x^2) + x\ddot{y} \times (1 - x^2).$$

### Example 5

Let the equation be sought whose root is

$$x + \tfrac{1}{4}x^2 + \tfrac{1}{9}x^3 + \tfrac{1}{16}x^4 + \&c.,$$

where the denominators are the squares of the numbers 1, 2, 3, etc. Take

$$y = x + \tfrac{1}{4}x^2 + \tfrac{1}{9}x^3 + \tfrac{1}{16}x^4 + \tfrac{1}{25}x^5 + \&c.$$

There will be

$$x\dot{y} = x + \tfrac{1}{2}x^2 + \tfrac{1}{3}x^3 + \tfrac{1}{4}x^4 + \tfrac{1}{5}x^5 + \&c.,$$

whose fluxion is

$$x\ddot{y} + \dot{y} = 1 + x + x^2 + x^3 + x^4 + \&c., \quad \text{that is} \quad x\ddot{y} + \dot{y} = \frac{1}{1-x}.$$

Similarly if the series is

$$x + \tfrac{1}{8}x^2 + \tfrac{1}{27}x^3 + \tfrac{1}{64}x^4 + \&c.$$

where the denominators are the cubes, you will find the equation to be

$$\dot{y} + 3x\ddot{y} + x^2\dddot{y} = \frac{1}{1-x}.$$

And if the series is

$$x + \tfrac{1}{16}x^3 + \tfrac{1}{81}x^3 + \tfrac{1}{256}x^4 + \&c.,$$

the denominators being the squares of the squares, the equation will come out as

$$\dot{y} + 6\ddot{y} + 7x^2\dddot{y} + x^3\ddddot{y} = \frac{1}{1-x}.$$

And so on.

(p.235)

## Scholion

Therefore when the relation of the coefficients is given, series are reduced either to fractions or to fluxions, and that with exactly the same ease in all cases; or series may be determined by a relation involving a greater or a smaller number of terms. For since by taking the fluxion of a series the terms are multiplied by the indices of the powers, which always have the same differences, and again by other quantities which have the same differences when second fluxions are taken, and so on, one may form in this way the relations of the terms, whatever these relations may be, by following the steps just described.

Now for the complete summation it is required that the fluxions be reduced to fluents; where this cannot be done, it has to be concluded that the series which is being considered is not one of those which can be summed. Now the fact that sums may be reduced to fluxions contributes much to the advancement of this doctrine, since the method for going back from fluxions to fluents, while it may be imperfect, is nevertheless better known and more developed than that of going back from differences to sums. But each provides assistance for the other: for if fluents can be found, they will express the sums, or conversely, when the sums have been found, they will give the fluents.

From the propositions which have now been presented and from others which can easily be deduced from the same principles one may find the root of any equation very accurately numerically provided it can be expressed in a simple series, even if this converges very slowly. But even now a difficulty remains, if neither does the series approximate rapidly, nor is the relation of the terms simple and suitable for defining by means of a difference equation. Indeed, if the root or its fluxion has degree more than one in the equation to be resolved, or if these are multiplied by each other, and members of this type cannot be eliminated, then in these cases the root cannot be reduced to a simple series made up of powers of the abscissa, at least by techniques presently known to me. But powers of the abscissa in the equation to be resolved, no matter how large these may be, in no way obstruct the simplicity of the series.

But since this opportunity has been presented, I will endeavour to remove the difficulty hitherto untouched, which is however very great in the reduction of fluxional equations. Therefore it has to be known that those fluxional equations which involve all the conditions of the problem, and so determine all the coefficients in its roots, can be resolved just like affected algebraic equations, and that by the rules previously given by *Newton*. But nevertheless it is usually the case that equations do not determine all the coefficients of the terms in their roots; that is because constant quantities vanish and are entirely lost from the equation on going from fluents to fluxions. But also they can vanish in different ways depending on how the fluxion is taken, which may be made clear by an example.

Consider the equation $y^2 = a^2 + bx + \frac{x^4}{c^2}$; and with $x$ flowing uniformly and one written for its fluxion, $2y\dot{y} = b + \frac{4x^3}{c^2}$ will be obtained by the direct method: the root of this extracted by the methods of *Newton* will be the same as the root of the fluent $y^2 = bx + \frac{x^4}{c^2}$, and that is because the quantity $a^2$ is now lost. And in this case the first term of the series, which certainly depends on the vanishing quantity $a^2$, is not determined.

Similarly, if the proposed equation is first divided by $x$, it will show $\frac{y^2}{x} = \frac{a^2}{x} + b + \frac{x^3}{c^2}$, whose fluxion is $\frac{2y\dot{y}}{x} - \frac{y^2}{x^2} = \frac{-a^2}{x^2} + \frac{3x^2}{c^2}$, or on multiplying by $x^2$,

$$2xy\dot{y} - y^2 = -a^2 + \frac{3x^4}{c^2}.$$

And the root of this taken simply is the same as that of the fluent $y^2 = a^2 + \frac{x^4}{c^2}$. In this case the second term of the series, which depends on the coefficient $b$ now lost, is not determined.

Finally, if the fluent equation is divided by $x^4$, then $\frac{y^2}{x^4} = \frac{a^2}{x^4} + \frac{b}{x^3} + \frac{1}{c^2}$ will be obtained, and thence the fluxion will be

$$\frac{2y\dot{y}}{x^4} - \frac{4y^2}{x^5} = \frac{-4a^2}{x^5} - \frac{3b}{x^4}, \quad \text{or} \quad 2xy\dot{y} - 4y^2 = -4a^2 - 3bx;$$

when this has been resolved by the common rules, it will give the same root as the fluent $y^2 = a^2 + bx$: and so the fifth term of the series, which depends on the coefficient $c^2$, is not determined. Hence it is clear that the fluxional equations

$$2y\dot{y} = b + \frac{4x^3}{c^2},$$

$$2xy\dot{y} - y^2 = -a^2 + \frac{3x^4}{c^2},$$

$$2xy\dot{y} - 4y^2 = -4a^2 - 3bx$$

result from the single fluent $y^2 = a^2 + bx + \frac{x^4}{c^2}$: if the first of these is used, the first term of the series is not determined; with the second, the first term is determined but not the second; finally, the first four terms of the series are determined by the third, but the fifth remains undetermined. Therefore various fluxional equations can result from the same fluent. But the method of resolution will not be completed before it has the power to produce all roots of the various fluents from which any proposed fluxion can result by any method. For it is necessary to consider quantities which either vanish in fact or which could vanish.

For one may not extract the root from a fluxional equation as if no quantity vanishes and then add a given quantity to the fluent found as in the quadrature of curves. In any case indeterminate terms are very often invariable quantities: and the addition of a given quantity to the root found is not equivalent to the addition of the quantity to the equation to be resolved. Thus the same fluxion $2y\dot{y} = b + \frac{4x^3}{c^2}$ can result from either of the fluents $y^2 = a^2 + bx + \frac{x^4}{c^2}$ or $y^2 = bx + \frac{x^4}{c^2}$; however the roots of these assign quite different forms, namely the former has $A + Bx + Cx^2 +$ &c., while the latter has $Ax^{1/2} + Bx^{3/2} + Cx^{5/2} +$ &c.

Moreover it can happen that a fluxional equation involves all the coefficients which the fluent involves, and this not withstanding, not all the terms are determined, as in the following example. The equation $y^2 = a^2 + bx + \frac{x^4}{c^2}$ produces directly the fluxion $2y\dot{y} = b + \frac{4x^3}{c^2}$: if this is multiplied by $y$, then

$2y^2\dot{y} = by + \dfrac{4x^3y}{c^2}$ will come out; and then for $y^2$ let its value $a^2 + bx + \dfrac{x^4}{c^2}$ be substituted, and finally $2a^2\dot{y} + 2bx\dot{y} + \dfrac{2x^4\dot{y}}{c^2} = by + \dfrac{4x^3y}{c^2}$ will be obtained. And this is a fluxional equation involving all the coefficients which its fluent involves; nevertheless, the first term of the series is not determined in this. The same will also happen everywhere unless there is present in the equation a member which involves neither the root nor any of its fluxions.

And there are very many difficulties of no lesser significance which, as anyone will easily appreciate, are greatly increased in fluxions of higher orders: for in second fluxions, two terms in no way dependent on each other can be indeterminate, and three in the case of third fluxions, and so on. But it happens not rarely that all terms are determined by an equation which also involves fluxions of higher orders. And all that has been said here about a fluxion whose fluent is known is also true about fluxions whose fluents are not known.

A root of any equation is a quantity which, when written for the letter denoting the root, will cause all the terms of the resulting equation to vanish. Now terms can vanish in only two ways, either as a result of opposite signs in like members, or also where a constant quantity is found in a fluent; for this leaves no trace in the fluxion. Thus if $y = Ax^n$, there will be $\dot{y} = nAx^{n-1}$, $\ddot{y} = (n^2 - n)Ax^{n-2}$; if $n = 0$, the value of the first fluxion will vanish; if $n^2 - n = 0$, that is, $n = 0$ or $n = 1$, the value of the second fluxion will vanish; and that is without other like members which can remove them. And these are the principles by means of which the universal difficulties which occur in the resolution of fluxional equations are to be unravelled.

Suppose that the equation to be resolved is $r^2\dot{y}^2 = r^2\dot{x}^2 - \dot{x}^2y^2$, or if $\dot{x}$ is put equal to one, $r^2\dot{y}^2 = r^2 - y^2$. By the parallelogram or other methods of *Newton*, you will find the root

$$y = x - \frac{x^3}{6r^2} + \frac{x^5}{120r^4} - \frac{x^7}{5040r^6} + \&c.$$

For if the square of this is written for $y^2$ and the square of the fluxion for $\dot{y}^2$ it will cause all members of the resulting equation to cancel each other out as a result of opposite signs. But let us see if there is another root which cannot be found in this way. To that end, suppose $Ax^n$ to be the first term of the series, or $y = Ax^n$ approximately; and there will be $\dot{y} = nAx^{n-1}$, and so $\dot{y}^2 = n^2A^2x^{2n-2}$ and $y^2 = A^2x^{2n}$; when these have been written in, there results $n^2r^2A^2x^{2n-2} = r^2 - A^2x^{2n}$, or on multiplying the whole expression by $x^2$,

$$n^2r^2A^2x^{2n} = r^2x^2 - A^2x^{2n+2}.$$

Here it is clear that the member $n^2r^2A^2x^{2n}$ vanishes when $n = 0$: therefore let 0 be substituted for $n$, and the equation will become

$$0 \times r^2A^2x^0 = r^2x^2 - A^2x^2;$$

therefore in this case the member $r^2A^2x^0$ in which $x$ is of the smallest power vanishes: and for that reason $Ax^0$, or the constant quantity $A$, will be the first term of the series, which converges more rapidly the smaller $x$ is. And by following up this calculation you will find by the common methods that

$$y = A \times \left(1 - \frac{x^2}{2r^2} + \frac{x^4}{24r^4} - \frac{x^6}{720r^6} + \&\text{c.}\right).$$

But the quantity $A$ is not determined by the fluxional equation. For if the fluxion of the equation $r^2\dot{y}^2 = r^2 - y^2$ is taken, $2r^2\dot{y}\ddot{y} = -2y\dot{y}$ will be obtained, or on dividing and transposing, $r^2\ddot{y} + y = 0$. Now let us put $y = Ax^n$; there will be $\dot{y} = nAx^{n-1}$ and $\ddot{y} = (n^2 - n)Ax^{n-2}$, and when these have been substituted, $(n^2 - n) \times r^2Ax^{n-2} + Ax^n = 0$ comes out; here it is clear that the index $n - 2$ cannot be compared with the index $n$; and consequently no root can be extracted by that method. But nevertheless, on taking the coefficient $n^2 - n = 0$, one will obtain $n = 0$ and $n = 1$; and so $A$ or $Ax$ can be the first term of the series, as we have already found.

The first series gives the sine and the second the cosine of a given arc $x$. And the coefficient $A$ in the second is equal to the radius $r$. For if you put $y$ for the sine or the cosine, you will always come up with the equation $r^2\dot{y}^2 = r^2 - y^2$, which is resolved by no root apart from the two which have just been produced. Now one may note from this example that the form of the series is always determined by the equation even if the coefficients are not determined. And where the coefficient of some term is not determined, the index of $x$ in it will always be found by setting some member of the resulting equation equal to zero. But where the coefficient of the term is determined, the index of $x$ in it is found by comparison of two indices in the resulting equation by means of *Newton's* rules.

Let $\ddot{y} + a^2y - x\dot{y} - x^2\ddot{y} = 0$ be the equation to be resolved, where $x$ flows uniformly, and one is written for its fluxion. Suppose $y = Ax^n$ approximately, and there will be $\dot{y} = nAx^{n-1}$, $\ddot{y} = (n^2 - n)Ax^{n-2}$: when these have been substituted into the equation, there results $(n^2 - n)Ax^{n-2} + (a^2 - n^2)Ax^n = 0$. Now set the coefficient $n^2 - n = 0$ and $n = 0$ or $n = 1$ will come out: when these values have been substituted for $n$ in the equation, they produce

$$0 \times Ax^{-2} + a^2Ax^0 = 0, \text{ and}$$
$$0 \times Ax^{-1} + (a^2 - 1)Ax = 0;$$

and since in both cases the member in which $x$ is of smallest power vanishes, I conclude that 0 or 1 is the index of $x$ in the first term of the series which converges more rapidly the smaller $x$ is: and so one may use

$$y = A + Bx^2 + Cx^4 + Dx^6 + \&\text{c.}$$
$$\text{or} \quad y = Ax + Bx^3 + Cx^5 + Dx^7 + \&\text{c.}$$

Now in the resulting equation $(n^2-n)Ax^{n-2}+(a^2-n^2)Ax^n=0$, set the coefficient $a^2-n^2=0$; and there will be $n=\pm a$: therefore write $+a$ and $-a$ for $n$, and

$$(a^2-a)Ax^{a-2}+0\times Ax^a=0$$
$$(a^2+a)Ax^{-a-2}+0\times Ax^{-a}=0$$

will come out. Therefore in both cases the member in which $x$ is of largest power vanishes; and for that reason $+a$ or $-a$ will be the index of the first term of the series which converges more rapidly the larger $x$ is. For we may take

$$y=Ax^a+Bx^{a-2}+Cx^{a-4}+Dx^{a-6}+\&c.$$
$$\text{or}\quad y=Ax^{-a}+Bx^{-a-2}+Cx^{-a-4}+Dx^{-a-6}+\&c.$$

When the coefficients of these four series have been determined they give the following values of the roots:

$$A+\frac{0-a^2}{1.2}Ax^2+\frac{4-a^2}{3.4}Bx^2+\frac{16-a^2}{5.6}Cx^2+\frac{36-a^2}{7.8}Dx^2+\&c.;$$

$$Ax+\frac{1-a^2}{2.3}Ax^2+\frac{9-a^2}{4.5}Bx^2+\frac{25-a^2}{6.7}Cx^2+\frac{49-a^2}{8.9}Dx^2+\&c.;$$

$$Ax^a-\frac{a(a-1)}{4(a-1)}\times\frac{A}{x^2}-\frac{(a-2)(a-3)}{8(a-2)}\times\frac{B}{x^2}-\frac{(a-4)(a-5)}{12(a-3)}\times\frac{C}{x^2}$$
$$-\frac{(a-6)(a-7)}{16(a-4)}\times\frac{D}{x^2}-\&c.;$$

$$\frac{A}{x^a}+\frac{a(a+1)}{4(a+1)}\times\frac{A}{x^2}+\frac{(a+2)(a+3)}{8(a+2)}\times\frac{B}{x^2}+\frac{(a+4)(a+5)}{12(a+3)}\times\frac{C}{x^2}$$
$$+\frac{(a+6)(a+7)}{16(a+4)}\times\frac{D}{x^2}+\&c.$$

The first two relate to the multiplication or division of a circular arc, while the latter pair are concerned with the area of a hyperbola. Let these examples suffice for the illustration of the rule provided here for the determination of the index. It is deduced from Proposition 5; and by means of it and the rules of *Newton*, the extraction of roots as infinite series is brought to a conclusion, so that anyone will easily understand, who has given some consideration to those things which authors have previously made known about this matter. For it will easily be demonstrated that an equation does not admit as a root any series which is not produced by this method. But here I do not speak about series which are made up of terms in which the indeterminate $x$ has indeterminate indices.

# PART TWO

(p.239) ## *On the Interpolation of Series*

Let there be an arbitrary straight line $PQ$ which is given in position, above which let any number of ordinates $A$, $B$, $C$, $D$ etc. be erected, which are parallel to each other and separated one from the other by equal intervals: moreover, let these ordinates represent the terms of a regular series, continually increasing or decreasing and having the same sign; and passing through the extremities of them all, there will be exactly one curve, which will in fact be defined by the given equation for the series, that is, from the given equation which expresses in general the relation between any two or more successive ordinates.

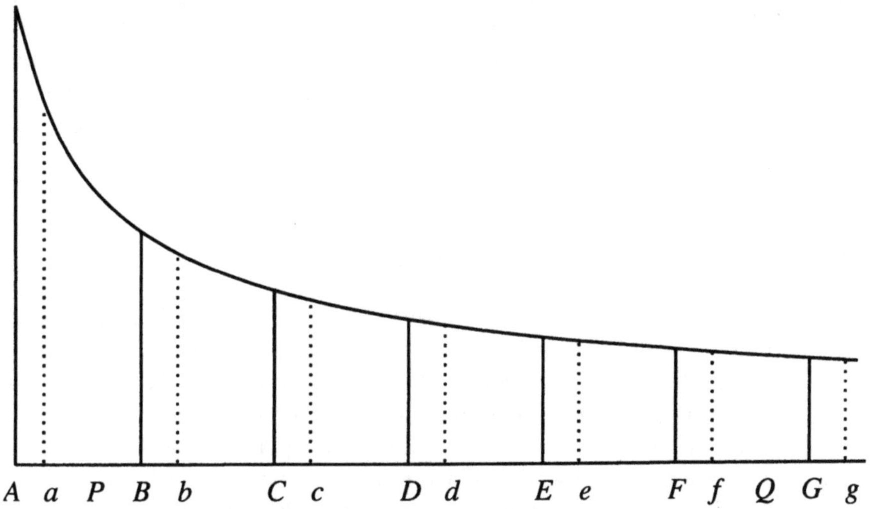


If the algebraic equation of this curve, that is to say the equation which defines the relation between abscissae and corresponding ordinates, can be extracted from that given difference equation, then any ordinate will be obtained when its abscissa is given by resolution of the affected equations, and so the complete interpolation of the series will be obtained; this in fact consists of assigning any term, principal or intermediate, when its position in the series has been given. But when the algebraic equation of the curve cannot be found, which is commonly the case, there is nothing further to be looked for apart from expressing the value of any term sought by means of a convergent series, or perhaps by the quadrature of curves.

Now here I am speaking about equidistant terms whose relations in fact come out on writing equidifferent numbers successively for the abscissa $z$ in the difference equation defining the series. For the common interval of the ordinates standing above the abscissa is proportional to the constant increment of the indeterminate $z$.

(p.239)

## Proposition 16

*If the common interval is the same for the principal terms and the intermediate terms, and if one of the intermediate terms is given, then they will all be determined by the given equation for the principal terms.*

In the figure above let $A$, $B$, $C$, $D$, etc. denote the principal terms and $a$, $b$, $c$, $d$, etc. denote intermediate terms; and let the intervals $AB$, $BC$, $CD$, etc. of the principal terms be equal to the intervals $ab$, $bc$, $cd$, etc. of the intermediate terms, each to each. I say that all the intermediate terms are given if any one of them is given along with the difference equation defining the relation of the principal terms.

For the equation which assigns the relation of the principal terms defines the curve passing through their extremities, and the equation which defines the relation of the intermediate terms also defines the curve passing through the extremities of those terms. But by the definition of the principal and intermediate terms, the same curve passes through the extremities of both; consequently, the same equation which defines the curve will define the relation of the terms in both series. And that equation is given by hypothesis; and so the law is given for continuing the intermediate terms, which thus will all be given if anyone of them is given. Q.E.D.

*Corollary.* Suppose that in any difference equation expressing the relation of the principal terms $A$, $B$, $C$, $D$, etc. the successive values of the abscissa $z$ are $PA$, $PB$, $PC$, $PD$, etc., where $P$ is any initial point of the abscissa: then the relations of the intermediate terms will be obtained by writing for $z$ successively $Pa$, $Pb$, $Pc$, $Pd$, etc. in the same equation. For the difference equation expresses in general the relation between any two ordinates located at a specific distance from each other, whether they occur in the series of principal terms or in that of the intermediate terms.

### Example 1

If in the geometric progression of terms $1, x, x^2, x^3, x^4$, etc. the terms standing in the middle between the principal terms are $a$, $b$, $c$, $d$, $e$, etc., then there will be $b = ax$, $c = bx$, $d = cx$, $e = dx$, etc., for which the relation is in fact the same as that for the principal terms.

### Example 2

If the principal terms are 1, 1, 2, 6, 24, 120, 720, etc., whose relations are $B = A$, $C = 2B$, $D = 3C$, $E = 4D$, etc., and the term located right in the middle between the first two terms 1 and 1 is $a$, then the rest will be given by the equations

$$b = \tfrac{3}{2}a, \quad c = \tfrac{5}{2}b, \quad d = \tfrac{7}{2}c, \quad e = \tfrac{9}{2}d, \quad \&c.,$$

these being located in the middle between any two principal terms, from which they are therefore equidistant.

But if $a$ denotes the term between the first and the second whose distance from the first is a third part of the common interval, put

$$b = \tfrac{4}{3}a, \quad c = \tfrac{7}{3}b, \quad d = \tfrac{10}{3}c, \quad e = \tfrac{13}{3}d, \quad \&c.$$

and $a$, $b$, $c$, $d$, $e$, etc. will be the intermediate terms, each of which is a third part of the common interval from the principal term before it, or two third parts of the same interval from the principal term immediately following it. It is established therefore that the relation of the intermediate terms is given by interpolation of the relation of the principal terms: and it can always be interpolated, since it is assigned by the equation for the series.

### Example 3

Consider the series 1, $\frac{1}{2}$, $\frac{3}{8}$, $\frac{5}{16}$, $\frac{25}{128}$, etc., which is produced by successive multiplication by the numbers $\frac{1}{2}$, $\frac{3}{4}$, $\frac{5}{6}$, $\frac{7}{8}$, etc., and let $a$ be the term in the middle between the first and the second, and put

$$b = \tfrac{2}{3}a, \quad c = \tfrac{4}{5}b, \quad d = \tfrac{6}{7}c, \quad e = \tfrac{8}{9}d, \quad \&c.$$

Then $b$, $c$, $d$, $e$, etc. will be the remaining intermediate terms which stand in the middle between any two principal terms.

### Scholion

If the equation for the series involves three terms, then two must be given in order that the remaining intermediate terms may be obtained; and three must be given if it involves four terms, and so on. Now *Newton's* Proposition 7 in the Treatise on the Quadrature of Curves is of this type; however, it applies not only to curves but also to any series. And this theorem comes into use whenever an intermediate term is sought which is located near the beginning of the series: for in that case its value comes out in a very slowly converging series; consequently, a corresponding intermediate term has first to be sought which is sufficiently removed from the beginning so that its value will come out in a series which approximates rapidly: then, once this is given, one has to go back from it to that first proposed by means of the relations of the terms, as in this proposition.

## (p.241) Proposition 17

*Every series can be interpolated, whose terms are made up of factors which can be interpolated.*

Let $A \times a \times \alpha$, $B \times b \times \beta$, $C \times c \times \gamma$, $D \times d \times \delta$, etc. be a series whose terms are made up from three factors: I say that it can be interpolated, if the three series of factors, namely $A$, $B$, $C$, $D$, etc., $a$, $b$, $c$, $d$, etc., and $\alpha$, $\beta$, $\gamma$, $\delta$, etc., can be interpolated.

For since the intermediate terms in the compound series are formed in the same way from the corresponding intermediate terms in the simple series as the principal terms are formed from the individual principal terms, these will be found by multiplying the respective intermediate terms of the simple series by each other. Thus if $T$ is a term between $B$ and $C$, if $t$ is the corresponding term between $b$ and $c$, and if $\tau$ is the corresponding term between $\beta$ and $\gamma$, then the corresponding term in the compound series, namely that between $B \times b \times \beta$ and $C \times c \times \gamma$, will be the product of those three, namely $T \times t \times \tau$. And the proposition will be demonstrated similarly if there are more or fewer factors. Q.E.D.

*Corollary.* Hence if two or more series of factors in a series which has to be interpolated can be interpolated exactly, they can be removed from the calculation: then the remaining terms have to be interpolated by methods which are to be presented below. For interpolation is not to be undertaken lightly; but before the task is begun one must enquire what is the simplest series on whose interpolation that of the proposed series depends. And this preparation is for the most part entirely necessary, if we are to arrive at neat and refined conclusions.

## Example 1

If the series to be interpolated is

$$1, \quad \tfrac{1}{6}x, \quad \tfrac{3}{40}x^2, \quad \tfrac{5}{112}x^3, \quad \tfrac{35}{1152}x^4, \quad \&c.,$$

I first resolve it into three series of simple factors in the following manner,

$$\begin{array}{cccccc} x^0, & x^1, & x^2, & x^3, & x^4, & \&c., \\ 1, & \frac{1}{3}, & \frac{1}{5}, & \frac{1}{7}, & \frac{1}{9}, & \&c., \\ 1, & \frac{1}{2}, & \frac{3}{8}, & \frac{5}{16}, & \frac{35}{128}, & \&c., \end{array}$$

and it is obvious that of these three series the first two can be interpolated, but not so with the third; however, since it is simpler, it will interpolated more easily than the proposed series. Thus, suppose that the term in the middle between the second and third terms of the compound series is required; it is clear that the corresponding terms in the first and second of the simple series are $x^{3/2}$ and $\frac{1}{4}$ respectively: let the corresponding term in the third be called $T$; then the product of these three $x^{3/2}$, $\frac{1}{4}$, and $T$, or $\frac{1}{4}Tx\sqrt{x}$, will be what is required. And so the interpolation of the compound series is reduced to the interpolation of a simpler series.

## Example 2

If a series of this type

$$1, \quad \frac{r}{p}A, \quad \frac{r+1}{p+1}B, \quad \frac{r+2}{p+2}C, \quad \frac{r+3}{p+3}D, \quad \&c.$$

is given, one may interpolate the series of numerators and denominators separately, that is, the series

$$1, \quad r, \quad r(r+1), \quad r(r+1)(r+2), \quad \&c.$$
$$1, \quad p, \quad p(p+1), \quad p(p+1)(p+2), \quad \&c.$$

Then any term in the series of numerators divided by the corresponding term in the series of denominators will give the corresponding term in the proposed series. If the difference between $r$ and $p$ is a small number, there is no need of this artifice. But when the said difference is large, it is necessary to interpolate the numerators and denominators separately.

## Scholion

Very many preparations of this type must also be applied to this. For example, suppose that for the series

$$\&c., \quad e, \quad d, \quad c, \quad b, \quad a, \quad A, \quad B, \quad C, \quad D, \quad E, \quad \&c.,$$

which goes off to infinity on both sides, the term in the middle between the middle two principal terms $a$ and $A$ is required. Multiply together the principal terms which are equidistant from the middle; that is, multiply $A$ by $a$, $B$ by $b$, etc., and a new series will be formed going off to infinity on both sides,

$$\&c., \quad Dd, \quad Cc, \quad Bb, \quad Aa, \quad Aa, \quad Bb, \quad Cc, \quad Dd, \quad \&c.;$$

the terms of this which are equidistant from the middle are equal to each other, and the term in the middle between $Aa$ and $Aa$ will be the square of the term sought which is located in the middle between $a$ and $A$ in the series previously proposed. Therefore that term which is sought can be extracted by interpolation of either series.

It is to be noted that the desired term can be located in various series, and from that consideration it can sometimes be found more easily. Thus, if some term sought lies in the middle between the first and the second in both of the following series

$$1, \quad r, \quad r(r+1), \quad r(r+1)(r+2), \quad \&c.,$$

$$1, \quad \frac{1}{p}, \quad \frac{1}{p(p+1)}, \quad \frac{1}{p(p+1)(p+2)}, \quad \&c.,$$

then by multiplying corresponding terms together a new series will be produced,

$$1, \quad \frac{r}{p}A, \quad \frac{r+1}{p+1}B, \quad \frac{r+2}{p+2}C, \quad \frac{r+3}{p+3}D, \quad \&c.;$$

the term of this series which has the middle position between the first and second terms is equal to the square of that first proposed.

Sometimes interpolation is also carried out successfully by means of logarithms, especially if the differences of the terms are very large. But these and similar matters are to be learned additionally by practical experience. For just as in common algebra the entire art of the analyst does not consist of the resolution of affected equations, but rather in reducing problems to these, so also in this analysis less skill is required for the resolution of difference equations or the interpolation of series; for there is far greater difficulty in finding series which determine unknown quantities and which are suitable for interpolation.

(p.241)

## Proposition 18

*Suppose that the terms of two series are formed by repeated multiplication by fractions whose numerators and denominators increase continuously by the addition of one and that the numerators are the same in both. I say that the term of one series whose distance from the beginning is the difference of the factors in the other is equal to the term of the second series whose distance from the beginning is the difference of the factors in the first series, provided the first terms are equal to each other.*

Let the two series whose first terms $A$ and $a$ are equal to each other and in which the numerators are the same be

$$B = \frac{r}{p}A, \quad C = \frac{r+1}{p+1}B, \quad D = \frac{r+2}{p+2}C, \quad E = \frac{r+3}{p+3}D, \quad \&c.,$$

$$b = \frac{r}{q}a, \quad c = \frac{r+1}{q+1}b, \quad d = \frac{r+2}{q+2}c, \quad e = \frac{r+3}{q+3}d, \quad \&c.$$

I say that the term of the former whose distance from the beginning is equal to the difference of the factors in the latter, namely $q-r$, is equal to the term of the latter whose distance from the beginning is $p-r$, namely the difference of the factors in the former series. And it is to be noted that, where $p-r$ or $q-r$ is a negative quantity, the terms which are being discussed are located before the first terms by these intervals.

With $A = 1$ let us take any term of the former series, for instance the seventh, namely

$$G = 1 \times \frac{r}{p} \times \frac{r+1}{p+1} \times \frac{r+2}{p+2} \times \frac{r+3}{p+3} \times \frac{r+4}{p+4} \times \frac{r+5}{p+5}.$$

And first of all if $p-r=0$, or $p=r$, there will be $p+1=r+1$, $p+2=r+2$, etc. and so all numerators and denominators cancel each other, and there remains $G=1$.

If $p-r=1$, there will be $p=r+1$, and thence $p+1=r+2$, $p+2=r+3$, etc., in which case all the numerators except the first and all the denominators except the last cancel out, so that

$$G = 1 \times \frac{r}{p+5}, \quad \text{or} \quad G = 1 \times \frac{r}{r+6},$$

on account of $p+5$ being equal to $r+6$.

If $p-r=2$, or $p=r+2$, there will be $p+1=r+3$, $p+2=r+4$, etc. and now all the numerators apart from the first two and all the denominators apart from the last two will cancel out, there remaining

$$G = 1 \times \frac{r}{p+4} \times \frac{r+1}{p+5}, \quad \text{or} \quad G = 1 \times \frac{r}{r+6} \times \frac{r+1}{r+7},$$

on account of $p+4=r+6$ and $p+5=r+7$.

And likewise, if $p-r=3$, or $p=r+3$, all the numerators apart from the first three and all the denominators apart from the last three will cancel out, and in that case there will be

$$G = 1 \times \frac{r}{r+6} \times \frac{r+1}{r+7} \times \frac{r+2}{r+8}.$$

And generally in the value of the term $G$ there will be as many numerators and as many denominators as there are units in $p-r$, as in the following table

$$p-r=0, \quad G=1,$$

$$p-r=1, \quad G = 1 \times \frac{r}{r+6},$$

$$p-r=2, \quad G = 1 \times \frac{r}{r+6} \times \frac{r+1}{r+7},$$

$$p-r=3, \quad G = 1 \times \frac{r}{r+6} \times \frac{r+1}{r+7} \times \frac{r+2}{r+8},$$

$$p-r=4, \quad G = 1 \times \frac{r}{r+6} \times \frac{r+1}{r+7} \times \frac{r+2}{r+8} \times \frac{r+3}{r+9},$$

&c.

And so if one puts $q=r+6$, or $q-r=6$, the term of this series 1, $\frac{r}{p}A$, $\frac{r+1}{p+1}B$, etc. whose interval from the beginning is $q-r$ or 6 is equal to the term of this series 1, $\frac{r}{q}a$, $\frac{r+1}{q+1}b$, etc. whose interval from the beginning is $p-r$. And the proposition will be obvious in other cases from the same reasoning. Q.E.D.

*Corollary.* Hence if the difference of the factors $p$ and $r$ is not very large, a term of the former series, however far removed from the beginning, will always be determined by a term in the latter series which is not far from the beginning. This will be clear from the following examples.

### Example 1

Let $r=3$, $p=5$, $q=10$; and when these values have been substituted, the two series come out as

$$1,\quad \frac{3}{5},\quad \frac{3.4}{5.6},\quad \frac{3.4.5}{5.6.7},\quad \frac{3.4.5.6}{5.6.7.8},\quad \&c.,$$

$$1,\quad \frac{3}{10},\quad \frac{3.4}{10.11},\quad \frac{3.4.5}{10.11.12},\quad \frac{3.4.5.6}{10.11.12.13},\quad \&c.$$

Now $q-r=7$, $p-r=2$; and so the term in the former series whose interval from the beginning is 7 will be equal to the term in the latter whose interval from the beginning is 2, or what is the same, the eighth term of the former series $\frac{3.4.5.6.7.8.9}{5.6.7.8.9.10.11}$ is equal to the third term $\frac{3.4}{10.11}$ of the latter series. And it is to be noted that where the difference between $p$ and $r$ is an integer, then any term of the former series is always equal to some principal term in the latter.

### Example 2

Let the first series be

$$1,\quad \tfrac{2}{1}A,\quad \tfrac{4}{3}B,\quad \tfrac{6}{5}C,\quad \tfrac{8}{7}D,\quad \&c.$$

and since the increment of the factors is two, divide the numerators and denominators by two, and the series

$$1,\quad \frac{1}{\frac{1}{2}}A,\quad \frac{2}{\frac{1}{2}+1}B,\quad \frac{3}{\frac{1}{2}+2}C,\quad \&c.$$

will result, where the increment of the factors is now one; and consequently this series can be compared with that in the theorem, with the result that $r=1$, $p=\frac{1}{2}$. Also let $m$ be the interval between the first term of the series and any other term, and there will be $m=q-r$, or $m=q-1$, and $q=m+1$; when this has been substituted, the second series will become

$$1,\quad \frac{a}{m+1},\quad \frac{2b}{m+2},\quad \frac{3c}{m+3},\quad \frac{4d}{m+4},\quad \&c.,$$

whose term which is removed from the beginning by the interval $p-r$, or $-\frac{1}{2}$, will be equal to the term of the first series whose interval from the beginning is any quantity $m$. That is to say, a term of the first series, however far removed from the beginning, namely by an arbitrarily large distance $m$, will always be equal to the term of the second series which is located before the first term by half the common interval.

Or if the series

$$1, \quad \tfrac{1}{2}A, \quad \tfrac{3}{4}B, \quad \tfrac{5}{6}C, \quad \tfrac{7}{8}D, \quad \&c.$$

is taken, whose terms are the reciprocals of the terms of the series which we have just been discussing, there will be $r = \frac{1}{2}$, $p = 1$; and if $m$ is the interval between the first term and any other term, there will be $m = q - r$, or $m = q - \frac{1}{2}$, and $q = m + \frac{1}{2}$, and the second series will become

$$1, \quad \frac{a}{2m+1}, \quad \frac{3b}{2m+3}, \quad \frac{5c}{2m+5}, \quad \frac{7d}{2m+7}, \quad \&c.$$

The term in this which is removed from the beginning by a distance $p - r$, or $\frac{1}{2}$, that is the term in the middle between the first and the second, will be equal to whatever term of the first series whose distance from the beginning is the arbitrary quantity $m$. Thus, if the thousand-and-first term of the series is required, whose interval from the beginning is in fact one thousand, there will be $m = 1000$, and the term of the series

$$1, \quad \frac{a}{2001}, \quad \frac{3b}{2003}, \quad \frac{5c}{2005}, \quad \frac{7d}{2007}, \quad \&c.$$

which is located in the middle between the first 1 and the second $\frac{1}{2001}$ will be equal to the thousand-and-first term of this series

$$1, \quad \tfrac{1}{2}A, \quad \tfrac{3}{4}B, \quad \tfrac{5}{6}C, \quad \tfrac{7}{8}D, \quad \&c.$$

The intermediate terms of this series are also found in the same way; for if $999\frac{1}{2}$ is written for $m$, the series

$$1, \quad \frac{a}{2000}, \quad \frac{3b}{2002}, \quad \frac{5c}{2004}, \quad \frac{7d}{2006}, \quad \&c.$$

will result, whose term in the middle between the first and the second is equal to the term of this series

$$1, \quad \tfrac{1}{2}A, \quad \tfrac{3}{4}B, \quad \tfrac{5}{6}C, \quad \tfrac{7}{8}D, \quad \&c.$$

which is located in the middle between the thousandth and the thousand-and-first terms.

**Example 3**

Suppose that the term of this series

$$1, \quad \tfrac{2}{1}A, \quad \tfrac{5}{4}B, \quad \tfrac{8}{7}C, \quad \tfrac{11}{10}D, \quad \&c.$$

is sought whose interval from the beginning is an arbitrary quantity $m$: first divide the numerators and the denominators by their common increment 3, and the series will become

$$1, \quad \frac{\frac{2}{3}}{\frac{1}{3}}A, \quad \frac{\frac{5}{3}}{\frac{4}{3}}B, \quad \&\text{c.},$$

and so there will be $r = \frac{2}{3}$, $p = \frac{1}{3}$, $q - r = q - \frac{2}{3} = m$, and thence $q = m + \frac{2}{3}$: hence the second series will become

$$1, \quad \frac{2a}{3m+2}, \quad \frac{5b}{3m+5}, \quad \frac{8c}{3m+8}, \quad \frac{11d}{3m+11}, \quad \&\text{c.},$$

whose term which is at a distance $p - r$, or $-\frac{1}{3}$, from the beginning, that is, the term which is located before the first term by a third part of the common interval, is equal to the term of the first series which is removed from the beginning by an arbitrarily large interval $m$.

(p.242)

## *On the Differences of Quantities*

Consider $a$, $b$, $c$, $d$, $e$, a series of any number of quantities, and if each term is taken away from the following term, the first differences $b - a$, $c - b$, $d - c$, $e - d$ will remain: then if each of these differences is likewise taken away from the next one, the second differences $c - 2b + a$, $d - 2c + b$, $e - 2d + c$ will be left; again the differences of these form the third differences $d - 3c + 3b - a$, $e - 3d + 3c - b$ of the quantities $a$, $b$, $c$, $d$, $e$. And the process continues in this way until it reaches the last difference as in the following table.

| | 1st | 2nd | 3rd | 4th |
|---|---|---|---|---|
| $a$ | | | | |
| | $b - a$ | | | |
| $b$ | | $c - 2b + a$ | | |
| | $c - b$ | | $d - 3c + 3b - a$ | |
| $c$ | | $d - 2c + b$ | | $e - 4d + 6c - 4b + a$ |
| | $d - c$ | | $e - 3d + 3c - b$ | |
| $d$ | | $e - 2d + c$ | | |
| | $e - d$ | | | |
| $e$ | | | | |

Let $1 - x$ be the binomial in which the coefficients $+1$ and $-1$ are the same as the coefficients in the first differences; then the coefficients of the square $1 - 2x + x^2$, namely $+1$, $-2$, $+1$, will be the coefficients in the second differences; likewise, the coefficients of the cube $1$, $-3$, $+3$, $-1$ will be the coefficients in the third differences: and in general the coefficients in any order of differences will be the coefficients in the corresponding power of the binomial. And now that these things have been noted, one may proceed in one step to any desired order of differences without consideration of the intermediate ones.

Two quantities have a first difference, three have a second, four have a third; and they cannot have further differences. But it sometimes happens that a certain order of differences forms an identical progression, in which case further differences are not obtained, however large the number of quantities may be. Thus in an arithmetic progression the first differences are equal, and so the second differences are not given. And in the series of squares 1, 4, 9, 16, 25, etc., whose roots are equidifferent, the first differences 3, 5, 7, 9, etc. are in arithmetic progression, the second differences are equal, and for that reason the third differences are zero. So also in the series of cubes 1, 8, 27, 64, 125, 216, etc. the first differences are 7, 19, 37, 61, 91, etc., the second 12, 18, 24, 30, etc., the third 6, 6, 6, etc., all being equal; and so the fourth differences are zero.

And generally let $A$, $B$, $C$, $D$, $E$ be any number of given quantities and $z$ a variable; then in the expression $A+Bz+Cz^2+Dz^3+Ez^4$ write successively for $z$ any equidifferent numbers; and the final differences of the resulting quantities will be determined by the highest power $z^4$, no account having been taken of the lower powers. Thus in this case the fourth difference is the last on account of the fact that the fourth power $z^4$ is here the highest.

Very often differences form a convergent series in cases where they do not terminate. Thus if $a$, $b$, $c$, $d$, $e$, etc. are almost equal to each other, and their first differences $b-a$, $c-b$, $d-c$, $e-d$, etc. are also almost equal to each other, and likewise also the second and subsequent differences are approximately equal in each case, then

$$a, \quad b-a, \quad c-2b+a, \quad d-3c+3b-a, \quad \&c.$$

will form a convergent series. Likewise the differences of terms whose relation is defined by the equation

$$T \times \left(z^\theta + az^{\theta-1} + bz^{\theta-2} + \&c.\right) = T' \times \left(z^\theta + cz^{\theta-1} + dz^{\theta-2} + \&c.\right),$$

when taken as before, will form a convergent series. But it is not to be expected that the differences of arbitrary quantities will either converge or terminate. This only happens in those quantities which increase or decrease exactly or approximately at the same rate as certain fixed powers of equidifferent numbers.

## (p.242) *On the Description of Curves Through Given Points*

After pointing out the technique for avoiding very complicated series in the quadrature of trinomial curves, *Newton* says in his letter to *Oldenburg* sent in the year 1676, "But I consider these things of less importance, because where simple series are not sufficiently manageable, I have another method not yet made known, by which one may come arbitrarily close to what is sought. Its basis is a convenient, ready and general solution of this problem, *To describe the geometric curve, which will pass through any number of given points. Euclid* showed how to draw the circle through three points. Also a conic section can be drawn through five given points: and a curve of three dimensions can be drawn through seven given points; (so that I am able to describe all curves of that order, which are determined by just seven points). These come about immediately by geometry with no calculation involved. But the above problem is of another type: and however intractable it may seem at first sight, the matter nevertheless turns out otherwise. For it is quite amongst the most beautiful things which I desired to solve."

*Newton* teaches the description of the conical parabola through four points in Proposition 60 of the *Arithmetica Universalis*: or rather he teaches a method for finding an equation for the parabola which will pass through four given points. And by the same method one may describe the curve of the third order through nine points, and the curve of the fourth order through fourteen points. And so on in the remaining cases. But our purpose does not require such a general solution; for it is enough to describe the parabolic figure through the extremities of an arbitrary number of ordinates which are parallel to each other and also to the axis of the curve. But the organic description of curves by the movement of angles, or any other method, is not of use for the present purpose: the matter comes back to the same whether the curve is actually drawn or it is conceived as having been drawn. For curves are in no way necessary here, except in so far as they are an aid to the mind for the correct understanding of the problem. For the description of the parabola through given points is exactly the same problem as the assignment of quantities when their differences are given; this is always achieved by algebra alone, namely by the resolution of simple equations.

(p.243)

## Proposition 19

*Let a series of equidistant ordinates going off to infinity on just one side be given, and let it be required to find the parabolic curve which will pass through the extremities of all of them.*

Let $A$, $A_1$, $A_2$, $A_3$, $A_4$, etc. denote equidistant ordinates standing as perpendiculars on the abscissa. Form their first differences $B$, $B_1$, $B_2$, $B_3$, etc., their second $C$, $C_1$, $C_2$, etc., third $D$, $D_1$, etc., and so on. Thus, $A$ is the first ordinate, $B$ is the first difference of the first two ordinates, $C$ is the second difference of the first three ordinates, $D$ is the third difference of the first four ordinates, and so on. Now the differences must be formed by taking the first from the second everywhere, that is, by putting

$$B = A_1 - A, \quad B_1 = A_2 - A_1, \quad \&c., \quad C = B_1 - B, \quad \&c.,$$

and so on with the rest; those which arise from the subtraction of a larger from a smaller quantity are to be taken as negative.

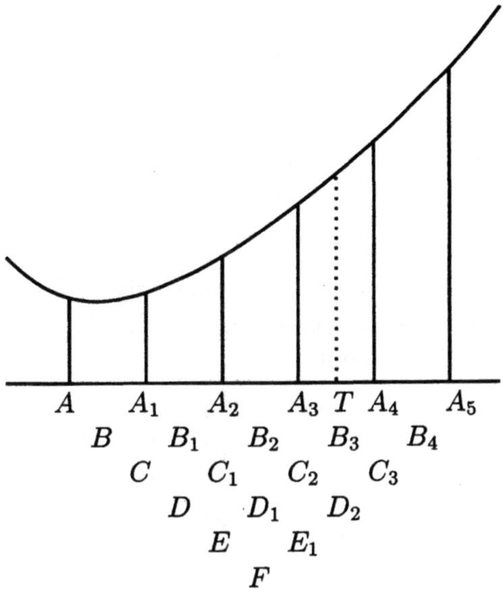


These things having been set down, let $T$ be any general ordinate, either principal or intermediate, whose distance from the first ordinate $A$, namely $AT$, is to the common interval of the equidistant ordinates as the indeterminate quantity $z$ is to one, and there will be

$$
\begin{aligned}
T = A &+ \\
&B \times \frac{z}{1} + \\
&C \times \frac{z}{1} \times \frac{z-1}{2} + \\
&D \times \frac{z}{1} \times \frac{z-1}{2} \times \frac{z-2}{3} + \\
&E \times \frac{z}{1} \times \frac{z-1}{2} \times \frac{z-2}{3} \times \frac{z-3}{4} + \\
&F \times \frac{z}{1} \times \frac{z-1}{2} \times \frac{z-2}{3} \times \frac{z-3}{4} \times \frac{z-4}{5} +
\end{aligned}
$$

&c.

This is the value of any ordinate $T$ lying on the same side of the first ordinate $A$ as the remaining ordinates lie: but if it lay on the other side, then the sign of the abscissa $z$ would have to be changed. For I take the abscissa to be positive if it is directed to the right among the later ordinates, and negative if it is extended on the other side. Now the proposition is proved as follows.

Imagine the ordinate $T$ to be taken away along the abscissa by parallel motion so that it takes up successively the positions of the rest. And since its distance from the first ordinate is taken so that it is to the common interval of the ordinates as $z$ is to one, then $z$ will be successively equal to 0, 1, 2, 3, 4, etc. and meanwhile $T$ will be equal to the ordinates $A$, $A_1$, $A_2$, $A_3$, etc. in turn, each in its place. Therefore, in order to extract the coefficients $A$, $B$, $C$, $D$, etc. which cause the parabola to pass through the extremities of the ordinates, in the equation for the figure

$$T = A + B\frac{z}{1} + C\frac{z}{1} \times \frac{z-1}{2} + D\frac{z}{1} \times \frac{z-1}{2} \times \frac{z-2}{3} + \&c.$$

write the ordinates $A$, $A_1$, $A_2$, $A_3$, etc. successively for $T$ and at the same time for $z$ the lengths of the abscissa in order, that is, 0, 1, 2, 3, etc., and the following equations will come out

$$\begin{aligned} A &= A, & \text{whence } A &= A, \\ A_1 &= A + B, & B &= A_1 - A, \\ A_2 &= A + 2B + C, & C &= A_2 - 2A_1 + A, \\ A_3 &= A + 3B + 3C + D, & D &= A_3 - 3A_2 + 3A_1 - A, \\ A_4 &= A + 4B + 6C + 4D + E, & E &= A_4 - 4A_3 + 6A_2 - 4A_1 + A, \\ &\&c. & &\&c. \end{aligned}$$

In fact the values of the coefficients $A$, $B$, $C$, etc. are extracted in turn from the values of the ordinates $A$, $A_1$, $A_2$, etc. It is clear from these that the first ordinate $A$ is the first coefficient; likewise the difference of the first two ordinates is the second coefficient; and the second difference of the first three is the third coefficient; and so on to infinity. Therefore when the values of the coefficients have been put in the solution, they make the parabola pass through the extremities of the ordinates. Q.E.D.

## *The Same Thing in Another Way*

Let us suppose that in general

$$T = A + B\frac{z}{1} + C\frac{z}{1} \times \frac{z-1}{2} + D\frac{z}{1} \times \frac{z-1}{2} \times \frac{z-2}{3} + \&c.,$$

where $A$, $B$, $C$, $D$, etc. are coefficients to be determined. Write the next values of the variables $T'$, $z+1$ in place of the preceding ones $T$, $z$, and

$$T' = A + B\frac{z+1}{1} + C\frac{z+1}{1} \times \frac{z}{2} + D\frac{z+1}{1} \times \frac{z}{2} \times \frac{z-1}{3} + \&c.,$$

will come out; if the value of $T$ is subtracted from this,

$$T' - T = B + C\frac{z}{1} + D\frac{z}{1} \times \frac{z-1}{2} + \&c.$$

will be obtained. On substituting $T''$, $T'$, $z+1$ for $T'$, $T$, $z$,

$$T'' - T' = B + C\frac{z+1}{1} + D\frac{z+1}{1} \times \frac{z}{2} + \&c.$$

will come out, and when the value of $T' - T$ has been subtracted from this it leaves

$$T'' - 2T' + T = C + D\frac{z}{1} + \&c.$$

And you will find similarly

$$T''' - 3T'' + 3T' - T = D + \&c.$$

Now let $T$ denote the first ordinate, and the corresponding value of the abscissa will be zero; when this has been substituted you will find

$$\begin{aligned} A &= T, \\ B &= T' - T, \\ C &= T'' - 2T' + T, \\ D &= T''' - 3T'' + 3T' - T, \\ &\&c. \end{aligned}$$

That is, the first coefficient $A$ is equal to the first ordinate $T$, the second $B$ is equal to the difference between the first two ordinates $T$ and $T'$, the third $C$ is equal to the second difference of the first three ordinates $T$, $T'$, $T''$, the fourth $D$ is equal to the third difference of the first four ordinates, and so on with the rest as has just been demonstrated.

### Example 1

Let five ordinates 1, 4, 2, 3, 9 be given, through the extremities of which it is required that the parabola should pass. Form their first differences 3, −2, 1, 6, their second differences −5, 3, 5, their third differences 8, 2, and their final difference −6. Then according to what has been written out in the solution of the proposition, the first ordinate and the first of each difference are to be taken respectively for $A$, $B$, $C$, etc.; that is,

$$\begin{array}{ccccccccc} 1 & & 4 & & 2 & & 3 & & 9 \\ & 3 & & -2 & & 1 & & 6 & \\ & & -5 & & 3 & & 5 & & \\ & & & 8 & & 2 & & & \\ & & & & -6 & & & & \end{array}$$

$$A = 1, \quad B = 3, \quad C = -5, \quad D = 8, \quad E = -6,$$

while $F$, $G$, etc. will be zero. When these values have been substituted, the equation for the parabola will come out as

$$T = 1 + 3\frac{z}{1} - 5\frac{z}{1} \times \frac{z-1}{2} + 8\frac{z}{1} \times \frac{z-1}{2} \times \frac{z-2}{3} - 6\frac{z}{1} \times \frac{z-1}{2} \times \frac{z-2}{3} \times \frac{z-3}{4},$$

which, having been brought back into order, is

$$T = \frac{12 + 116z - 111z^2 + 34z^3 - 3z^4}{12}.$$

And to check the working, write 0, 1, 2, 3, 4, successively for the abscissa $z$, and the five proposed ordinates will result for $T$.

But the ordinates can be taken in reverse order, provided the signs of the differences are changed in alternate orders: then we have to put

$$A = 9, \quad B = -6, \quad C = 5, \quad D = -2, \quad E = -6;$$

when these have been written in and the resulting equation has been brought back into order, we will obtain finally

$$T = \frac{108 - 92z + 9z^2 + 14z^3 - 3z^4}{12};$$

if 0, 1, 2, 3, 4 are substituted for $z$ in this, the proposed ordinates will come out in reverse order. And here two equations are obtained for the same parabola, since the abscissa is drawn starting from the first ordinate in one case, but from the last in the other.

### Example 2

Now let it be required to find an equation for the parabola which passes through the extremities of the six equidistant ordinates 5, 3, 7, 23, 57, 115. Form their first differences, and then the rest until we have arrived at the final ones, as at the side. And you will find $A = 5$, $B = -2$, $C = 6$, $D = 6$. When these have been substituted, there arises

| | | | | | | | | | | |
|---|---|---|---|---|---|---|---|---|---|---|
| 5 | | 3 | | 7 | | 23 | | 57 | | 115 |
| | −2 | | 4 | | 16 | | 34 | | 58 | |
| | | 6 | | 12 | | 18 | | 24 | | |
| | | | 6 | | 6 | | 6 | | | |

$$T = 5 - 2\frac{z}{1} + 6\frac{z}{1} \times \frac{z-1}{2} + 6\frac{z}{1} \times \frac{z-1}{2} \times \frac{z-2}{3},$$

which after reduction is $T = 5 - 3z + z^3$. And if you write for $z$ successively 0, 1, 2, 3, 4, 5, the six proposed ordinates will come out.

A straight line passes through two points, a conical parabola through three, a cubic through four, a biquadratic through five, and so on to infinity. Now it sometimes happens that a curve of lower order passes through more points as in the last example. Moreover, the order of the parabola is always denoted by the final order of differences. But if the number of ordinates is infinite, and a progression of equal differences does not arise, I say that in that case the curve will be of infinitely many dimensions, the value of $T$ expanding into an infinite series.

### Scholion

In this solution we have taken the common distance of the ordinates to be one; now if an arbitrary quantity $n$ had been used for this, there would have resulted

$$T = A + B \times \frac{z}{n} + C \times \frac{z}{n} \times \frac{z-n}{2n} + D \times \frac{z}{n} \times \frac{z-n}{2n} \times \frac{z-2n}{3n} + \&c.$$

Now set

the second ordinate $A_1 = A + \dot{A}n$,
the third ordinate $A_2 = A + 2\dot{A}n + \ddot{A}n^2$,
the fourth ordinate $A_3 = A + 3\dot{A}n + 3\ddot{A}n^2 + \dddot{A}n^3$,
the fifth ordinate $A_4 = A + 4\dot{A}n + 6\ddot{A}n^2 + 4\dddot{A}n^3 + \ddddot{A}n^4$,
&c.

And you will find in turn

$$B = \dot{A}n, \quad C = \ddot{A}n^2, \quad D = \dddot{A}n^3, \quad E = \ddddot{A}n^4, \quad \text{\&c.};$$

when these values have been substituted for $B$, $C$, $D$, $E$, etc., there will arise

$$T = A + \dot{A}\frac{z}{1} + \ddot{A}\frac{z}{1} \times \frac{z-n}{2} + \dddot{A}\frac{z}{1} \times \frac{z-n}{2} \times \frac{z-2n}{3} + \text{\&c.}$$

Now let the common interval $n$ become zero, and $\dot{A}$, $\ddot{A}$, $\dddot{A}$, etc. will become the fluxions of the first ordinate $A$, provided that the fluxion of the abscissa $z$ is one; and

$$T = A + \dot{A}z + \tfrac{1}{2}\ddot{A}z^2 + \tfrac{1}{6}\dddot{A}z^3 + \tfrac{1}{24}\ddddot{A}z^4 + \text{\&c.}$$

will result. Therefore when the equidistant ordinates coincide, we come upon the series in which the coefficients of the terms are the fluxions of the first ordinate divided respectively by the numbers 1, 2, 6, 24, etc., which are generated by repeated multiplication of these numbers 1, 2, 3, 4, etc. And Mr *Taylor* was the first to discover this in his *Methodus Incrementorum*, and afterwards *Herman* gave it in the Appendix to his *Phoronomia*.

Hence suppose that the ordinate of any curve is resolved into a series of this form $A+Bz+Cz^2+Dz^3+$&c., where the exponents of the abscissa $z$ are positive integers; the first term $A$ is the first ordinate, namely the one which passes through the beginning of the abscissa; the first two $A + Bz$ denote the straight line which passes through two coincident points of the curve and which consequently is tangential to the curve; the first three terms $A + Bz + Cz^2$ define the conical parabola which passes through three coincident points of the curve and which for that reason is tangential to the curve and has the same curvature at the point through which the first ordinate passes; the first four terms $A + Bz + Cz^2 + Dz^3$ define the cubic parabola which passes through four coincident points of the curve, that is, which is tangential to the curve and has the same curvature and variation of curvature at the point of contact. Finally, the whole series $A + Bz + Cz^2 + Dz^3 +$&c. is the ordinate of the parabola of infinitely many dimensions which is tangential to the curve, and at the point of contact has the same curvature, variation of curvature, variation of variation, and so on to infinity, as is explained by *Newton* in Proposition 10 of the second book of the *Principia*. Or what comes back to the same, the whole series is the ordinate of the parabola passing through infinitely many ordinates of the curve, which are equidistant and coincident with the first ordinate.

Hence we have an idea of the analogy which there is between the method of differences and the common method of series; the latter proceeds by means of fluxions or ultimate ratios of differences, while the former generally uses differences of arbitrary magnitude.

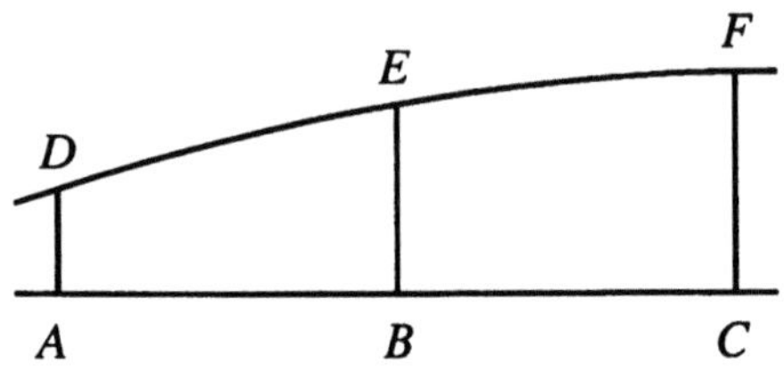


Let $DEF$ denote any curve, whose abscissa $AC$ meets the equidistant ordinates $AD$, $BE$, $CF$ at right angles. And let $AB = z$, $AD = A$; and from the above there will be

$$BE = A + \dot{A}z + \tfrac{1}{2}\ddot{A}z^2 + \tfrac{1}{6}\dddot{A}z^3 + \tfrac{1}{24}\ddddot{A}z^4 + \&c.,$$

that is to say, this value of BE is the ordinate of the parabolic curve which coincides with the other curve at the point D: therefore for the area of the curve one may use the area of the same parabola, which by the inverse method of fluxions produces

$$ABED = Az + \tfrac{1}{2}\dot{A}z^2 + \tfrac{1}{6}\ddot{A}z^3 + \tfrac{1}{24}\dddot{A}z^4 + \tfrac{1}{120}\ddddot{A}z^5 + \&c.$$

And in exactly the same manner, if $BE$ is called $y$ and $AB = BC = z$, the area $BCFE$ will be

$$BCFE = yz + \tfrac{1}{2}\dot{y}z^2 + \tfrac{1}{6}\ddot{y}z^3 + \tfrac{1}{24}\dddot{y}z^4 + \tfrac{1}{120}\ddddot{y}z^5 + \&c.$$

If the sign of the abscissa $z$ is changed in this, the area $BEDA$ will be obtained, expressed as a negative number, that is to say, by changing the sign of the abscissa the area lying on the other side of the ordinate will be obtained. But that area expressed positively becomes

$$BEDA = yz - \tfrac{1}{2}\dot{y}z^2 + \tfrac{1}{6}\ddot{y}z^3 - \tfrac{1}{24}\dddot{y}z^4 + \tfrac{1}{120}\ddddot{y}z^5 - \&c.$$

And this is the series of Mr *Johann Bernoulli* expressing the area in terms of the last ordinate and its fluxions; and we have now given this area in terms of fluxions of the first ordinate. But it has to be noted that the former series does not extend to cases in which the first ordinate is tangential to the curve, and the series of Mr *Bernoulli* does not extend to those in which the last ordinate is tangential to the curve. For the parabola whose area is used for the area of the curve whose quadrature is required can be tangential to none of the ordinates; and so it cannot coincide with the other curve tangential to its ordinate. For expressions of this type for the area and ordinate of curves presuppose the form of the series to be $A + Bz + Cz^2 + \&c.$ in which the exponents of the abscissa $z$ are positive integers.

(p.244)

## Proposition 20

*Let a series of equidistant ordinates going off to infinity on both sides be given, and let it be required to find the parabolic curve which will pass through the extremities of them all.*

### *First Case*

Let $a$ denote the ordinate in the middle of them all, and let $a_2$, $a_4$, $a_6$, $a_8$, etc. be the ordinates on one side, and ${}_2a$, ${}_4a$, ${}_6a$, ${}_8a$ etc. those on the other, the progression going off to infinity on both sides. Form their first differences ${}_7B$, ${}_5B$, ${}_3B$, ${}_1B$, $B_1$, $B_3$, $B_5$, $B_7$, their second differences ${}_6b$, ${}_4b$, ${}_2b$, $b$, $b_2$, $b_4$, $b_6$, their third differences ${}_5C$, ${}_3C$, ${}_1C$, $C_1$, $C_3$, $C_5$, their fourth differences ${}_4c$, ${}_2c$, $c$, $c_2$, $c_4$, and so on with the rest, by always taking preceding from subsequent as in the previous proposition.

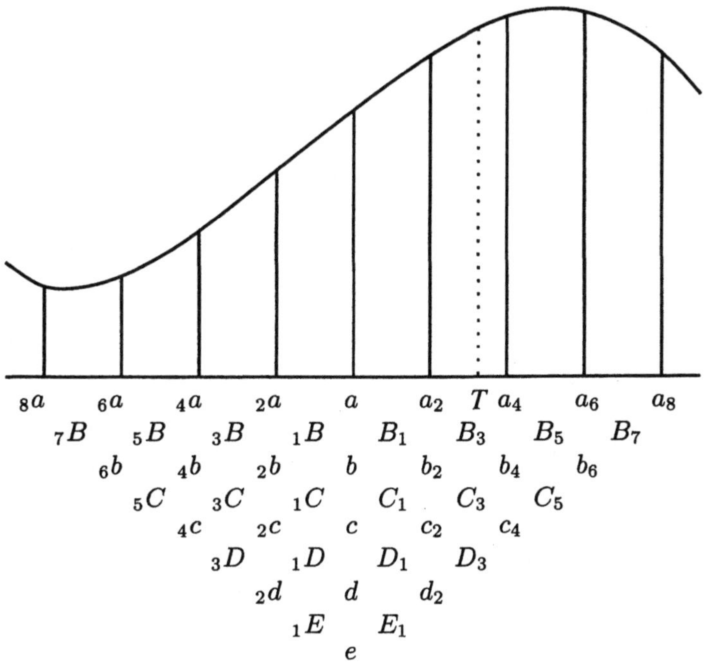


Now let $a$, $b$, $c$, $d$, $e$, etc. be the middle ordinate and the differences in alternate orders respectively. And let ${}_1B$ and $B_1$, ${}_1C$ and $C_1$, ${}_1D$ and $D_1$, ${}_1E$ and $E_1$, etc. be the two middle differences in the other orders; put

$$B = {}_1B + B_1, \quad C = {}_1C + C_1, \quad D = {}_1D + D_1, \quad E = {}_1E + E_1, \quad \&c.$$

And let the interval between an arbitrary ordinate $T$ and the middle one $a$ be to the common interval of the equidistant ordinates as $z$ is to one; and the ordinate $T$ will be

$$\begin{aligned} T = a + {} & \frac{Bz + bz^2}{1.2} + \\ & \frac{2Cz + cz^2}{1.2} \times \frac{z^2 - 1}{3.4} + \\ & \frac{3Dz + dz^2}{1.2} \times \frac{z^2 - 1}{3.4} \times \frac{z^2 - 4}{5.6} + \\ & \frac{4Ez + ez^2}{1.2} \times \frac{z^2 - 1}{3.4} \times \frac{z^2 - 4}{5.6} \times \frac{z^2 - 9}{7.8} + \\ & \frac{5Fz + fz^2}{1.2} \times \frac{z^2 - 1}{3.4} \times \frac{z^2 - 4}{5.6} \times \frac{z^2 - 9}{7.8} \times \frac{z^2 - 16}{9.10} + \\ & \&c. \end{aligned}$$

Here it is to be noted that the abscissa $z$ is negative when the desired ordinate $T$ lies on the opposite side of the middle ordinate.

## *Second Case*

Now let ${}_1A$ and $A_1$ be the two middle ordinates, and $A_3$, $A_5$, $A_7$, $A_9$, etc. those on one side, ${}_3A$, ${}_5A$, ${}_7A$, ${}_9A$, etc. those on the other. Form their first differences ${}_8a$, ${}_6a$, ${}_4a$, ${}_2a$, $a$, $a_2$, $a_4$, $a_6$, $a_8$, their second differences ${}_7B$, ${}_5B$, ${}_3B$, ${}_1B$, $B_1$, $B_3$, $B_5$, $B_7$, their third differences ${}_6b$, ${}_4b$, ${}_2b$, $b$, $b_2$, $b_4$, $b_6$, and so on, by taking the former from the latter everywhere.

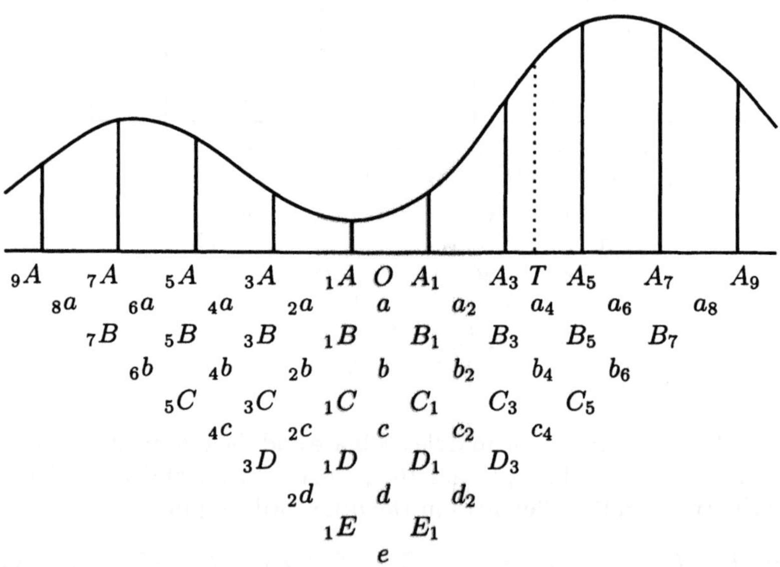


&c.

Now take out the middle differences $a$, $b$, $c$, $d$, $e$, etc. and also the middle two in the other orders, namely ${}_1A$ and $A_1$, ${}_1B$ and $B_1$, ${}_1C$ and $C_1$, ${}_1D$ and $D_1$, ${}_1E$ and $E_1$ etc. and put

$$A = {}_1A + A_1,\ B = {}_1B + B_1,\ C = {}_1C + C_1,\ D = {}_1D + D_1,\ E = {}_1E + E_1,$$

Let $O$ be the point in the middle between the two middle ordinates ${}_1A$ and $A_1$. And let the distance of an arbitrary ordinate T from the middle point, namely $OT$, be to the common interval of the equidistant ordinates as $z$ is to two: and there will be

$$\begin{aligned} T = {} & \frac{A+az}{2} + \\ & \frac{3B+bz}{2} \times \frac{z^2-1}{4.6} + \\ & \frac{5C+cz}{2} \times \frac{z^2-1}{4.6} \times \frac{z^2-9}{8.10} + \\ & \frac{7D+dz}{2} \times \frac{z^2-1}{4.6} \times \frac{z^2-9}{8.10} \times \frac{z^2-25}{12.14} + \\ & \frac{9E+ez}{2} \times \frac{z^2-1}{4.6} \times \frac{z^2-9}{8.10} \times \frac{z^2-25}{12.14} \times \frac{z^2-49}{16.18} + \\ & \&c. \end{aligned}$$

And also in this case $z$ is positive when $T$ lies on the same side of the middle point $O$ as in the diagram, and it is negative when it lies on the opposite side. Now both cases are very easily demonstrated by the method of the previous proposition.

**Example 1**

Let five ordinates $-3$, $-8$, $1$, $12$, $37$ be given, through the extremities of which a parabola is to be drawn. Determine their first differences $-5$, $9$, $11$, $25$, their second differences $14$, $2$, $14$, their third differences $-12$, $12$, and their final difference $24$. Then since the number of ordinates is odd, proceed according to the first case. And starting from the middle ordinate, continue to the middle differences in alternate orders, putting

| | | | | | | | | |
|---|---|---|---|---|---|---|---|---|
| −3 | | −8 | | 1 | | 12 | | 37 |
| | −5 | | 9 | | 11 | | 25 | |
| | | 14 | | 2 | | 14 | | |
| | | | −12 | | 12 | | | |
| | | | | 24 | | | | |

$$a = 1, \quad b = 2, \quad c = 24 \quad \text{and then} \quad B = 9 + 11 = 20, \quad C = -12 + 12 = 0.$$

And when these have been substituted, there results

$$T = 1 + \frac{20z + 2z^2}{1.2} + \frac{24z^2}{1.2} \times \frac{z^2-1}{3.4}, \quad \text{or} \quad T = 1 + 10z + z^4.$$

And this is the ordinate of the parabola of four dimensions passing through the extremities of the five proposed ordinates, as will be confirmed by writing the numbers $-2, -1, 0, 1, 2$ successively for $z$. In this case the curve crosses the base, since some of the ordinates are negative and some positive.

## Example 2

Now let six ordinates 1, 5, 10, 10, 5, 1 be given through the extremities of which it is required to draw a parabola.

Determine their differences as at the side: and since the number of ordinates is even, the second case is applied. Then beginning from the two middle ordinates, and proceeding to the two middle differences, there will be

$$\begin{array}{ccccccccccc}
1 & & 5 & & 10 & & 10 & & 5 & & 1 \\
& 4 & & 5 & & 0 & & -5 & & -4 & \\
& & 1 & & -5 & & -5 & & 1 & & \\
& & & -6 & & 0 & & 6 & & & \\
& & & & 6 & & 6 & & & & \\
& & & & & 0 & & & & &
\end{array}$$

$$A = 10 + 10 = 20,\ B = -5 - 5 = -10,\ C = 6 + 6 = 12;$$
$$\text{then } a = 0,\ b = 0,\ c = 0.$$

When these values have been substituted, there will be

$$T = \frac{20}{2} - \frac{30}{2} \times \frac{z^2 - 1}{4.6} + \frac{60}{2} \times \frac{z^2 - 1}{4.6} \times \frac{z^2 - 9}{8.10};$$

when this equation has been brought back into order it becomes

$$T = \frac{689 - 50z^2 + z^4}{64}.$$

And to test the operation, write $-5, -3, -1, 1, 3, 5$ successively for $z$, and they will produce the proposed ordinates. For in the second case of the proposition the common interval of the ordinates, or, what is equivalent, the increment of the abscissa, is equal to two.

In this example powers of the abscissa of odd dimensions are lacking, since the ordinates on both sides at equal distances from the beginning of the abscissa have the same sign and are equal to each other. For in this case the equation for the parabola remains the same, even if the sign of the abscissa is changed. But if the proposed ordinates had been $+1, -5, +10, -10, +5, -1$, or $+1, +5, +10, -10, -5, -1$, where the ordinates equidistant from the middle are still equal, but have different signs, then I say that in that case the powers of the abscissa of even dimensions would be lacking.

## Scholion

After *Newton* several celebrated geometers have dealt with the description of the curve of parabolic type through any number of given points. But all their solutions are the same as those which have just been shown; indeed these differ scarcely from *Newton's* solutions, as will be confirmed by the fifth Lemma of the third Book of the *Principia* and the *Methodus Differentialis* which was edited by Mr *Jones*. Indeed *Newton* describes the parabola through given points; others have studied the assignation of terms from given differences; but in whatever way it may be perceived, in whatever form it may be expressed, the problem is the same. And of course the discovery of the forms which the values of the ordinate $T$ have is extremely ingenious and worthy of the most distinguished author: but after the forms are obtained, the investigation of the problem is easy, for which nothing other than the resolution of simple equations is required.

But it is to be noted that the form of the ordinate $A + Bz + Cz^2 + Dz^3 + \&c.$ made up of powers, which *Newton* applied for demonstrating the foundation of his method, is ill-suited to the present purpose. For the value of any coefficient comes out in an infinite series; but anyone who applies the forms used here, will arrive at the previous conclusions with very little effort.

(p.245)

## Proposition 21

*Given a series of principal terms to find the intermediate terms which are not far removed from the beginning.*

Let ordinates which are respectively equal to the principal terms be erected at right angles above a straight line which is given in position and equidistant from each other; then let the parabolic line which passes through the extremities of all of them be determined by means of the two previous propositions; this will also pass through the extremities of the intermediate terms, which will therefore be given once the equation for the parabola is given. Q.E.I.

### Example 1

Let the series to be interpolated be

$$1,\ \tfrac{1}{2},\ \tfrac{3}{8},\ \tfrac{5}{16},\ \tfrac{35}{128},\ \tfrac{63}{256},\ \&c.$$

whose terms are produced by repeated multiplication by the numbers $\frac{1}{2}$, $\frac{3}{4}$, $\frac{5}{6}$, $\frac{7}{8}$, etc. Determine the differences of the terms and the differences of the differences as follows:

$$\begin{array}{ccccccccccc}
1 & & \frac{1}{2} & & \frac{3}{8} & & \frac{5}{16} & & \frac{35}{128} & & \frac{63}{256}\ \&c. \\
& -\frac{1}{2} & & -\frac{1}{8} & & -\frac{1}{16} & & -\frac{5}{128} & & -\frac{7}{256} & \\
& & \frac{3}{8} & & \frac{1}{16} & & \frac{3}{128} & & \frac{3}{256} & & \\
& & & -\frac{5}{16} & & -\frac{5}{128} & & -\frac{3}{256} & & & \\
& & & & \frac{35}{128} & & \frac{7}{256} & & & & \\
& & & & & -\frac{63}{256} & & & & &
\end{array}$$

Since this series goes off to infinity on only one side, the interpolation has to be carried out by means of Proposition 19. And if the first term 1 is used for the first ordinate, there will be

$$A = 1, \quad B = -\tfrac{1}{2}, \quad C = +\tfrac{3}{8}, \quad D = -\tfrac{5}{16}, \quad E = +\tfrac{35}{128}, \quad \&c.$$

When these have been substituted,

$$T = 1 - \frac{1}{2} \times \frac{z}{1} + \frac{3}{8} \times \frac{z}{1} \times \frac{z-1}{2} - \frac{5}{16} \times \frac{z}{1} \times \frac{z-1}{2} \times \frac{z-2}{3} + \&c.$$

comes out, that is,

$$T = 1 - \frac{1}{2} A \frac{z}{1} - \frac{3}{4} B \frac{z-1}{2} - \frac{5}{6} C \frac{z-2}{3} - \frac{7}{8} D \frac{z-3}{4} - \&c.,$$

where $A$, $B$, $C$, $D$, etc. now denote the terms of this series in the *Newtonian* manner. But any principal term can be used for the first ordinate; for example, let the second be taken, and there will be

$$A = \tfrac{1}{2}, \quad B = -\tfrac{1}{8}, \quad C = +\tfrac{1}{16}, \quad D = -\tfrac{5}{128}, \quad E = +\tfrac{7}{256}, \quad \&c.$$

and so

$$T = \frac{1}{2} - \frac{1}{8} \times \frac{z}{1} + \frac{1}{16} \times \frac{z}{1} \times \frac{z-1}{2} - \&c.,$$

that is,

$$T = \frac{1}{2} - \frac{1}{4} A \frac{z}{1} - \frac{3}{6} B \frac{z-1}{2} - \frac{5}{8} C \frac{z-2}{3} - \frac{7}{10} D \frac{z-3}{4} - \&c.$$

But it has to be realised that $z$ is the distance between the term sought and that term which is used for the first ordinate. Thus if the desired term $T$ stands in the middle between the first and the second, put $+\frac{1}{2}$ for $z$ in the first expression for $T$ and $-\frac{1}{2}$ for it in the second expression; and for the same term $T$ you will have the following two series

$$1 - \tfrac{1}{4}A + \tfrac{3}{16}B + \tfrac{15}{36}C + \tfrac{35}{64}D + \tfrac{63}{100}E + \&c.$$

$$\tfrac{1}{2} + \tfrac{1}{8}A + \tfrac{9}{24}B + \tfrac{25}{48}C + \tfrac{49}{80}D + \tfrac{81}{120}E + \&c.$$

And since these series converge extremely slowly, they are to be summed by means of the theorem in the scholion of Proposition 11. And it should be understood that the values of the terms turn out to be very simple when the term which is located nearest to the desired intermediate term is used for the first ordinate. But when the term sought is very far removed from the beginning, Proposition 18 comes into use, as will be shown in what follows.

### Example 2

Let the series to be interpolated be 1, 1, 2, 6, 24, 120, 720, etc. whose terms are generated by repeated multiplication of the numbers 1, 2, 3, 4, 5, etc.

Since these terms increase very rapidly, their differences will form a divergent progression, as a result of which the ordinate of the parabola does not approach the true value. Therefore in this and in similar cases I interpolate the logarithms of the terms, whose differences can in fact form a rapidly convergent series, even if the terms themselves increase very rapidly as in the present example.

Now I propose to find the term which stands in the middle between the first two 1 and 1. And since the logarithms of the initial terms have slowly convergent differences, I first seek the term standing in the middle between two terms which are sufficiently far removed from the beginning, for example, that between the eleventh term 3628800 and the twelfth term 39916800: and when this is given, I may go back to the term sought by means of Proposition 16. And since there are some terms located on both sides of the intermediate term which is to be determined first, I set up the operation by means of the second case of Proposition 20. For where the calculation is in numbers rather than symbols, one may proceed by this method as long as a sufficiently large number of terms located on both sides of the term sought is given, even if the series to be interpolated does not actually go off to infinity on both sides.

Now I extract from the table the logarithms of the twelve terms, the first of which is the sixth term 120, so that there are six before and just as many after the term which is sought. Then since that desired term is located right in the middle of them all, the abscissa $z$ in the second case of Proposition 20 will be $z = 0$; and consequently the first, third, and the rest of the odd-order differences, which are multiplied by $z$, will not enter into the calculation; therefore I only collect together the second, fourth, and the other even-order differences as you see:

| Logarithms | 2nd diff. | 4th diff. | 6th diff. | 8th | 10th |
|---|---|---|---|---|---|
| 2.0791812460 | | | | | |
| 2.8573324964 | 669467896 | | | | |
| 3.7024305364 | 579919470 | 21154180 | | | |
| 4.6055205234 | 511525224 | 14443928 | 2568588 | | |
| 5.5597630329 | 457574906 | 10302264 | 1446210 | 541511 | |
| 6.5597630329 | 413926852 | 7606810 | 865343 | 259252 | 156590 |
| 7.6011557180 | 377885608 | 5776699 | 543728 | 133583 | 65082 |
| 8.6803369641 | 347621063 | 4490316 | 355696 | 72996 | |
| 9.7942803164 | 321846834 | 3559629 | 240660 | | |
| 10.9404083521 | 299632234 | 2869602 | | | |
| 12.1164996111 | 280287236 | | | | |
| 13.3206195938 | | | | | |

And now I take out the two middle logarithms and the two middle differences in each case; and I put their sums equal to $A$, $B$, $C$, $D$, etc. respectively as you see:

$$\begin{array}{r} 6.5597630329 \\ 7.6011557180 \\ \hline A = 14.1609187509, \end{array} \quad \begin{array}{r} 413926852 \\ 377885608 \\ \hline B = 791812460, \end{array} \quad \begin{array}{r} 7606810 \\ 5776699 \\ \hline C = 13383509, \end{array}$$

$$\begin{array}{r} 865343 \\ 543728 \\ \hline D = 1409071, \end{array} \quad \begin{array}{r} 259252 \\ 133583 \\ \hline E = 392835, \end{array} \quad \begin{array}{r} 156590 \\ 65082 \\ \hline F = 221672. \end{array}$$

Then in the second case of Proposition 20 I substitute 0 for $z$, and I have

$$T = \tfrac{1}{2}A - \tfrac{1}{16}B + \tfrac{3}{256}C - \tfrac{5}{2048}D + \tfrac{35}{65536}E - \tfrac{63}{524288}F + \&c.$$

Now write in the values just found for $A$, $B$, $C$, $D$, $E$, $F$, and the calculation will be as follows:

$$\begin{array}{r} 7.08045937545 \\ 1568380 \\ 2098 \\ \hline + \ 7.08047508023 \end{array} \qquad \begin{array}{r} 494882787 \\ 34401 \\ 266 \\ \hline - \ 494917454 \end{array}$$

And on subtracting the sum of the negative terms from that of the positive terms, there will remain $T = 7.07552590569$. And this is the logarithm of the number 11899423.08, which indeed stands in the middle between the terms 3628800 and 39916800.

Now just as the principal terms are formed by multiplying the first repeatedly by the numbers 1, 2, 3, 4, etc., so by Proposition 16 the intermediate terms are generated by multiplying the intermediate term between the first and the second repeatedly by the numbers $1\frac{1}{2}$, $2\frac{1}{2}$, $3\frac{1}{2}$, $4\frac{1}{2}$, etc. For example, the product of the ten factors $1\frac{1}{2}$, $2\frac{1}{2}$, $3\frac{1}{2}$, $4\frac{1}{2}$, $5\frac{1}{2}$, $6\frac{1}{2}$, $7\frac{1}{2}$, $8\frac{1}{2}$, $9\frac{1}{2}$, $10\frac{1}{2}$ and the term which stands in the middle between the first and the second is equal to the intermediate term just found, 11899423.08, whose place is indeed in the middle between the eleventh and twelfth terms. Therefore if that intermediate term is divided by $10\frac{1}{2}$, and the quotient by $9\frac{1}{2}$, and the new quotient by $8\frac{1}{2}$, and so on up to the divisor $1\frac{1}{2}$, then the last quotient will be equal to the term in the middle between 1 and 1. Now the intermediate terms coming out by that division are

| Principal | Intermediate |
|---|---|
| 39916800 | |
| | 11899423.08 |
| 3628800 | |
| | 1133278.389 |
| 362880 | |
| | 119292.4620 |
| 40320 | |
| | 14034.40729 |
| 5040 | |
| | 1871.254305 |
| 720 | |
| | 287.8852777 |
| 120 | |
| | 52.34277777 |
| 24 | |
| | 11.63172839 |
| 6 | |
| | 3.323350969 |
| 2 | |
| | 1.329340388 |
| 1 | |
| | .8862269251 |
| 1 | |
| | 1.7724538502 |

From this it is established that the term between 1 and 1 is .8862269251, whose square is .7853... etc., namely the area of a circle whose diameter is one. And twice that term, 1.7724538502, namely the term which stands before the first principal term by half the common interval, is equal to the square root of the number 3.1415926... etc., which denotes the circumference of a circle whose diameter is one. For if the squares of the principal terms form a new series 1, 1, 4, 36, 576, 1440, etc. the term in the middle between the first and the second will be to one as the area of the circle is to the circumscribed square: and the term which stands before the first by half the common interval will be to one as the circumference of a circle is to its diameter. But it will be shown in what follows how series of this type can be interpolated without logarithms.

### Example 3

Suppose that quadrature of the curve whose ordinate is $x^{\theta-1} \times (e+fx^{\eta})^{\lambda}$ is required. Write 0, 1, 2, 3, etc. successively for the index $\lambda$, and the series of equidistant ordinates

$$x^{\theta-1} \times (e+fx^{\eta})^{0}, \quad x^{\theta-1} \times (e+fx^{\eta})^{1}, \quad x^{\theta-1} \times (e+fx^{\eta})^{2}, \quad \text{\&c.},$$

will come out, between which the proposed ordinate will be located at an interval $\lambda$ from the beginning. Therefore the area sought will have the same position among the areas of those ordinates, which form the following progression of equidistant terms:

$$\frac{1}{\theta}x^{\theta}, \quad \frac{e}{\theta}x^{\theta}+\frac{f}{\theta+\eta}x^{\theta+\eta}, \quad \frac{e^2}{\theta}x^{\theta}+\frac{2ef}{\theta+\eta}x^{\theta+\eta}+\frac{f^2}{\theta+2\eta}x^{\theta+2\eta}, \quad \text{\&c.}$$

Now if these areas were to be interpolated by means of Proposition 19, the same series would result for the area sought as that obtained by the common method of resolving the ordinate into a convergent series in order that the fluent may be found from it. But suppose the terms in the series of areas are first divided respectively by the terms of this geometric progression, namely

$$(e+fx^{\eta})^0, \quad (e+fx^{\eta})^1, \quad (e+fx^{\eta})^2, \quad (e+fx^{\eta})^3, \quad \text{\&c.}$$

That is, suppose we put

$$A = x^{\theta} \times \frac{\frac{1}{\theta}}{(e+fx^{\eta})^0},$$

$$A_1 = x^{\theta} \times \frac{\frac{e}{\theta}+\frac{fx^{\eta}}{\theta+\eta}}{(e+fx^{\eta})^1},$$

$$A_2 = x^{\theta} \times \frac{\frac{e^2}{\theta}+\frac{2efx^{\eta}}{\theta+\eta}+\frac{f^2x^{2\eta}}{\theta+2\eta}}{(e+fx^{\eta})^2},$$

$$A_3 = x^{\theta} \times \frac{\frac{e^3}{\theta}+\frac{3e^2fx^{\eta}}{\theta+\eta}+\frac{3ef^2x^{2\eta}}{\theta+2\eta}+\frac{f^3x^{3\eta}}{\theta+3\eta}}{(e+fx^{\eta})^3},$$

&c.

Let the calculation be set up according to Proposition 19, and the differences will be found to be

$$B = \frac{-\eta f x^{\theta+\eta}}{\theta(\theta+\eta)(e+fx^{\eta})}, \quad C = \frac{+2\eta^2 f^2 x^{\theta+2\eta}}{\theta(\theta+\eta)(\theta+2\eta)(e+fx^{\eta})^2},$$

$$D = \frac{-6\eta^3 f^3 x^{\theta+3\eta}}{\theta(\theta+\eta)(\theta+2\eta)(\theta+3\eta)(e+fx^{\eta})^3}, \quad \text{\&c.}$$

When these have been substituted for $A$, $B$, $C$, $D$, etc. and $\lambda$ for $z$, you will find the term at distance $\lambda$ from the beginning to be

$$\frac{x^{\theta}}{\theta}-\frac{\lambda\eta f x^{\theta+\eta}}{\theta(\theta+\eta)(e+fx^{\eta})}+\frac{\lambda(\lambda-1)\eta^2 f^2 x^{\theta+2\eta}}{\theta(\theta+\eta)(\theta+2\eta)(e+fx^{\eta})^2}$$

$$-\frac{\lambda(\lambda-1)(\lambda-2)\eta^3 f^3 x^{\theta+3\eta}}{\theta(\theta+\eta)(\theta+2\eta)(\theta+3\eta)(e+fx^{\eta})^3}+\text{\&c.}$$

But because the terms to be interpolated had been divided by the powers of $e + fx^\eta$, namely each by the power whose index was the distance of the term from the beginning, now, inversely, multiply the term just found by the power of the stated binomial, whose index is $\lambda$, in fact its distance from the beginning, and for the area of the curve

$$\frac{x^\theta}{\theta}\times(e+fx^\eta)^\lambda\times\left(1-\frac{\lambda\eta fx^\eta}{(\theta+\eta)(e+fx^\eta)}+\frac{\lambda(\lambda-1)\eta^2f^2x^{2\eta}}{(\theta+\eta)(\theta+2\eta)(e+fx^\eta)^2}-\text{\&c.}\right)$$

will be obtained. Or on putting

$$y=\frac{fx^\eta}{e+fx^\eta},\quad r=\frac{\theta+\eta}{\eta},$$

and writing the series in the *Newtonian* manner,

$$\frac{x^\theta}{\theta}\times(e+fx^\eta)^\lambda-\frac{\lambda}{r}Ay-\frac{\lambda-1}{r+1}By-\frac{\lambda-2}{r+2}Cy-\frac{\lambda-3}{r+3}Dy-\text{\&c.}$$

will come out for the area of the curve whose ordinate is, generally, $x^{\theta-1}\times(e+fx^\eta)^\lambda$. And when this series has been transformed by means of Proposition 7, it will change into *Newton's* series for the quadrature of binomial curves. It terminates when the index $\lambda$ is a positive integer, but after the necessary preparation of the ordinate it will always terminate when quadrature of the curve can be effected. But its principal usefulness is that it expresses the areas in a series which is quite simple. If the coefficients $e$, $f$ have opposite signs, *Newton's* series is to be preferred, and ours where the signs are the same.

### Example 4

Let it be required to assign the binomial coefficients given the middle coefficient in a power whose index is an even number. If $n$ denotes the index of the power and the middle coefficient is multiplied repeatedly by the fractions

$$\frac{n}{n+2},\quad\frac{n-2}{n+4},\quad\frac{n-4}{n+6},\quad\text{\&c.},$$

the products will be the remaining coefficients located on both sides of the middle one:

$$a_6 = a \times \frac{n}{n+2} \times \frac{n-2}{n+4} \times \frac{n-4}{n+6},$$
$$a_4 = a \times \frac{n}{n+2} \times \frac{n-2}{n+4},$$
$$a_2 = a \times \frac{n}{n+2},$$
$$a = a,$$
$${}_2a = a \times \frac{n}{n+2},$$
$${}_4a = a \times \frac{n}{n+2} \times \frac{n-2}{n+4},$$
$${}_6a = a \times \frac{n}{n+2} \times \frac{n-2}{n+4} \times \frac{n-4}{n+6}.$$

And if $a$ denotes the middle coefficient, then $a_2$, $a_4$, $a_6$, etc. on one side and ${}_2a$, ${}_4a$, ${}_6a$, etc. on the other will denote the remaining coefficients. Then, beginning the calculation according to the first case of Proposition 20, you will find

$$a - \frac{r^2}{2(n+2)}A - \frac{r^2-4}{4(n+4)}B - \frac{r^2-16}{6(n+6)}C - \frac{r^2-36}{8(n+8)}D - \text{\&c.}$$

to be the term of the series which has to be interpolated, whose distance from the middle term $a$ is to the common interval of the principal terms as $r$ is to two. For example, in the twelfth power the coefficients are 1, 12, 66, 220, 495, 792, 924, 792, etc., the middle coefficient being $a = 924$. And if that coefficient is required which is at distance three from the middle, there will be $r = 6$; and when this has been substituted along with 12 for $n$,

$$924 - \tfrac{36}{2.14}A - \tfrac{32}{4.16}B - \tfrac{20}{6.18}C$$

comes out for the coefficient sought, the series terminating: and these terms with the fractions removed are $924 - 1188 + 594 - 110$, the sum of which according to their own signs is 220, which is the value of the coefficient sought.

And in the same way, if the index $n$ is an odd number and $A$ is either of the middle coefficients, then the coefficient whose distance from the intermediate point between the two middle coefficients is to the common interval as $r$ is to two will be

$$A - \frac{r^2-1}{2(n+3)}A - \frac{r^2-9}{4(n+5)}B - \frac{r^2-25}{6(n+7)}C - \frac{r^2-49}{8(n+9)}D - \text{\&c.}$$

And in a very large power these series will converge provided that the interval between the middle coefficient and the one sought is very small compared with the index of the power.

### Scholion

After the series to be interpolated has been prepared by means of Proposition 17 as required, even if it goes off to infinity on both sides, one may proceed by means of Proposition 19 except where the terms on both sides at equal distances from the middle are equal to each other; where this happens, the first case of Proposition 20 may be used, if by a certain rule some principal term can claim the place in the middle of all the terms: or if two terms can claim the middle place by the same rule, the second case of the same proposition may be used. And in other cases one may proceed more or less at will.

(p.251)

## Proposition 22

*Given a series of equidistant terms, to find any term, principal or intermediate, no matter how far distant from the beginning of the series.*

If the term sought is far removed from the beginning, then by means of Proposition 18 determine another series in which that desired term forms a term near the beginning; then proceed as in the above proposition.

### Example 1

Let it be proposed to find any term at an arbitrarily large interval $m$ from the beginning for this series

$$1, \quad \tfrac{2}{1}A, \quad \tfrac{4}{3}B, \quad \tfrac{6}{5}C, \quad \tfrac{8}{7}D, \quad \&c.$$

By Proposition 18 the term of the series

$$1, \quad \frac{a}{m+1}, \quad \frac{2b}{m+2}, \quad \frac{3c}{m+3}, \quad \frac{4d}{m+4}, \quad \&c.$$

which is located before the first term by half the common interval, will be equal to the term of the former series whose interval from the beginning is $m$. Now it is established by Example 2 of Proposition 21 that the term which is at a distance half of the common interval before the first term in the series of numerators 1, 1.1, 1.1.2, 1.1.2.3, etc., that is, in this series 1, 1, 2, 6, 24, 120, etc., is the square root of the number 3.1415926 ... etc. Consequently, I only interpolate the denominators, namely

$$1, \quad \frac{1}{m+1}, \quad \frac{1}{(m+1)(m+2)}, \quad \frac{1}{(m+1)(m+2)(m+3)}, \quad \&c.$$

And since this series can be continued to infinity on both sides, it will in fact be continued, and will become

$$\&c., \quad (m-2)(m-1)m, \quad (m-1)m, \quad m, \quad 1, \quad \frac{1}{m+1}, \quad \frac{1}{(m+1)(m+2)}, \quad \&c.$$

Here the term sought is located in the middle between the two middle terms $m$ and 1: but because the differences of these terms are very large, let those which are at the same distance on both sides of the middle be multiplied by each other; that is, $m$ by 1, $(m-1)m$ by $\frac{1}{m+1}$, and so on; and the new series

$$\&c., \quad \frac{m-2}{m+2}\cdot\frac{m-1}{m+1}m, \quad \frac{m-1}{m+1}m, \quad m, \quad m, \quad m\frac{m-1}{m+1}, \quad m\frac{m-1}{m+1}\cdot\frac{m-2}{m+2}, \quad \&c.$$

will result, which goes off to infinity on both sides and has equal those terms which are the same distance from the middle. But also the term located in the middle between the two middle principal terms $m$ and $m$ is the square of the term in the middle between $m$ and 1 in the former series. Therefore that term between $m$ and $m$ in the last series may be sought by means of the second case of Proposition 20, and this will be found to be

$$m + \frac{m}{4(m+1)} + \frac{9m}{32(m+1)(m+2)} + \frac{75m}{128(m+1)(m+2)(m+3)} + \&c.$$

When this has been multiplied by the circumference of the circle, namely the square of the corresponding term in the series of numerators, you will have for the square of the term sought

$$3.14159\ldots\&c. \times \left(m + \frac{A}{4(m+1)} + \frac{9B}{8(m+2)} + \frac{25C}{12(m+3)} + \frac{49D}{16(m+4)} + \&c.\right)$$

Therefore the term of the proposed series 1, $\frac{2}{1}A$, $\frac{4}{3}B$, $\frac{6}{5}C$, etc. at a distance $m$ from the beginning is equal to the mean proportional between the circumference of the circle and that series, which converges more rapidly the larger $m$ is, that is to say, the further the term sought is from the beginning.

For example, let $m = 100$, and the first term of the series will be the circumference of the circle multiplied by 100, or $A = 314.15927$; then there will be

$$B = \frac{A}{4 \times 101} = .77762, \quad C = \frac{9B}{8 \times 102} = .00858, \quad D = \frac{25C}{12 \times 103} = .00017;$$

and the sum of these four terms is 314.94564, whose square root 17.7467079 is the hundred-and-first term of the series to be interpolated, or the product $\frac{2}{1} \times \frac{4}{3} \times \frac{6}{5} \times \frac{8}{7}$ etc. with one hundred factors. And in the same way one may find any intermediate term: for if $99\frac{1}{2}$ is written for $m$, the term in the middle between the hundredth and hundred-and-first terms will be obtained. Or if $99\frac{1}{3}$ is substituted for $m$, the term located after the hundredth term by a third part of the common interval will be obtained.

$$\begin{array}{r} 314.15927 \\ 77762 \\ 858 \\ 17 \\ \hline 314.94564 \end{array}$$

It should be noted that the reciprocals of the terms of some series can be interpolated: thus the reciprocals of the terms in the last series form the series

$$1, \quad \tfrac{1}{2}A, \quad \tfrac{3}{4}B, \quad \tfrac{5}{6}C, \quad \tfrac{7}{8}D, \quad \&c.$$

in which the term removed by an arbitrary interval equal to $\frac{1}{2}m$ will be equal to the term of the series

$$1, \quad \frac{a}{m+1}, \quad \frac{3b}{m+3}, \quad \frac{5c}{m+5}, \quad \frac{7d}{m+7}, \quad \&c.$$

which is located in the middle between the first and second terms, and which consequently will be found to be the mean proportional between the following series

$$\frac{1}{m+1} + \frac{A}{2(m+3)} + \frac{9B}{4(m+5)} + \frac{25C}{6(m+7)} + \frac{49D}{8(m+9)} + \&c.$$

and the number .6366197723676, which is equal to one divided by the semi-circumference of the circle: this will be confirmed by following the steps of the first part of this example.

### Example 2

Suppose that for this series

$$1, \quad \tfrac{2}{1}A, \quad \tfrac{5}{4}B, \quad \tfrac{8}{7}C, \quad \tfrac{11}{10}D, \quad \&c.$$

a term which is arbitrarily far removed from the beginning, namely by an interval $m$, is sought: and by Proposition 18 that term will be equal to the term of this series

$$1, \quad \frac{2a}{3m+2}, \quad \frac{5b}{3m+5}, \quad \frac{8c}{3m+8}, \quad \frac{11d}{3m+11}, \quad \&c.$$

which is located before the first term by a third part of the common interval. Therefore the numerators and denominators may be interpolated separately as in Example 2 of Proposition 21, namely by logarithms; and the term sought will be obtained.

### Scholion

Hence it is established that very remote terms of series can be determined no less accurately than intermediate terms near the beginning. But in the series for interpolation

$$1, \quad \frac{r}{p}A, \quad \frac{r+1}{p+1}B, \quad \&c.$$

if the difference between $p$ and $r$ is great, likewise the work involved in finding an arbitrary term will be great. Now the easiest case of all is where $p - r$

is equal to $\pm\frac{1}{2}$, as in Example 1, except where that difference is a whole number, in which case the series will be exactly interpolable.

(p.254)

## Proposition 23

*To find the ratio which the middle coefficient has to the sum of all the coefficients in any power of the binomial.*

### *First Solution*

If the index is an even number, let it be called $n$; but if it is odd, let it be called $n-1$; and as the middle coefficient is to the sum of all the coefficients of the same power, so one will be to the mean proportional between the semicircumference of the circle and either of the following series:

$$n + \frac{A}{2(n+2)} + \frac{9B}{4(n+4)} + \frac{25C}{6(n+6)} + \frac{49D}{8(n+8)} + \frac{81E}{10(n+10)} + \&c.$$

or

$$n + 1 - \frac{A}{2(n-1)} - \frac{9B}{4(n-3)} - \frac{25C}{6(n-5)} - \frac{49D}{8(n-7)} - \frac{81E}{10(n-9)} - \&c.$$

For example, if the ratio of the middle coefficient to the sum of all the coefficients in the hundredth or ninety-ninth power is required, there will be $n = 100$; when this has been multiplied by the semicircumference of the circle, it produces the first term $A = 157.079632679$; then there will be

$$B = \frac{A}{204}, \quad C = \frac{9B}{416}, \quad D = \frac{25C}{636}, \quad \&c.$$

and by carrying out the calculation as at the side, the sum of the terms will be found to be 157.866984459, whose square root 12.5645129018 is to one as the sum of all the coefficients is to the middle coefficient in the hundredth or ninety-ninth power. Now this calculation has been done using the former series: for although the difference is very small, I prefer the one in which the terms are all of the same sign.

| |
|---:|
| 157.079632679 |
| 769998199 |
| 16658615 |
| 654820 |
| 37137 |
| 2734 |
| 246 |
| 26 |
| 3 |
| 157.866984459 |

### *Second Solution*

With $n$ remaining as before, the sum of all the coefficients will be to the middle one as the square root of the ratio of the semicircumference of the circle to either of the following series:

$$\frac{1}{n+1} + \frac{A}{2(n+3)} + \frac{9B}{4(n+5)} + \frac{25C}{6(n+7)} + \frac{49D}{8(n+9)} + \frac{81E}{10(n+11)} + \&c.$$

or

$$\frac{1}{n} - \frac{A}{2(n-2)} - \frac{9B}{4(n-4)} - \frac{25C}{6(n-6)} - \frac{49D}{8(n-8)} - \frac{81E}{10(n-10)} - \&c.$$

Or which comes back to the same, put $a = .6366197723676$, namely the quantity which results on dividing one by the semicircumference of the circle; and the mean proportional between the number $a$ and either of those series will be to one as the middle coefficient is to the sum of them all.

Thus if the index is $n = 100$, as in the calculation above, there will be according to the first series

$$A = \frac{a}{101}, \ B = \frac{A}{206}, \ C = \frac{9B}{420}, \ D = \frac{25C}{642}, \ \&c.$$

Now from the calculation at the side it appears that the sum of the terms is .00633444670787, whose square root .0795892373872 is to one as the middle coefficient is to the sum of all the coefficients in the ninety-ninth or hundredth power.

| |
|---:|
| .00630316606305 |
| 3059789351 |
| 65566914 |
| 2553229 |
| 143473 |
| 10469 |
| 934 |
| 98 |
| 12 |
| 2 |
| .00633444670787 |

And so altogether there are four series of the same simplicity for the solution of this problem. But in practice there is no need to revert to series: for it suffices to take the mean proportional between the semicircumference of the circle and $n + \frac{1}{2}$; for this will always approximate more closely than the first two terms of the series, of which even the first alone suffices for the most part. For example, if $n = 100$, there will be $n + \frac{1}{2} = 100\frac{1}{2}$, which multiplied by the semicircumference of the circle produces 157.865, whose square root is 12.5644, less by one in the last figure.

Now the same approximation may be expressed otherwise and in a manner much more convenient in practice as follows. Let $c$ be to one as the square of the diameter is to the circle; that is, let $c = 1.2732395447352$; and the sum of the coefficients will be to the middle coefficient as one is to $\sqrt{\frac{c}{2n+1}}$ approximately, the error being an excess of about $\frac{1}{16n^2}\sqrt{\frac{c}{2n+1}}$. If $n = 100$, there will be $\frac{c}{2n+1} = .006334525$, and its square root .07958973 is accurate in the sixth decimal: in fact, when this has been divided by $16n^2$, that is by 160000, it will give the correction .00000050; and when this has been subtracted from the approximation, it leaves the number sought .07958923, which is accurate in the last figure.

Likewise if $n = 900$, there will be $\frac{c}{2n+1} = .000706962545$, whose square root .026588767 exceeds the true value by two in the last figure. But if the correction is computed and subtracted from the approximation, the number sought will be obtained accurate in the thirteenth decimal.

But here is an equally easy and more accurate approximation. Let the difference between the logarithms of the numbers $n+2$ and $n-2$ be divided by 16, let the quotient be added to half the logarithm of the index $n$, and then let the constant logarithm .0980599385151, namely half the logarithm of the semicircumference of the circle, be added to this sum; the final sum will be the logarithm of the number which is to one as the sum of the coefficients is to the middle one. For example, if $n = 900$, the calculation will be

| | |
|---|---|
| $\frac{1}{2}l, 900$ | 1.4771212547 |
| 16) Diff. of Log. 902 & Log. 898 | .0001206376 |
| Constant Log. | .0980599385 |
| Sum | 1.5753018308 |

And this sum exceeds the true value by two in the last figure, and it is the logarithm of the number 37.6098698 which is to one as the sum of the coefficients is to the middle coefficient in the power 900 or 899. And if you wish the reciprocal of that number, take the complement of the logarithm, namely $-2.4246981692$, and the number corresponding to this will be .0265887652 which gives the ratio of the middle coefficient to the sum of all the coefficients in the powers already stated.

## Demonstration

The powers of the binomial whose indices are even numbers have a unique middle coefficient; but those whose indices are odd have two middle coefficients. And hence two cases of the problem arise. In the first, where the index is even, divide the sums of the coefficients 1, 4, 16, 64, 256, 1024, etc. by the corresponding middle coefficients 1, 2, 6, 20, 70, 252, etc. and the quotients

$$1,\ 2,\ \tfrac{8}{3},\ \tfrac{16}{5},\ \tfrac{128}{35},\ \tfrac{256}{63},\ \text{\&c. or } 1,\ \tfrac{2}{1}A,\ \tfrac{4}{3}B,\ \tfrac{6}{5}C,\ \tfrac{8}{7}D,\ \text{\&c.}$$

will be to one as the sums of the coefficients are to the middle coefficients in the various powers.

Likewise, if the sums of the coefficients in the odd powers, namely 2, 8, 32, 128, 512, &c. are divided by the corresponding middle coefficients 1, 3, 10, 35, 126, &c. the quotients will again turn out the same, being

$$2,\ \tfrac{8}{3},\ \tfrac{16}{5},\ \tfrac{128}{35},\ \text{\&c.}$$

For there is the same relation between the sum of the coefficients and the middle coefficient in any even power as there is between the sum of the coefficients and the middle coefficient in the odd power immediately below. And so the interpolation of the series 1, $\frac{2}{1}A$, $\frac{4}{3}B$, $\frac{6}{5}C$, $\frac{8}{7}D$, etc. as in the first Example of the twenty-second Proposition solves both cases of the problem. But here we will give an investigation of these series without the method of differences.

## *Analysis of the First Solution*

The series to be interpolated 1, $\frac{2}{1}A$, $\frac{4}{3}B$, etc. is defined by the equation $T' = \dfrac{n+2}{n+1}T$, where $n$ is a variable quantity, and its successive values are 0, 2, 4, 6, 8, etc., namely the indices of the powers when they are even, or the indices augmented by one when they are odd. Square both sides of the equation to be resolved, and

$$T'^2 = \frac{n^2+4n+4}{n^2+2n+1}T^2$$

will be obtained, or what is the same

$$2T'^2 + (n+2)(T^2 - T'^2) - \frac{T'^2}{n+2} = 0.$$

Now take

$$T^2 = An + \frac{Bn}{n+2} + \frac{Cn}{(n+2)(n+4)} + \frac{Dn}{(n+2)(n+4)(n+6)} + \&c.$$

And after the required reduction according to rules already presented, you will find

$$T^2 = An + B + \frac{C-2B}{n+2} + \frac{D-4C}{(n+2)(n+4)} + \&c.$$

In this write the subsequent values of the variables for the preceding ones, that is, $T'^2$ for $T^2$ and $n+2$ for $n$, and

$$T'^2 = A(n+2) + B + \frac{C-2B}{n+4} + \frac{D-4C}{(n+4)(n+6)} + \&c.$$

will emerge. Then by taking the difference of these values and multiplying this by $n+2$,

$$(n+2)(T^2 - T'^2) = -2A(n+2) \quad * \quad + \frac{2C-4B}{n+4} + \frac{4D-16C}{(n+4)(n+6)} + \&c.$$

will result. But if $n+2$ is written for $n$ in the expression previously taken for $T^2$

$$T'^2 = (n+2)\times\left(A + \frac{B}{n+4} + \frac{C}{(n+4)(n+6)} + \frac{D}{(n+4)(n+6)(n+8)} + \&c.\right)$$

will be obtained. And on dividing by $n+2$,

$$\frac{T'^2}{n+2} = A + \frac{B}{n+4} + \frac{C}{(n+4)(n+6)} + \frac{D}{(n+4)(n+6)(n+8)} + \&c.$$

Now substitute in the equation to be resolved the values of

$$T'^2, \quad (n+2)(T^2 - T'^2), \quad \frac{T'^2}{n+2},$$

reduced to the same form, and

$$2B - A + \frac{4C - 9B}{n+4} + \frac{6D - 25C}{(n+4)(n+6)} + \frac{8E - 49D}{(n+4)(n+6)(n+8)} + \&c. = 0$$

will result. Finally by setting the numerators equal to zero, you will have

$$2B - A = 0, \quad 4C - 9B = 0, \quad 6D - 25C = 0, \quad 8E - 49D = 0, \quad \&c.$$

And these are the relations of the coefficients in the first series. And the latter series in the first solution arises in the same way.

## *Analysis of the Second Solution*

The second solution is accomplished by interpolation of the series 1, $\frac{1}{2}a$, $\frac{3}{4}b$, $\frac{5}{6}c$, $\frac{7}{8}d$, etc. whose terms are the reciprocals of those in the first: in fact it is defined by the equation $T' = \frac{n+1}{n+2}T$, in which the successive values of $n$ are 0, 2, 4, 6, 8, etc. as before. And on squaring

$$T'^2 = \frac{n^2 + 2n + 1}{n^2 + 4n + 4} T^2$$

results, that is,

$$(n+1)(n+3)(T^2 - T'^2) - 2(n+1)T^2 - T'^2 = 0.$$

Let us now form

$$T^2 = \frac{A}{n+1} + \frac{B}{(n+1)(n+3)} + \frac{C}{(n+1)(n+3)(n+5)} + \frac{D}{(n+1)(n+3)(n+5)(n+7)} + \&c.$$

And by substituting $n+2$ for $n$,

$$T'^2 = \frac{A}{n+3} + \frac{B}{(n+3)(n+5)} + \frac{C}{(n+3)(n+5)(n+7)} + \frac{D}{(n+3)(n+5)(n+7)(n+9)} + \&c.$$

will come forth, and so

$$(n+1)(n+3)(T^2 - T'^2) = 2A + \frac{4B}{n+5} + \frac{6C}{(n+5)(n+7)} + \frac{8D}{(n+5)(n+7)(n+9)} + \&c.,$$

which after reduction to the required form is

$$2A + \frac{4B}{n+3} + \frac{6C-8B}{(n+3)(n+5)} + \frac{8D-24C}{(n+3)(n+5)(n+7)} + \&c.$$

Now write these values in the equation to be resolved, and

$$\frac{2B-A}{n+3} + \frac{4C-9B}{(n+3)(n+5)} + \frac{6D-25C}{(n+3)(n+5)(n+7)}$$

$$+ \frac{8E-49D}{(n+3)(n+5)(n+7)(n+9)} + \&c. = 0$$

will result. When the numerators have been equated to zero, they will give the relation of the coefficients of the former series in the second solution.

That moreover the coefficient $A$ is the semicircumference of the circle in one case and its reciprocal in the other is demonstrated thus. By the first series

$$T^2 = An \times \left(1 + \frac{1}{2(n+2)} + \&c.\right).$$

Now the larger $n$ is, the closer the equation $T^2 = An$ approaches the truth, since the latter terms eventually become infinitely smaller than the former. Therefore if in the equation $T^2 = An$, or $A = \dfrac{T^2}{n}$, for $n$ are written successively its values 2, 4, 6, 8, 10, etc. and at the same time the squares of the corresponding terms for $T^2$, the following equations will arise, which continuously approximate to the true value of $A$:

$$A = 2,$$
$$A = 2 \times \frac{8}{9},$$
$$A = 2 \times \frac{8}{9} \times \frac{24}{25},$$
$$A = 2 \times \frac{8}{9} \times \frac{24}{25} \times \frac{48}{49},$$
$$\&c.$$

Hence the value of $A$ is the product of all terms

$$2 \times \frac{8}{9} \times \frac{24}{25} \times \frac{48}{49} \times \frac{80}{81} \times \&c.$$

up to infinity; this is equal to the semicircumference of the circle according to Wallis's *Arithmetica Infinitorum*.

p.262)

## Proposition 24

*If the whole numbers 0, 1, 2, 3, 4, etc. are written successively for $z$ in the ordinate of the curve $x^{r+z-1} \times (1-x)^{p-r-1}$, I say that there is the same* *relation between the areas of the resulting ordinates as there is between the terms of the series*

$$a, \quad \frac{r}{p}a, \quad \frac{r+1}{p+1}b, \quad \frac{r+2}{p+2}c, \quad \frac{r+3}{p+3}d, \quad \&c.$$

*provided the abscissa $x$ is equal to one.*

For let the areas and corresponding ordinates be

| Areas | Ordinates |
|---|---|
| $A$ | $x^{r-1} \times (1-x)^{p-r-1}$ |
| $B$ | $x^{r} \times (1-x)^{p-r-1}$ |
| $C$ | $x^{r+1} \times (1-x)^{p-r-1}$ |
| $D$ | $x^{r+2} \times (1-x)^{p-r-1}$ |
| $E$ | $x^{r+3} \times (1-x)^{p-r-1}$ |
| &c. | &c. |

Then on comparing these areas by means of Proposition 7 of *Newton's De Quadratura Curvarum*, you will find

$$B = \frac{rA - x^r \times (1-x)^{p-r}}{p},$$

$$C = \frac{(r+1)B - x^{r+1} \times (1-x)^{p-r}}{p+1},$$

$$D = \frac{(r+2)C - x^{r+2} \times (1-x)^{p-r}}{p+2},$$

$$E = \frac{(r+3)D - x^{r+3} \times (1-x)^{p-r}}{p+3},$$

&c.

Now let $x$ be equal to one, as is assumed in the theorem; and there will be $1 - x = 0$, in which case the relation of the areas is

$$B = \frac{r}{p}A, \quad C = \frac{r+1}{p+1}B, \quad D = \frac{r+2}{p+2}C, \quad E = \frac{r+3}{p+3}D, \quad \&c.$$

And so there is the same relation between the areas of these curves and between the terms of the proposed series when the abscissa $x$ is one. Q.E.D.

*Corollary.* Hence in the series

$$a, \quad \frac{r}{p}a, \quad \frac{r+1}{p+1}b, \quad \frac{r+2}{p+2}c, \quad \&c.,$$

if $z$ denotes the interval between the first term $a$ and any other principal or intermediate term $T$, the first term $a$ will be to any other principal or intermediate term which is distanced from the beginning by an interval $z$ as the area of the curve whose ordinate is $x^{r-1} \times (1-x)^{p-r-1}$ is to the area of the curve whose ordinate is $x^{r+z-1} \times (1-x)^{p-r-1}$.

### Example 1

Let the series

$$1, \quad \tfrac{1}{2}a, \quad \tfrac{3}{4}b, \quad \tfrac{5}{6}c, \quad \tfrac{7}{8}d, \quad \&c.$$

be given for interpolation. Since the common difference of both the numerators and the denominators is 2, divide these by two, so that this difference becomes one as in the theorem, and the series will become

$$1, \quad \frac{\frac{1}{2}}{1}a, \quad \frac{\frac{3}{2}}{2}b, \quad \&c.$$

which, when compared with that in the proposition, gives $p = 1$, $r = \frac{1}{2}$; when these have been substituted, we will have that the first term of the series, or one, is to any other principal or intermediate term distanced by an interval $z$ from the beginning as the area of the ordinate $x^{-1/2} \times (1-x)^{-1/2}$ is to the area of the ordinate $x^{z-1/2} \times (1-x)^{-1/2}$, that is, as the area of this ordinate $\dfrac{1}{\sqrt{x-x^2}}$ is to the area of this ordinate $\dfrac{x^z}{\sqrt{x-x^2}}$.

Thus if the required term is located in the middle between the first and the second, there will be $z = \frac{1}{2}$, in which case the latter ordinate is $\dfrac{x^{1/2}}{\sqrt{x-x^2}}$, or $\dfrac{1}{\sqrt{1-x}}$: and so, one is to the term in the middle between the first and the second terms as the area of the ordinate $\dfrac{1}{\sqrt{x-x^2}}$ is to the area of the ordinate $\dfrac{1}{\sqrt{1-x}}$, that is, as the circumference of the circle 3.1415926... etc. is to 2, which by this means produces .63661977... etc.

If the hundred-and-first term of the same series is sought, there will be $z = 100$; and so, one is to the proposed term as 3.1415... etc. is to the area of the ordinate $\dfrac{x^{100}}{\sqrt{x-x^2}}$. And similarly on putting $z = 100\frac{1}{2}$, the term in the middle between the hundred-and-first and the hundred-and-second terms will be determined by the circumference of the circle and the area of the ordinate $\dfrac{x^{100}}{\sqrt{1-x}}$: throughout, however, the parts of the areas which lie above the abscissa equal to one have to be taken.

The reciprocals of the terms can also be interpolated, and that is sometimes more convenient than interpolation of the terms themselves. The reciprocals of the terms in the last series are $1$, $\frac{2}{1}a$, $\frac{4}{3}b$, etc., and so $r = 1$, $p = \frac{1}{2}$; and thence, the first term is to the term distanced by an interval $z$ from the beginning as the area of the ordinate $x^0 \times (1-x)^{-3/2}$ is to the area of the ordinate $x^z \times (1-x)^{-3/2}$, that is, as 2 is to the area of the ordinate $\dfrac{x^z}{(1-x)\sqrt{1-x}}$; for quadrature can be effected exactly for the first curve.

### Example 2

Let the series to be interpolated be

$$1, \quad \tfrac{1}{2}a, \quad \tfrac{4}{5}b, \quad \tfrac{7}{8}c, \quad \tfrac{10}{11}d, \quad \text{\&c.}$$

Divide the numerators and the denominators by their increment 3; and you will find $p = \frac{2}{3}$, $r = \frac{1}{3}$; hence let the first ordinate be $x^{-2/3} \times (1-x)^{-2/3}$, or $\dfrac{1}{\sqrt[3]{x^2 - 2x^3 + x^4}}$; and let the second ordinate be $\dfrac{x^z}{\sqrt[3]{x^2 - 2x^3 + x^4}}$. Then, the first term of the series is to any other whose distance from the beginning is $z$ as the area of the former is to the area of the latter.

(p.263)

## Proposition 25

*If the whole numbers 0, 1, 2, 3, 4, etc. are written successively for $z$ in the ordinate of the curve $x^{p-z} \times (1-x)^{r+z-1}$, there will be the same relation between the areas of the resulting ordinates as there is between the terms of the series*

$$a, \quad \frac{r}{p}a, \quad \frac{r+1}{p-1}b, \quad \frac{r+2}{p-2}c, \quad \frac{r+3}{p-3}d, \quad \text{\&c.},$$

*where the numerators increase continually, while the denominators decrease. And here I also put the abscissa $x$ equal to one.*

This proposition is demonstrated as above.

*Corollary.* Hence in the series

$$a, \quad \frac{r}{p}a, \quad \frac{r+1}{p-1}b, \quad \frac{r+2}{p-2}c, \quad \frac{r+3}{p-3}d, \quad \text{\&c.},$$

as the first term $a$ is to any other term which is distanced from the beginning by an interval $z$, so the area of the curve whose ordinate is $x^p \times (1-x)^{r-1}$ is to the area of the curve whose ordinate is $x^{p-z} \times (1-x)^{r+z-1}$.

### Example 1

Let the series

$$1, \quad \frac{n}{1}a, \quad \frac{n-1}{2}b, \quad \frac{n-2}{3}c, \quad \frac{n-3}{4}d, \quad \text{\&c.}$$

be given for interpolation, whose terms are the binomial coefficients in the power whose index is $n$. Since this series does not fall directly under this proposition, I interpolate the reciprocals of the terms

$$1, \quad \frac{1}{n}a, \quad \frac{2}{n-1}b, \quad \frac{3}{n-2}c, \quad \&c.,$$

by which means there will be $r = 1$, $p = n$; and so, one is to the term in the latter series distanced from the beginning by an interval $z$ as the area of the ordinate $x^n \times (1-x)^0$ is to the area of the ordinate $x^{n-z} \times (1-x)^z$: or, one is to the term in the first series removed from the beginning by an interval $z$ as the area of the ordinate $x^{n-z} \times (1-x)^z$ is to $\dfrac{1}{n+1}$.

Thus if the coefficient of the fifth term in the ninth power is required, there will be $n = 9$, $z = 4$; when these have been written in, the area of the ordinate $x^5 \times (1-x)^4$ will be to $\frac{1}{10}$ as one is to the coefficient sought. Now the expanded ordinate is $x^5 - 4x^6 + 6x^7 - 4x^8 + x^9$, and its area is $\frac{1}{6} - \frac{4}{7} + \frac{6}{8} - \frac{4}{9} + \frac{1}{10}$, or $\frac{1}{1260}$; then, as $\frac{1}{1260}$ is to $\frac{1}{10}$ so one is to 126, which is the proposed coefficient.

### Example 2

If the term which is located in the middle between the two coefficients 1 and 1 in the simple power of the binomial is required, the index of the binomial will be $n = 1$ and $z = \frac{1}{2}$; and hence one is to the term between the coefficients 1 and 1 as the area of the ordinate $x^{1/2} \times (1-x)^{1/2}$ is to $\frac{1}{2}$, that is, as the area of the circle is to the circumscribed square.

### Scholion

When curves whose quadratures are required are of very many dimensions, some of their ordinates have to be found by means of a table of logarithms; the areas will be given by these by means of *Newton's* parabola. However, if the relation of the terms in the series to be interpolated involves several terms, the interpolation will be completed by comparison with other curves. But passing over this material, let me add certain things concerning other methods of interpolation.

p.265)

## Proposition 26

*Let the series to be interpolated be*

$$1, \quad \frac{r}{p}a, \quad \frac{r+1}{p+1}b, \quad \frac{r+2}{p+2}c, \quad \frac{r+3}{p+3}d, \quad \&c.$$

*and put* $n = r - p$, *and*

$$B = \frac{n}{1} \cdot \frac{n-1}{2} \times A,$$

$$C = \frac{n-1}{2} \cdot \frac{n-2}{2} \times \left(B + \frac{n}{3}A\right),$$

$$D = \frac{n-2}{3} \cdot \frac{n-3}{2} \times \left(C + \frac{n-1}{3}B + \frac{n}{3} \cdot \frac{n-1}{4}A\right),$$

$$E = \frac{n-3}{4} \cdot \frac{n-4}{2} \times \left(D + \frac{n-2}{3}C + \frac{n-1}{3} \cdot \frac{n-2}{4}B + \frac{n}{3} \cdot \frac{n-1}{4} \cdot \frac{n-2}{5}A\right),$$

$$F = \frac{n-4}{5} \cdot \frac{n-5}{2} \times \left(E + \frac{n-3}{3}D + \frac{n-2}{3} \cdot \frac{n-3}{4}C + \frac{n-1}{3} \cdot \frac{n-2}{4} \cdot \frac{n-3}{5}B + \frac{n}{3} \cdot \frac{n-1}{4} \cdot \frac{n-2}{5} \cdot \frac{n-3}{6}A\right),$$

&c.

*Then the principal or intermediate term of the series to be interpolated, whose distance from the beginning is $z - p$, will be*

$$z^n \times \left(A + \frac{B}{z} + \frac{C}{z^2} + \frac{D}{z^3} + \frac{E}{z^4} + \&c.\right).$$

It is to be noted that by Proposition 18 the coefficient $A$ is equal to the term in the series of numerators $1, ra, (r+1)b, (r+2)c$, etc. which is distanced by an interval $p - r$ from the beginning; and that is determined by means of Example 2 of Proposition 21.

## Demonstration

The proposed series is defined by the difference equation $T' = \frac{z+n}{z}T$, where $n = r - p$ as in the theorem and the successive values of the indeterminate $z$ are $p, p+1, p+2$, etc. Now suppose that

$$T = z^n \times \left(A + \frac{B}{z} + \frac{C}{z^2} + \frac{D}{z^3} + \frac{E}{z^4} + \&c.\right),$$

then write for $T$ and $z$ their next values $T'$ and $z+1$, respectively, and

$$T' = (z+1)^n \times \left(A + \frac{B}{z+1} + \frac{C}{(z+1)^2} + \frac{D}{(z+1)^3} + \frac{E}{(z+1)^4} + \&c.\right)$$

will come out. Or on developing the powers

$$T' = z^n \times \left(A + \frac{nA+B}{z} + \frac{(nA+2B)(n-1)+2C}{2z^2} + \frac{(nA+3B)(n-1)(n-2)+6(n-2)C+6D}{6z^3} + \&\text{c.}\right).$$

The equation to be resolved $T' = \dfrac{z+n}{z}T$ may be written in the following manner, $T'z - Tz - Tn = 0$; let the expressions for $T$ and $T'$ be substituted in this, and

$$z^n \times \left(\frac{n(n-1)A - 2B}{2z} + \frac{(nA+3B)(n-1)(n-2) - 12C}{6z^2} + \&\text{c.}\right) = 0$$

will result. Now let the numerators be set equal to zero, and

$$B = \frac{n}{1}\cdot\frac{n-1}{2} \times A, \quad C = \frac{n-1}{2}\cdot\frac{n-2}{2} \times \left(B + \frac{n}{3}A\right), \quad \&\text{c.}$$

will be obtained. And on continuing the calculation the rest of the coefficients will come out as in the theorem. Q.E.D.

**Example 1**

Consider the series

$$1, \quad \tfrac{1}{2}a, \quad \tfrac{3}{4}b, \quad \tfrac{5}{6}c, \quad \tfrac{7}{8}d, \quad \&\text{c.},$$

which is defined by the equation $T' = \dfrac{z-\frac{1}{2}}{z}T$, in which the successive values of $z$ are 1, 2, 3, 4, etc. When this has been compared with the equation $T' = \dfrac{z-n}{z}T$, it gives $n = -\frac{1}{2}$, and hence

$$\begin{aligned}
B &= \frac{3}{8} \times A & &= A \times \frac{3}{8},\\
C &= \frac{15}{16} \times \left(B - \frac{1}{6}A\right) & &= A \times \frac{25}{128},\\
D &= \frac{35}{24} \times \left(C - \frac{3}{6}B + \frac{1}{16}A\right) & &= A \times \frac{105}{1024},\\
E &= \frac{63}{32} \times \left(D - \frac{5}{6}C + \frac{5}{16}B - \frac{1}{32}A\right) & &= A \times \frac{1659}{32768},\\
F &= \frac{99}{40} \times \left(E - \frac{7}{6}D + \frac{35}{48}C - \frac{7}{32}B + \frac{7}{384}A\right) & &= A \times \frac{6237}{262144},
\end{aligned}$$

&c.

And so

$$T = \frac{A}{\sqrt{z}} \times \left(1 + \frac{3}{8z} + \frac{25}{128z^2} + \frac{105}{1024z^3} + \frac{1659}{32768z^4} + \frac{6237}{262144z^5} + \&c.\right).$$

Now the quantity $A$ in examples of this type can be determined as follows. Find a principal term sufficiently far from the beginning by means of the relation of the terms to be interpolated, for example, the sixteenth, which produces .144464448... etc. here. Write this for $T$ and at the same time for $z$ its corresponding value, namely 16; and you will have

$$.144464448 = \frac{A}{4} \times \left(1 + \frac{3}{8.16} + \frac{25}{128.16.16} + \&c.\right),$$

or, on collecting the terms into one sum,

$$.144464448 = \frac{A}{4} \times 1.02422627,$$

from which $A = .564189583548$ comes out: when this has been given, $T$ will be given in any desired case by a very few of the terms in its expression.

**Example 2**

Let it be required to interpolate the series

$$1, \quad \tfrac{2}{1}a, \quad \tfrac{5}{4}b, \quad \tfrac{8}{7}c, \quad \tfrac{11}{10}d, \quad \&c.$$

which is defined by the equation $T' = \dfrac{z + \frac{1}{3}}{z} T$, the successive values of the abscissa $z$ being $\frac{1}{3}, \frac{4}{3}, \frac{7}{3}, \frac{10}{3}$, etc. Having been compared with the equation in the theorem, this gives $n = \frac{1}{3}$; when this value has been substituted,

$$T = A\sqrt[3]{z} \times \left(1 - \frac{1}{9z} \;*\; + \frac{10}{2187z^3} + \frac{11}{19683z^4} - \frac{77}{59049z^5} + \&c.\right)$$

will be obtained.

Now to extract the coefficient $A$, I seek the fourteenth term of the series to be interpolated, which comes out as 4.652136: then I write this value for $T$ and for $z$ its fourteenth value $\frac{40}{3}$, and I have

$$4.652136 = A\sqrt[3]{\frac{40}{3}} \times \left(1 - \frac{1}{120} + \frac{1}{518400} + \&c.\right).$$

Or on extracting the cube root of the number $\frac{40}{3}$ and collecting the terms into one sum, I obtain $4.652136 = A \times 2.351506$, and so $A = 1.978364$. Now that $A$ has been given, any other term will be found with the greatest ease. Let the term which is located before the thousand-and-first term by a third part of the common interval be sought: for $z$ write its corresponding value 1000 and the expression for $T$ will come out as

$$10A \times \left(1 - \frac{1}{9000}\right), \quad \text{or} \quad T = 19.78144.$$

For where the required term is far removed from the beginning and the calculation does not have to be extended to a great number of figures, a very few terms in the expression for $T$ are amply sufficient.

### Scholion

In the same way as the root is extracted from the equation $T' = \frac{z+n}{z}T$ in this proposition, it is also extracted from any other which is contained in the following form:

$$T \times \left(z^\theta + az^{\theta-1} + bz^{\theta-2} + \&c.\right) = T' \times \left(z^\theta + cz^{\theta-1} + dz^{\theta-2} + \&c.\right).$$

For the index $n = a - c$ according to Proposition 6; and then the form of the series to be taken for $T$ will be

$$T = Az^n + Bz^{n-1} + Cz^{n-2} + \&c.$$

Indeed in series of this type which are roots of difference equations, the indices of $z$ have one for their decrement, except in certain cases which are very special. Therefore when the index of $z$ in the first term has been obtained, the form of the series to be taken for the root $T$ is determined: then on writing $z + 1$ for $z$ and $T'$ for $T$

$$T' = A(z+1)^n + B(z+1)^{n-1} + C(z+1)^{n-2} + \&c.$$

will come out. Then by bringing this expression back to the form of $T$, as has been shown above, and by multiplying both expressions by the quantities which the equation to be resolved already contains, the coefficients taken will be given by combining like members in the resulting equation.

## (p.267) Proposition 27

*If the equation for the series is* $T' = \frac{z^2}{z^2+r}T$, *the root will be*

$$T = A + \frac{r}{z}A + \frac{r+1}{2(z+1)}B + \frac{r+4}{3(z+2)}C + \frac{r+9}{4(z+3)}D + \frac{r+16}{5(z+4)}E + \&c.$$

Let the equation to be resolved $T' = \frac{z^2}{z^2+r}T$ be written in this form $z^2T - z^2T' - rT' = 0$, and let $T$ be taken as

$$T = A + \frac{B}{z} + \frac{C}{z(z+1)} + \frac{D}{z(z+1)(z+2)} + \frac{E}{z(z+1)(z+2)(z+3)} + \&c.$$

And there will be

$$T' = A + \frac{B}{z+1} + \frac{C}{(z+1)(z+2)} + \frac{D}{(z+1)(z+2)(z+3)} + \frac{E}{(z+1)(z+2)(z+3)(z+4)} + \&c.$$

Hence

$$T - T' = \frac{B}{z(z+1)} + \frac{2C}{z(z+1)(z+2)} + \frac{3D}{z(z+1)(z+2)(z+3)} + \frac{4E}{z(z+1)(z+2)(z+3)(z+4)} + \&c.$$

And on multiplying by $z^2$

$$z^2T - z^2T' = \frac{Bz}{z+1} + \frac{2Cz}{(z+1)(z+2)} + \frac{3Dz}{(z+1)(z+2)(z+3)} + \frac{4Ez}{(z+1)(z+2)(z+3)(z+4)} + \&c.$$

And as a result of reduction

$$z^2T - z^2T' = B + \frac{2C - B}{z+1} + \frac{3D - 4C}{(z+1)(z+2)} + \frac{4E - 9D}{(z+1)(z+2)(z+3)} + \&c.$$

Write this expression for $z^2T - z^2T'$ and for $T'$ the expression previously found, and

$$B - rA + \frac{2C - (r+1)B}{z+1} + \frac{3D - (r+4)C}{(z+1)(z+2)} + \frac{4E - (r+9)D}{(z+1)(z+2)(z+3)} + \&c. = 0$$

will result. And by comparing like members,

$$B = \frac{r}{1}A, \quad C = \frac{r+1}{2}B, \quad D = \frac{r+4}{3}C, \quad E = \frac{r+9}{4}D, \quad \&c.$$

These are the values of the coefficients: but if $A$, $B$, $C$, etc. denote whole terms, the value of $T$ which has now been assigned will come out. Q.E.D.

## Example

*Wallis* found the ultimate term of this series

$$1, \quad \tfrac{8}{9}A, \quad \tfrac{24}{25}B, \quad \tfrac{48}{49}C, \quad \tfrac{80}{81}D, \quad \&c.$$

to be the area of a circle whose diameter is one; here the denominators are the squares of the odd numbers and are one more than the numerators. But let us see here what is the ultimate term of this series

$$1, \quad \tfrac{4}{3}A, \quad \tfrac{16}{15}B, \quad \tfrac{36}{35}C, \quad \tfrac{64}{63}D, \quad \&c.,$$

where the numerators are the squares of the even numbers and are one more than the denominators. The equation for this series will be $T' = \frac{z^2}{z^2 - \frac{1}{4}} T$, the successive values of the abscissa $z$ being 1, 2, 3, 4, etc.: therefore on comparing this equation with that in the proposition, there will be $r = -\frac{1}{4}$, and when this has been substituted, it produces

$$T = A - \frac{A}{4z} + \frac{3B}{8(z+1)} + \frac{15C}{12(z+2)} + \frac{35D}{16(z+3)} + \frac{63E}{20(z+4)} + \&c.$$

And for the determination of the coefficient $A$, seek the tenth term of the series, namely 1.5300.1727.35; substitute this for $T$ and at the same time for $z$ its corresponding value 10; and you will obtain

$$1.5300.1727.35 = A \times \left(1 - \frac{1}{4.10} - \frac{1.3}{4.8.10.11} - \frac{1.3.15}{4.8.12.10.11.12} - \&c.\right)$$

That is, on collecting the terms into one,

$$1.5300.1727.35 = A \times .9740.3924.54, \quad \text{and hence} \quad A = 1.57079633,$$

in fact the semicircumference of the circle: when this has been given, any principal or intermediate term of the series to be interpolated will be given very easily. Moreover, it is established that its ultimate term, or the product of all the factors

$$\frac{4}{3} \times \frac{16}{15} \times \frac{36}{35} \times \frac{64}{63} \times \frac{100}{99} \times \&c.,$$

is equal to the first coefficient $A$, and so to the semicircumference of the circle.

(p.269)

## Proposition 28

*To find the sum of any number of logarithms, whose arguments are in arithmetic progression.*

Let $x+n$, $x+3n$, $x+5n$, $x+7n$, ..., $z-n$ denote an arbitrary collection of numbers in arithmetic progression, the first of which is $x+n$, the last $z-n$, and whose common difference is $2n$. Moreover, let $l,z$ and $l,x$ denote the tabular logarithms of the numbers $z$ and $x$; and let $a = .43429.44819.03252$, namely the reciprocal of the natural logarithm of ten. And the sum the logarithms set forth will be equal to the difference between the following two series

$$\frac{z\,l,z}{2n} - \frac{az}{2n} - \frac{an}{12z} + \frac{7an^3}{360z^3} - \frac{31an^5}{1260z^5} + \frac{127an^7}{1680z^7} - \frac{511an^9}{1188z^9} + \&c.$$

$$\frac{x\,l,x}{2n} - \frac{ax}{2n} - \frac{an}{12x} + \frac{7an^3}{360x^3} - \frac{31an^5}{1260x^5} + \frac{127an^7}{1680x^7} - \frac{511an^9}{1188x^9} + \&c.$$

Moreover, these series are continued to infinity as follows: put

$$-\frac{1}{3.4} = A,$$
$$-\frac{1}{5.8} = A + 3B,$$
$$-\frac{1}{7.12} = A + 10B + 5C,$$
$$-\frac{1}{9.16} = A + 21B + 35C + 7D,$$
$$-\frac{1}{11.20} = A + 36B + 126C + 84D + 9F,$$

&c.,

where the numbers which multiply $A$, $B$, $C$, $D$ etc. in the various identities are the alternate coefficients in the odd powers of the binomial. These things having been stated, the coefficient of the third term will be $-\frac{1}{12} = A$, that of the fourth $+\frac{7}{360} = B$, of the fifth $-\frac{31}{1260} = C$, and so on.

## Demonstration

Let the variable $z$ be reduced by its decrement $2n$; or what is the same, let $z - 2n$ be substituted for $z$ in the series

$$\frac{z\,l,z}{2n} - \frac{az}{2n} - \frac{an}{12z} + \frac{7an^3}{360z^3} - \frac{31an^5}{1260z^5} + \&\text{c.}$$

and its successive value

$$\frac{(z-2n)\,l,(z-2n)}{2n} - \frac{a}{2n} \times (z-2n)$$
$$- \frac{an}{12(z-2n)} + \frac{7an^3}{360(z-2n)^3} - \frac{31an^5}{1260(z-2n)^5} + \&\text{c.}$$

will result. Subtract this from the former value, the terms having been reduced first to the same form by division, and

$$l,z - \frac{an}{z} - \frac{an^2}{2z^2} - \frac{an^3}{3z^3} - \frac{an^4}{4z^4} - \&\text{c.}$$

will be left, that is, the logarithm of the number $z - n$. And so in general the decrement of two successive values of the series equates to the logarithm of $z - n$; this represents in general any one of the logarithms which had to be summed. Therefore the series will be the sum of the logarithms set forth provided the second series is subtracted from it. For sums just as areas sometimes have to be corrected in order that they evaluate correctly.

### Example 1

Let it be proposed to find the sum of the logarithms of the ten numbers 101, 103, 105, 107, 109, 111, 113, 115, 117, 119; on comparison with $x+n$, $x+3n$, $x+5n$, ..., $z-n$ these give common difference $2n = 2$ and so $n = 1$; and the first $x+1 = 101$, the last $z-1 = 119$, so that $x = 100$, $z = 120$. Now when these have been substituted, as well as .43429.44819.03252 for $a$ and the logarithms of 100 and 120 for $l,x$ and $l,z$ respectively, the values of the two series will be found to be 78.28491.40012.1 and 98.69290.42601.6, whose difference gives 20.40799.02589.5 for the required sum of logarithms.

### Example 2

Now let the sum of the logarithms of the numbers 11, 12, 13, ..., 1000 be required, the first of which is 11 and the last 1000, and the common difference is one. Therefore $n = \frac{1}{2}$, $x + \frac{1}{2} = 11$, $z - \frac{1}{2} = 1000$; whence $x = \frac{21}{2}$, $z = \frac{2001}{2}$; when these have been written in, as well as the logarithms of $\frac{21}{2}$ and $\frac{2001}{2}$ for $l,x$ and $l,z$, the numbers 2567.20555.42879 and 6.16067.30987 result for the values of the series, whose difference leaves 2561.04488.11892 for the required sum of logarithms.

Further, if you wish the sum of the logarithms of the natural numbers 1, 2, 3, 4, 5, etc., no matter how many, let $z-n$ be the last of the numbers with $n = \frac{1}{2}$, and three or four terms of this series

$$z\,l,z - az - \frac{a}{24z} + \frac{7a}{2880z^3} - \&c.$$

added to half of the logarithm of the circumference of the circle whose radius is one, that is, to 0.39908.99341.79, will give the desired sum, and that with less effort the more logarithms there are to be summed. Thus if you put $z - \frac{1}{2} = 1000$, or $z = \frac{2001}{2}$, the value of the series will be 2567.20555.42879 as before, which, when added to the constant logarithm, produces 2567.60464.42221 for the sum of the logarithms of the first thousand numbers of the series 1, 2, 3, 4, 5, etc.

### Example 3

Suppose that it is required to find the five-hundredth coefficient in the thousandth power of the binomial. It follows from Newton's Theorem for expanding the binomial that that coefficient is equal to the product of the four hundred and ninety-nine factors

$$\frac{1000}{1}, \quad \frac{999}{2}, \quad \frac{998}{3}, \quad \frac{997}{4}, \quad \frac{996}{5}, \quad \ldots, \quad \frac{502}{499},$$

the first of which is $\frac{1000}{1}$ and the last $\frac{502}{499}$; both the numerators and the denominators are in arithmetic progression. In order to find the sum of the logarithms of the numerators 1000, 999, 998, 997, ..., 502, put the common difference $1 = 2n$, the greatest of them $1000 = z - \frac{1}{2}$, the least $502 = x + \frac{1}{2}$; then there will be $n = \frac{1}{2}$, $z = 1000\frac{1}{2}$, $x = 501\frac{1}{2}$, and when these have been substituted, 2567.20555.42879 will result for the value of the former series and 1136.38715.63268 for the value of the latter; then the difference of these, 1430.81839.79611, is equal to the sum of the logarithms of the numerators. Then in order to obtain the sum of the logarithms of the denominators 1, 2, 3, 4, ..., 499, put $n = \frac{1}{2}$, $z - \frac{1}{2} = 499$, or $z = 499\frac{1}{2}$, and when these have been written in the former series, its value 1130.98834.85966 will result, to which the logarithm .39908.99342 is added according to the rule in the previous example; and you will have 1131.38743.85308 for the sum of the logarithms of the denominators. Finally, when this has been subtracted from the sum of the logarithms of the numerators, 299.43095.94303 will remain, which is in fact the logarithm of the desired coefficient.

(p.274)

## Scholion

Series of this type

$$1, \quad \frac{r}{p}A, \quad \frac{r+1}{p+1}B, \quad \frac{r+2}{p+2}C, \quad \&c.$$

are interpolated by means of Proposition 26, when the difference between $r$ and $p$ is very small, and in general by this proposition if no account is taken of that difference. And in exactly the same way one can find the sum of logarithms of numbers which are made up in a much more complicated fashion than by equal differences; and in that way one can determine the terms of series whose interpolation is considered to be very difficult. The areas of curves whose ordinates are of this type $(1-x)^{1000}$, where the index of the binomial is very large, are also found by this problem, but only in that case when the part of the area sought lies above the part of the abscissa equal to one.

And indeed almost all problems concerning interpolations may be treated by this analysis, even if three or more terms of the series to be interpolated appear in the difference equation: for I have the resolution of these in my power. And it is appropriate to note here that series which arise by Newton's parabola also arise by our method. For suppose that it is proposed to interpolate the series

$$a, \quad \frac{r+n}{r}a, \quad \frac{r+n+1}{r+1}b, \quad \frac{r+n+2}{n+2}c, \quad \frac{r+n+3}{r+3}d, \quad \&c.,$$

which is defined by the equation

$$T' = \frac{z+n+r}{z+r}T,$$

in which the successive values of the abscissa $z$ are 0, 1, 2, 3, etc. Let it be written in the following way:

$$(z+r)(T'-T)-nT=0.$$

Let $T$ be expressed in the form

$$T=A+Bz+Cz(z-1)+Dz(z-1)(z-2)+Ez(z-1)(z-2)(z-3)+\&c.$$

And on substituting $T'$ for $T$ and $z+1$ for $z$

$$T'=A+B(z+1)+C(z+1)z+D(z+1)z(z-1)+E(z+1)z(z-1)(z-2)+\&c.$$

will be obtained. Whence

$$T'-T=B+2Cz+3Dz(z-1)+4Ez(z-1)(z-2)+\&c.$$

Now when these values have been substituted into the equation which has to be resolved and the terms have been reduced to the same form,

$$\left.\begin{array}{l}+rB\\-nA\end{array}\right\}\ \left.\begin{array}{l}+2(r+1)C\\-\ (n-1)B\end{array}\right\}z\ \ \left.\begin{array}{l}+3(r+2)D\\-\ (n-2)C\end{array}\right\}z(z-1)$$
$$\left.\begin{array}{l}+4(r+3)E\\-\ (n-3)D\end{array}\right\}z(z-1)(z-2)+\&c.=0$$

will result. Finally, on setting like terms equal to zero, the following identities will be obtained:

$$B=\frac{n}{r}A,\ C=\frac{1}{2}\times\frac{n-1}{r+1}B,\ D=\frac{1}{3}\times\frac{n-2}{r+2}C,\ E=\frac{1}{4}\times\frac{n-3}{r+3}D,\ \&c.$$

And so

$$T=A+A\times\frac{n}{r}\times\frac{z}{1}+A\times\frac{n}{r}\times\frac{z}{1}\times\frac{n-1}{r+1}\times\frac{z-1}{r+2}+\&c.$$

That is,

$$T=A+\frac{n}{r}A\frac{z}{1}+\frac{n-1}{r+1}B\frac{z-1}{2}+\frac{n-2}{r+2}C\frac{z-2}{3}+\frac{n-1}{r+3}D\frac{z-3}{4}+\&c.,$$

where $A$, $B$, $C$, $D$, etc. no longer denote coefficients, but whole terms. And the first coefficient $A$, which is not determined by the equation, is equal to the term of the series to be interpolated which passes through the beginning of the abscissa $z$. And this is that very value of the term $T$ which would have been produced by Proposition 19. And also the series which Proposition 20 provides will be found through resolution of the difference equation by taking the required form of the root.

(p.275)

## Proposition 29

*Let a series of ordinates be given where the ordinates are distanced from each other by arbitrary intervals and the series goes off to infinity on just one side; and let it be required to find the parabolic line which goes through the extremities of them all.*

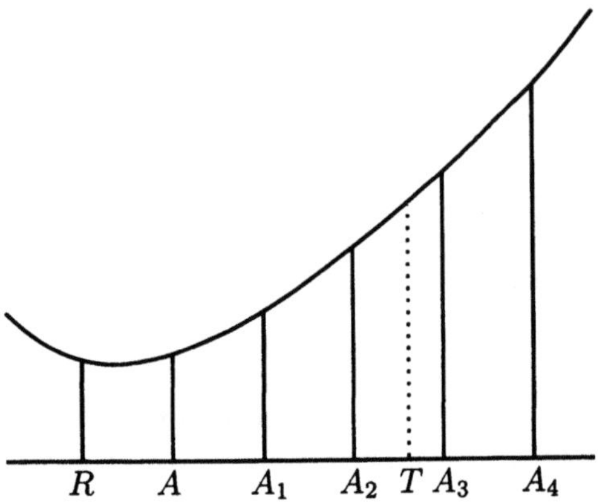


Let $A$, $A_1$, $A_2$, $A_3$, etc. be ordinates standing at right angles on the abscissa; and let $R$ be any point on the abscissa; and put

$$a = RA, \quad b = RA_1, \quad c = RA_2, \quad d = RA_3, \quad e = RA_4, \quad \&c.,$$

that is to say, let $a$, $b$, $c$, $d$, $e$, etc. be the intervals between the ordinates and the point $R$ respectively. And let $T$ denote in general any ordinate whose distance from the point $R$ is $z$. Then put

$$B = \frac{A_1 - A}{b - a}, \quad C = \frac{B_1 - B}{c - a}, \quad D = \frac{C_1 - C}{d - a}, \quad E = \frac{D_1 - D}{e - a}, \quad \&c.$$

$$B_1 = \frac{A_2 - A_1}{c - b}, \quad C_1 = \frac{B_2 - B_1}{d - b}, \quad D_1 = \frac{C_2 - C_1}{e - b}, \quad \&c.$$

$$B_2 = \frac{A_3 - A_2}{d - c}, \quad C_2 = \frac{B_3 - B_2}{e - c}, \quad \&c.$$

$$B_3 = \frac{A_4 - A_3}{e - d}, \quad \&c.$$

&c.

And the ordinate will be

$$\begin{aligned} T = A &+ \\ &B \times (z - a) + \\ &C \times (z - a) \times (z - b) + \\ &D \times (z - a) \times (z - b) \times (z - c) + \\ &E \times (z - a) \times (z - b) \times (z - c) \times (z - d) + \\ &F \times (z - a) \times (z - b) \times (z - c) \times (z - d) \times (z - e) + \\ &\qquad \&c. \end{aligned}$$

It is to be noted that the beginning of the abscissa, namely the point $R$, is to be taken arbitrarily, either between the ordinates or outside of them all as in the diagram, as long as account is taken of + and − signs. Now the proposition is demonstrated by substituting the ordinates $A$, $A_1$, $A_2$, etc. successively for $T$, and at the same time for $z$ its successive lengths in order $a$, $b$, $c$, etc. For by taking differences of the resulting equations and dividing these by the intervals of the ordinates, the values of the coefficients assigned above will come out.

### Example 1

Let the intervals of the ordinates from the beginning of the abscissa be $a = 2$, $b = 3$, $c = 5$, $d = 6$; let the ordinates themselves be

$$\begin{array}{llll} A = 2, & B = 1, & C = 0, & D = \frac{1}{2}, \\ A_1 = 3, & B_1 = 1, & C_1 = 2, & \\ A_2 = 5, & B_2 = 7, & & \\ A_3 = 12. & & & \end{array}$$

And by setting up the calculation according to the instructions of the theorem, it will be found that $B = 1$, $C = 0$, $D = \frac{1}{2}$: and when these have been substituted along with 2 for $A$, it will be found that

$$T = 2 + (z - 2) + \tfrac{1}{2} \times (z - 2) \times (z - 3) \times (z - 5),$$

which is

$$T = \frac{z^3 - 10z^2 + 33z - 30}{2}$$

after it has been brought back into order. For if 2, 3, 5, 6 are written in it for $z$, the proposed ordinates 2, 3, 5, 12 will come out.

### Example 2

Let it be required to determine the time of the solstice given some meridian altitudes of the sun about the same time. Let the ordinates denote the altitudes of the sun and let their intervals denote the times between the observations; then let the parabola pass through the extremities of the ordinates, and its abscissa which corresponds to the least ordinate, whether it be one of the given ordinates or some intermediate ordinate, will determine the moment of time at which the sun enters the Tropic. For example, in the year 1500 *B. Walther* of *Nürnberg* observed the distances of the sun from its zenith as follows:

| | | |
|---|---|---|
| 44975 | 8th | of June. |
| 44934 | 9th | |
| 44883 | 12th | |
| 44990 | 16th | |

Now let the observed distance on the eighth day be the first ordinate, and let the beginning of the abscissa be at the same point; and there will be $a = 0$, $b = 1$, $c = 4$, $d = 8$; and the calculation will be

$$\begin{array}{llll} A = 44975, & B = -41, & C = +6, & D = +\frac{1}{32}, \\ A_1 = 44934, & B_1 = -17, & C_1 = +\frac{175}{28}, & \\ A_2 = 44883, & B_2 = +\frac{107}{4}, & & \\ A_3 = 44990. & & & \end{array}$$

And on substituting these values for $A$, $B$, $C$, it will be found that

$$T = 44975 - 41z + 6z(z-1) + \tfrac{1}{32}z(z-1)(z-4),$$

that is,

$$T = 44975 - \tfrac{375}{8}z + \tfrac{187}{32}z^2 + \tfrac{1}{32}z^3.$$

Now since the abscissa sought will correspond to the least ordinate, let the fluxion of $T$ be set equal to zero, and

$$3z^2 + 374z = 1500,$$

will be obtained, whose root 3.889355 expresses the number of days which have elapsed between the meridian of the eighth of June and the moment of the solstice; and so this took place at approximately 21 hours $20\frac{2}{3}$ minutes after the meridian of the eleventh, according to these observations. The time of the solstice can also be determined by means of more observations and a parabola of more dimensions, or by means of three observations using the conical parabola as *Halley* instructed. But it is necessary that the differences between the observed altitudes are significantly greater than the errors which can be made in the course of observation, otherwise nothing can be concluded with certainty.

(p.276)

## Scholion

*Newton* uses this proposition for determining the position of a comet which falls between some positions known by observations. Certainly if one observes a number of longitudes which are denoted by just as many ordinates whose intervals are proportional to the times between the observations and the parabola is described through the extremities of the ordinates, the intermediate ordinates of this figure will denote the intermediate longitudes of the comet for times which are proportional to the abscissae. And by the same method the latitude will be given for any time if some latitudes are given. Now if the longitude and latitude have been given, the path of the comet in the heavens is given. And in this way very many things which are difficult to observe can be determined with sufficient accuracy from some previous and some subsequent observations.

This proposition is also applicable for the resolution of pure or affected equations. For by writing for the root in the equation to be resolved numbers which are not much different from it, their intervals will come out, which, having been interpolated, will produce the root. But after *Halley's* resolution of equations a more compendious solution is to be hoped for in vain.

In the case when the intervals of the ordinates are infinitely reduced this problem will give the root of a fluxional equation, even if neither the root nor the other indeterminate flows uniformly, and this is achieved by nothing more than substitution of fluxions of the root for differences of the ordinates, and of fluxions of the abscissa for their intervals. For just as the case of equidistant ordinates corresponds to uniformly increasing fluxions of the abscissa, so this proposition corresponds to fluxions which vary according to an arbitrary law. And the resolution of a fluxional equation in which both indeterminates flow according to arbitrary laws is not a corollary of this proposition, but its simplest case of all: it is appropriate to mention this here in passing, in order that it may be understood that the Method of Differences embraces in a very general way the universal doctrine of series, a fact which perhaps some have not realised.

(p.276)

## Proposition 30

*To find the asymptote of the hyperbola of logarithmic type given some of its equidistant ordinates.*

Let the equidistant ordinates be $A$, $B$, $C$, $D$, $E$, etc. which stand on the abscissa $PL$ at right angles to it; and let $QR$ be the asymptote of the curve, which is parallel to the abscissa and at distance $PQ$ from it. In fact, let any abscissa $AL$ be called $z$ and let the corresponding ordinate $Lm$ be $y$. Now the logarithmic hyperbola $HmK$ is defined by an equation of this form

$$y = a - br^z - cr^{2z} - dr^{3z} - er^{4z} - \&c.;$$

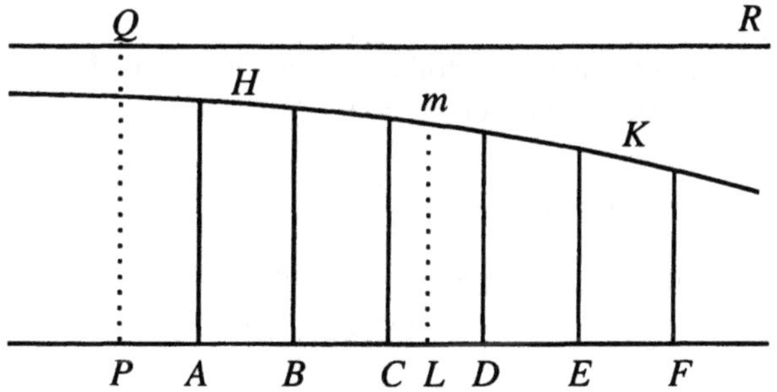


here $r$, $a$, $b$, $c$, $d$, $e$, etc. are constant quantities; and $PQ$, the distance between the abscissa and the asymptote, will be

$$
\begin{aligned}
PQ = A + {} & \\
& \frac{A-B}{r-1} + \\
& \frac{rA-(r+1)B+C}{(r-1)(r^2-1)} + \\
& \frac{r^3A-(r^3+r^2+r)B+(r^2+r+1)C-D}{(r-1)(r^2-1)(r^3-1)} + \\
& \frac{r^6-(r^6+r^5+r^4+r^3)B+(r^5+r^4+2r^3+r^2+r)C-(r^3+r^2+r+1)D+E}{(r-1)(r^2-1)(r^3-1)(r^4-1)} + \\
& \&\text{c.}
\end{aligned}
$$

The coefficients of the letters $A$, $B$, $C$, $D$, etc. in the various terms are formed by repeated multiplication of the numbers

$$1, \quad \frac{n-1}{r-1}, \quad \frac{n-r}{r^2-1}, \quad \frac{n-r^2}{r^3-1}, \quad \frac{n-r^3}{r^4-1}, \quad \&\text{c.}$$

Furthermore, in the first term $n = r^0$, in the second $n = r^1$, in the third $n = r^2$, in the fourth $n = r^3$, and so on. For example, in the fourth term the coefficients taken in reverse order and neglecting the signs are

$$1, \ 1 + r + r^2, \ r + r^2 + r^3, \ r^3;$$

in fact

$$1 \times \frac{r^3-1}{r-1} = 1 + r + r^2, \ (1 + r + r^2) \times \frac{r^3-r}{r^2-1} = r + r^2 + r^3,$$

and finally

$$(r + r^2 + r^3) \times \frac{r^3-r^2}{r^3-1} = r^3;$$

and in this way the coefficients are found. However, in this proposition I exclude the case in which $r = \pm 1$; for then the hyperbola degenerates to a straight line.

The series is investigated as follows. Let the sign of the abscissa $z$ be changed in the equation assumed, $y = a - br^z - cr^{2z} - \&\text{c.}$, and there will result

$$y = a - \frac{b}{r^z} - \frac{c}{r^{2z}} - \frac{d}{r^{3z}} - \&\text{c.}$$

Now let $z$ become infinitely large, and in the value of the ordinate $y$ all the terms after the first will vanish so long as $r$ is greater than one; in that way there will be $y = a$, that is, the ordinate removed to infinite distance, or the distance between the abscissa and the asymptote $PQ$, is equal to the first term $a$: for at infinite distance the curve coincides with its asymptote. Moreover, the quantity $a$ is investigated as follows in the equation first taken, $y = a - br^z - cr^{2z} - dr^{3z} - er^{4z} - \&\text{c.}$ Write the equidistant ordinates $A$, $B$, $C$, $D$, $E$, etc. successively for $y$ and correspondingly 0, 1, 2, 3, 4, etc. for the abscissa $z$; and the following equations will result:

$$\begin{aligned}
A &= a - b - c - d - e - \&\text{c.}\\
B &= a - br - cr^2 - dr^3 - er^4 - \&\text{c.}\\
C &= a - br^2 - cr^4 - dr^6 - er^8 - \&\text{c.}\\
D &= a - br^3 - cr^6 - dr^9 - er^{12} - \&\text{c.}\\
E &= a - br^4 - cr^8 - dr^{12} - er^{16} - \&\text{c.}\\
&\&\text{c.}
\end{aligned}$$

There are therefore as many equations as there are unknowns $a$, $b$, $c$, $d$, $e$, etc. from which $a$ is obtained by ordinary algebra, and its value will come out the same as has already been assigned to $PQ$. Q.E.D.

*Corollary.* Hence if some initial terms in an infinite series are given, whose differences are approximately in geometric progression, the last one of all, namely the one which is at infinite distance from the beginning, will be given. For if $A$, $B$, $C$, $D$, etc. taken in reverse order denote terms whose differences are almost in geometric proportion as $r^0$, $r^1$, $r^2$, $r^3$, etc. the last one of all will be equal to $PQ$, the interval between the abscissa and the asymptote: or, in the terminology of *Gregory*, the termination of the series will be given.

## Example

Given some regular polygons inscribed in a circle, to find the last of the polygons or the area of the circle. Let them be

| | | |
|---|---|---|
| 4 | 2.00000.00000.0000 = $F$ | 3.14033.11569.5475 |
| 8 | 2.82842.71247.4619 = $E$ | 126.08888.0294 |
| 16 | 3.06146.74589.2072 = $D$ | 6072.7439 |
| 32 | 3.12144.51522.5805 = $C$ | 5.5652 |
| 64 | 3.13654.84905.4594 = $B$ | 119 |
| 128 | 3.14033.11569.5475 = $A$ | |
| | | 3.14159.26535.8979 |

Let the last polygon be called $A$, the penultimate $B$, the antepenultimate $C$, and so on backwards. And since their differences $A - B$, $B - C$, $C - D$, etc. are approximately as the terms 1, 4, 16, 64, 256, etc., that is, as the powers of four, there will be $r = 4$; when this has been substituted, the general series becomes

$$A + \frac{A-B}{3} + \frac{4A - 5B + C}{3.15} + \frac{64A - 84B + 21C - D}{3.15.63} + \&\text{c.}$$

In this write for $A$, $B$, $C$, etc. their values, and the first five terms will give the area of the circle to fifteen places of figures, as is clear from the adjacent calculation. And the matter is carried out similarly by means of circumscribed polygons.

Moreover, any series can be summed by this method. For if the equidistant ordinates denote successive sums, the value of the whole series will be equal to the distance between the asymptote and the abscissa. If the series to be summed is of this type $a + bx + cx^2 + dx^3 + \&c.$, where $x$, $x^2$, $x^3$, etc. denote parts of the terms which are in geometric progression, there will be $r = x$; and the further the successive sums, which are denoted by $A$, $B$, $C$, $D$, etc. are from the beginning the more rapidly the value of $PQ$ will converge. But in the case where $r = \pm 1$, the hyperbola which is defined by this type of equation

$$y = \frac{1}{z^n} \times \left(a + \frac{b}{z} + \frac{c}{z^2} + \frac{d}{z^3} + \&c.\right)$$

is to be taken in place of the hyperbolic logarithm; and the index $n$ will be determined from the nature of the series to be summed.

It is to be noted that infinite series can equally be summed by means of *Newton's* parabola as by means of these hyperbolas. For, if the ordinates which are equidistant in the hyperbolas are set up at certain fixed distances, they will produce by means of the parabola the same expressions for the values of the series.

The number of figures which are true in the polygon $A$ is doubled by two terms of the series, tripled by three, and so on. Thus in the example given there are three true figures 314 in polygon $A$, and from this five terms of the series have given the area of the circle to fifteen places of figures. And these are approximations of the type which *James Gregory* and *Huygens* have previously found: indeed the latter trebled the true figures, while the former quadrupled them, quintupled, and in fact he extended them without limit, as may be seen in the Appendix to the *Vera Circuli et Hyperbolae Quadratura.*

## (p.280) Proposition 31

*To find the area of any curve very closely given some of its equidistant ordinates.*

Describe the parabolic figure through the extremities of the ordinates, and its area, which is found by known methods, will be approximately equal to the area of the proposed curve. Q.E.I.

### Scholion

Since it would be laborious to go back always to the parabola, I have computed the following table which shows the area of a curve directly given some of its equidistant ordinates.

*Table of Areas*

| | |
|---|---|
| 3 | $\frac{A+4B}{6}R$ |
| 5 | $\frac{7A+32B+12C}{90}R$ |
| 7 | $\frac{41A+216B+27C+272D}{840}R$ |
| 9 | $\frac{989A+5888B-928C+10496D-4540E}{28350}R$ |

*Table of Corrections*

| | |
|---|---|
| 3 | $\frac{P-4A+6B}{180}R$ |
| 5 | $\frac{P-6A+15B-20C}{470}R$ |
| 7 | $\frac{P-8A+28B-56C+70D}{930}R$ |
| 9 | $\frac{P-10A+45B-120C+210D-252E}{1600}R$ |

In these tables $A$ is the sum of the first and the last ordinates, $B$ is the sum of the second and the penultimate ordinates, $C$ is the sum of the third and the antepenultimate ordinates, and so on until the ordinate in the middle of them all has been reached, which is represented by the last of the letters $A$, $B$, $C$, etc. $R$ is the base above which the area lies, or that part of the abscissa intercepted between the first and the last ordinates. $P$ is the sum of the two ordinates one of which is located before the first ordinate, the other after the last ordinate, at distances equal to the common interval of the other ordinates. Now the number of ordinates, which here is odd, is indicated at the sides of the tables. The expressions in the Table of Areas are the areas contained by the base, the curve, and the extreme ordinates at both ends. And those in the Table of Corrections are approximately of the same magnitude as the differences between the true areas and those produced by means of the table: and so if the first figure of the correction is found, then let it be added if the correction is negative, or subtracted when this is positive; one may safely conclude that the area so corrected is true in that place of decimals in which the first figure of the correction appears, and not beyond it. And so the area found is corrected by means of the Table of Corrections, and at the same time the number of true figures is discerned.

**Example**

Let $\dfrac{1}{x+1}$ be the ordinate of an equilateral hyperbola, and let its area which lies above the abscissa equal to one be sought. Write for $x$ successively

$$\frac{0}{8},\ \frac{1}{8},\ \frac{2}{8},\ \frac{3}{8},\ \frac{4}{8},\ \frac{5}{8},\ \frac{6}{8},\ \frac{7}{8},\ \frac{8}{8},$$

and the nine ordinates

$$\frac{8}{8},\ \frac{8}{9},\ \frac{8}{10},\ \frac{8}{11},\ \frac{8}{12},\ \frac{8}{13},\ \frac{8}{14},\ \frac{8}{15},\ \frac{8}{16}$$

will come out. Therefore

$$A = \frac{8}{8} + \frac{8}{16} = \frac{3}{2}, \quad B = \frac{8}{9} + \frac{8}{15} = \frac{64}{45}, \quad C = \frac{8}{10} + \frac{8}{14} = \frac{48}{35},$$

$$D = \frac{8}{11} + \frac{8}{13} = \frac{192}{143}, \quad E = \frac{8}{12} = \frac{2}{3};$$

and when these have been substituted in the last expression for the areas along with one for $R$, the area arises as .69314721. Then in the ordinate $\dfrac{1}{1+x}$ write $-\frac{1}{8}$ and $\frac{9}{8}$ successively for $x$, and the two ordinates $\frac{8}{7}$ and $\frac{8}{17}$ will come out, of which the former stands before the first ordinate and the latter after the last ordinate; and so

$$P = \frac{8}{7} + \frac{8}{17} = \frac{192}{119};$$

when this has been substituted for $P$ and for $A$, $B$, $C$, $D$, $E$ their values, the correction for nine ordinates will give +.00000003, which, since it is positive, is subtracted from the value of the area previously found, and there will remain .69314718, which is accurate in the last figure.

I had computed these tables further, but the expressions for eleven or more ordinates are not suitable for application because of the huge magnitude of the numerical coefficients. But if nine ordinates do not give the area sufficiently accurately, let the base be divided into two or more parts, and then the area will be divided into just as many parts, and if each of these is obtained separately by means of nine ordinates, you will obtain the whole area as accurately as you wish. But sometimes it is also convenient to find part of an area by means of an infinite series, especially if the the curve crosses the base at right angles. And these things having been noted beforehand, any area will be obtained sufficiently accurately by means of the table now presented.

But also areas of curves can be expressed, not inconveniently, by means of differences of equidistant ordinates as follows.

*Table of Areas by Means of Differences of Ordinates*

| | |
|---|---|
| 1 | $A$ |
| 3 | $A+\frac{1}{6}B$ |
| 5 | $A+\frac{2}{3}B+\frac{7}{90}C$ |
| 7 | $A+\frac{3}{2}B+\frac{11}{20}C+\frac{41}{840}D$ |
| 9 | $A+\frac{8}{3}B+\frac{86}{45}C+\frac{92}{189}D+\frac{3}{86}E$ |
| 11 | $A+\frac{25}{6}B+\frac{175}{36}C+\frac{3445}{1512}D+\frac{4045}{9072}E+\frac{94}{3503}F$ |
| 13 | $A+6B+\frac{103}{10}C+\frac{158}{21}D+\frac{1833}{700}E+\frac{4813}{11550}F+\frac{66}{3050}G$ |

In this table $A$ is the ordinate in the middle of them all, $B$ is the second difference of the three ordinates in the middle, $C$ is the fourth difference of the five ordinates in the middle, and so on up to the last of the letters $A$, $B$, $C$, $D$, $E$, $F$, $G$, which is the last difference of all the ordinates. Thus if there are five ordinates $a$, $b$, $c$, $d$, $e$, there will be

$$A=c, \quad B=b-2c+d, \quad C=a-4b+6c-4d+e.$$

And it is similar in other cases. Now when these expressions have been multiplied by the base of the curve, or the part of the abscissa contained between the first and the last ordinates, they give the areas according to the given number of ordinates, which is indicated at the side. It is to be noted that the final terms in the expressions for nine, eleven and thirteen ordinates are not the true ones but are simpler than these and are in fact sufficiently close. For the middle ordinate $A$ and the differences $B$, $C$, $D$, $E$, etc. form a convergent series; and so it is not required that the coefficients of the final terms which enter into the calculation are absolutely accurate. Now it is discerned from the convergence of the series $A$, $B$, $C$, $D$, etc. to how many figures the area will be expressed accurately; and so this table does not require a table of corrections. But also the numerical coefficients are much smaller than those in the table above, and for that reason this table is to be preferred, especially where there is a large number of ordinates.

Again let the area of the hyperbola whose ordinate is $\dfrac{1}{1+x}$ be sought; for $x$ write $0$, $\frac{1}{4}$, $\frac{2}{4}$, $\frac{3}{4}$, $\frac{4}{4}$; and the five equidistant ordinates $1$, $\frac{4}{5}$, $\frac{2}{3}$, $\frac{4}{7}$, $\frac{1}{2}$ will come out; therefore there will be $A=\frac{2}{3}$, in fact the middle ordinate, $B=\frac{4}{105}$, that is to say, the second difference of the three ordinates in the middle $\frac{4}{5}$, $\frac{2}{3}$, $\frac{4}{7}$, and finally $C=\frac{1}{70}$, that is, it is equal to the last difference of them all. And when these values have been substituted in the expression for five ordinates, they give

$$\frac{2}{3}+\frac{8}{315}+\frac{1}{900}=\frac{4367}{6300},$$

which, when multiplied by the base, or one, and expressed in decimals, results in .69317 for the area sought. And this area is exact at least in the fourth decimal place, that is, where the first figure of the last term $\frac{1}{900}$ appears.

But in order that the differences may be more readily and easily found: let $a$ be the middle ordinate, $b$ the sum of the two ordinates on both sides which are nearest to the middle, then $c$ the sum of the two which follow on both sides, and so on; then put

$$\begin{aligned} A &= a, \\ B &= b - 2A, \\ C &= c - 2A - 4B, \\ D &= d - 2A - 9B - 6C, \\ E &= e - 2A - 16B - 20C - 8D, \\ F &= f - 2A - 25B - 50C - 35D - 10E, \\ G &= g - 2A - 36B - 105C - 112D - 54E - 12F, \\ &\text{\&c.} \end{aligned}$$

And $A$, $B$, $C$, $D$, etc. will be respectively the middle ordinate, the second difference of the three middle ordinates, the fourth difference of the five middle ordinates, and so on with the rest.

## (p.282) Proposition 32

*Let a, b, c, d, e, f, etc. denote equidistant terms which are tending continuously to the ratio of equality, and the following equations will approximate to their relations.*

| | |
|---|---|
| 2 | $a - b = 0$ |
| 3 | $a - 2b + c = 0$ |
| 4 | $a - 3b + 3c - d = 0$ |
| 5 | $a - 4b + 6c - 4d + e = 0$ |
| 6 | $a - 5b + 10c - 10d + 5e - f = 0$ |
| 7 | $a - 6b + 15c - 20d + 15e - 6f + g = 0$ |
| 8 | $a - 7b + 21c - 35d + 35e - 21f + 7g - h = 0$ |
| 9 | $a - 8b + 28c - 56d + 70e - 56f + 28g - 8h + i = 0$ |
| 10 | $a - 9b + 36c - 84d + 126e - 126f + 84g - 36h + 9i - k = 0$ |

&c.

This table is to be kept for use, so that it may be consulted as often as there is need. Now it is clear that the numerical coefficients are the coefficients of the powers of the binomial. And the demonstration is clear: for since the terms are supposed to tend continuously to the ratio of equality, their differences $a-b$, $b-c$, $c-d$, $d-e$, etc. will be small; then the differences of the differences $a-2b+c$, $b-2c+d$, $c-2d+e$, etc. will be less than the first differences; and the third differences $a-3b+3c-d$, $b-3c+3d-e$, etc. will be less than the second; and the fourth $a-4b+6c-4d+e$, etc. will be less than the third, and so on to infinity. Therefore when the first, second, third differences and the rest are set equal to zero as in the proposition, they will approximate continuously to the true relation of the terms. Q.E.D.

*Corollary.* Hence if any term is lacking in a series of equidistant terms, it can be found by means of this proposition. Thus if there are five terms $a$, $b$, $c$, $d$, $e$, their relation will be $a-4b+6c-4d+e=0$; and any one of them will be given very closely by this equation if the others have been given.

And it is to be noted that, other things being equal, the nearer a term is to the middle of all the terms, the more accurately it is determined: and errors from the true value are approximately as the reciprocals of the numerical coefficients of the terms sought. Therefore let the term sought be located either in the middle of all the terms or as close as possible to this.

## Example

Let the logarithm of the number 53 be sought given the logarithms of some preceding numbers. Put $a$ for the logarithm sought, and there will be

$$l,52=b=1.71600.33436,$$
$$l,51=c=1.70757.01761,$$
$$l,50=d=1.69897.00043,$$
$$l,49=e=1.69019.06800,$$
$$l,48=f=1.68124.12374,$$
$$l,47=g=1.67209.78579.$$

Then the relation between the seven terms $a$, $b$, $c$, $d$, $e$, $f$, $g$ will be

$$a-6b+15c-20d+15e-6f+g=0, \quad \text{and so} \quad a=6b-15c+20d-15e+6f-g:$$

on substituting in this for $b$, $c$, $d$, $e$, $f$, $g$ their values, 1.72427.58726 will be obtained for $a$, or the logarithm of the number 53, the error being .00000.00030 in excess. But if six logarithms are given, of which three are located before and three after the one which is sought, I say that in that case the logarithm sought will be determined very accurately.

Therefore let $A$, $B$, $C$, $D$, etc. denote the sums of the given terms which are at the same distance on both sides of the term sought, and its values will be as in the following table:

| | |
|---|---|
| 2 | $\frac{A}{2}$ |
| 4 | $\frac{4A-B}{6}$ |
| 6 | $\frac{15A-6B+C}{20}$ |
| 8 | $\frac{56A-28B+8C-D}{70}$ |
| 10 | $\frac{210A-120B+45C-10D+E}{252}$ |

&c.

For example, suppose that the logarithms of the numbers 50, 51, 52, 54, 55, 56 are given, and let it be required to find the logarithm of the number 53. Put

$$l,52 + l,54 = 3.44839.71034,$$
$$l,51 + l,55 = 3.44793.28656,$$
$$l,50 + l,56 = 3.44715.80313.$$

Then substitute these values for $A$, $B$, $C$ in the expression for six terms, and the number 1.72427.58695 will come out for the logarithm of the number 53, the error being one in the last figure. Hence if perhaps some term is missing in logarithmic, trigonometric, astronomical or other tables of this type, it can be deduced by means of this proposition; or if there is a suspicion that some term is erroneous, it can be corrected by the same method. For the expressions shown here are general and certainly do not depend upon the nature of any particular table.

(p.283)

## Proposition 33

*Let*

$$\&\text{c.}, \ \epsilon, \ \delta, \ \gamma, \ \beta, \ \alpha, \ a, \ b, \ c, \ d, \ e, \ \&\text{c.}$$

*denote successive terms in a series going off to infinity on both sides, and put*

$$A = a+\alpha, \ B = b+\beta, \ C = c+\gamma, \ D = d+\delta, \ E = e+\epsilon, \ \textit{and so on};$$

*and the term located in the middle between $a$ and $\alpha$ will be equal to*

$$\begin{aligned} &\tfrac{1}{2} \times A + \\ &\tfrac{1}{16} \times (A-B) + \\ &\tfrac{3}{256} \times (2A-3B+C) + \\ &\tfrac{5}{2048} \times (5A-9B+5C-D) + \\ &\tfrac{35}{65536} \times (14A-28B+20C-7D+E) + \end{aligned}$$

&c.

The numerical coefficients of the letters $A$, $B$, $C$, $D$, etc. are the differences of the coefficients in various powers of the binomial: and the coefficients which multiply the whole terms, namely $\frac{1}{2}$, $\frac{1}{16}$, $\frac{3}{256}$, etc., are generated by repeated multiplication of the numbers $\frac{1}{4\times 2}$, $\frac{3}{4\times 4}$, $\frac{5}{4\times 6}$, $\frac{7}{4\times 8}$, etc. Now that these things have been noted in advance, the series is continued at will.

And the series is investigated by putting $z = 0$ in the second case of Proposition 20; for then the ordinate or the term which is located in the middle of them all will be obtained.

Now let the terms of the series be collected into one sum as you see here:

$$\begin{array}{c|c} 2 & \dfrac{A}{2} \\ 4 & \dfrac{9A - B}{16} \\ 6 & \dfrac{150A - 25B + 3C}{256} \\ 8 & \dfrac{1225A - 245B + 49C - 5D}{2048} \\ 10 & \dfrac{39690A - 8820B + 2268C - 405D + 35E}{65536} \end{array}$$

&c.

The first expression is the first term of the series, the second expression is the sum of the first and the second terms, the third expression is the sum of the first three terms, and so on. Thus if the successive terms are given, the intermediate terms will be given rapidly by means of this table or of the series itself. The first expression $\dfrac{A}{2}$ suffices when the second term of the series is less than anything which may come into the calculation. And it is similar with the rest; for the terms of the series are the differences between the expressions and the true value very closely: and so one may always know which expression is sufficient for what has been proposed.

For example, suppose that the logarithm of the number 53 is required and that those of the numbers 46, 48, 50, 52, 54, 56, 58, 60 are given; put

$$l,52 + l,54 = A = 3.44839.71035,$$
$$l,50 + l,56 = B = 3.44715.80313,$$
$$l,48 + l,58 = C = 3.44466.92309,$$
$$l,46 + l,60 = D = 3.44090.90820.$$

And when these values have been written in the series, or in the expression for eight terms, 1.72427.58696 will be obtained for the logarithm of the number 53. And in the same way one may find any of the other intermediate logarithms. Therefore in the construction of tables it is enough to obtain first of all some terms at the required distances; for the rest can be inserted by this method. For the terms first found are to be interpolated repeatedly, until the process has reached the final terms which appear in the table. And it is to be noted that it is necessary to calculate all the terms about the beginning of the table because of the large differences; then step by step one may omit alternate terms with decreasing differences, and then groups of three and groups of seven, where the differences are smaller. And this is the method which *Newton* taught: but special rules deduced from the nature of the table to be constructed are to be preferred; for these for the most part will complete the task with less work.

*THE END*

# Notes

**Stirling's Preface (pp. 17–18).** This requires little comment since Stirling expands on many of the points he makes in the main body of the text. Concerning James Gregory see the note on Proposition 30. It is curious that Stirling refers to Newton's *first* letter to Oldenburg for his quotation (... ex Epistola ejus priori ad Oldenburgum ...). In fact it comes from the *epistola posterior*, which Stirling would have seen on pp. 67–86 (see p. 79) of the first edition or on pp. 142–190 (see p. 177) of the second edition of Collins's *Commercium Epistolicum* [12]; see also my note on the *Description of Curves Through Given Points* (pp. 242–243). Concerning De Moivre see the notes on Propositions 13, 23 and 28, and the Appendix.

In the final sentence of the Preface Stirling refers to Sir Alexander Cuming and the connection is elaborated in Stirling's letter to De Moivre which is translated in the Appendix. It has already been noted that Sir Alexander Cuming (ca 1690–1775) was one of Stirling's sponsors at the Royal Society (see my Introduction, footnote 7). He had a fascinating life, during which he was a baronet, a member of the Scottish bar, a Fellow of the Royal Society, a Cherokee chief, and an alchemist, but died in poverty (see [59]).

**Stirling's Introduction (pp. 19–32).** The first two sections, *On the Relation of Terms* and *On Difference Equations Which Define Series*, are quite clear and require little comment. In the first section Stirling discusses the MacLaurin expansion of $(r + sx + tx^2)^{-1}$ (on the assumption that 0 is not a root of the quadratic) and its Laurent expansion valid for $|x|$ sufficiently large. In connection with the reference to De Moivre see the end of the note on Proposition 13. In the second section Stirling introduces his notation for the terms of a series (sequence): if $T$ stands for $a_n$ then $T'$, $T''$, $T'''$, $T^{\mathrm{iv}}, \ldots$ stand for $a_{n+1}, a_{n+2}, a_{n+3}, a_{n+4}, \ldots$. Strictly, $T'$, $T''$, $T'''$ should be $T^{\mathrm{i}}$, $T^{\mathrm{ii}}$, $T^{\mathrm{iii}}$ but dashes are printed in the original and are retained here. The third section, *On the Form and Reduction of Series*, is much more substantial. While many of the particular rules which are stated and illustrated in it are not used in the sequel, the underlying ideas do pervade the rest of the work.

Integral powers of a variable $z$ appear naturally in analysis and are the familiar building blocks for polynomials, rational functions, power series and Taylor and Laurent expansions. However, they are not well-suited to numerical work since, for example, the difference $(z + 1)^n - z^n$ cannot be put in any simpler related form. The factorial expressions $z(z-1)\dots(z-n+1)$ and $(z(z+1)\dots(z+n))^{-1}$ are more suited for differencing: for example,

$$(z+1)z(z-1)\dots(z-n+2)-z(z-1)\dots(z-n+1)=nz(z-1)\dots(z-n+2).$$

For this and other purposes, Stirling discusses the representation of integral powers of $z$ in terms of such factorial expressions and provides two tables along with attendant rules to facilitate such processes. We will consider these in turn.

It is clear that for $m$ a positive integer there is a uniquely determined set of coefficients $S(m,1)$, ..., $S(m,m)$ such that

$$z^m = S(m,1)z + S(m,2)z(z-1) + \dots + S(m,m)z(z-1)\dots(z-m+1) \quad (1)$$

for all real or complex $z$. These coefficients are known nowadays as the *Stirling numbers of the second kind* and are important in combinatorial theory: $S(m,k)$ can be characterized as the number of partitions of a set of $m$ elements into $k$ nonempty classes. Stirling presents these numbers for $m = 1,2,\dots,9$ in his *First Table*; according to the rule which he states just before the table, the numbers in the rows are the coefficients in the expansions

$$\frac{1}{n-1} = \frac{1}{n}\left(1-\frac{1}{n}\right)^{-1} = \frac{1}{n} + \frac{1}{n^2} + \frac{1}{n^3} + \dots \quad (n>1),$$

$$\frac{1}{(n-1)(n-2)} = \frac{1}{n^2}\left(1-\frac{1}{n}\right)^{-1}\left(1-\frac{2}{n}\right)^{-1}$$

$$= \frac{1}{n^2} + \frac{3}{n^3} + \frac{7}{n^4} + \dots \quad (n>2),$$

$$\frac{1}{(n-1)(n-2)(n-3)} = \frac{1}{n^3}\left(1-\frac{1}{n}\right)^{-1}\left(1-\frac{2}{n}\right)^{-1}\left(1-\frac{3}{n}\right)^{-1}$$

$$= \frac{1}{n^3} + \frac{6}{n^4} + \frac{25}{n^5} + \dots \quad (n>3),$$

and so on. We can explain this as follows. Note that for any positive integer $k$ the Laurent expansion of $((z-1)(z-2)\dots(z-k))^{-1}$ for $|z|>k$ is of the form $\sum_{m=k}^{\infty} \alpha_m z^{-m}$ and the coefficient of $z^{-m}$ in this expansion is the coefficient of $z^{-1}$ in the corresponding expansion of

$$\frac{z^m}{z(z-1)(z-2)\dots(z-k)} = \frac{S(m,1)}{(z-1)\dots(z-k)} + \dots + \frac{S(m,k)}{z-k}$$
$$+ S(m,k+1) + S(m,k+2)(z-k-1) + \dots$$
$$+ S(m,m)(z-k-1)\dots(z-m+1). \quad (2)$$

But the only term on the right-hand side which can contribute to the term in $z^{-1}$ is

$$\frac{S(m,k)}{z-k} = \frac{S(m,k)}{z}\left(1-\frac{k}{z}\right)^{-1} = S(m,k)\sum_{r=0}^{\infty}\frac{k^r}{z^{r+1}} \quad (|z|>k).$$

Thus the required coefficient is $S(m,k)$. This shows that the $k$-th row of the table consists of the numbers $S(m,k)$ $(m=k,k+1,\ldots)$ and the $m$-th column contains $S(m,k)$ $(k=1,2,\ldots,m)$.

It is clear from (1) that

$$S(m,1) = S(m,m) = 1 \qquad (m=1,2,\ldots). \tag{3}$$

Furthermore, from the above discussion $S(m+1,k)$ will be the coefficient of $z^{-1}$ in the Laurent expansion of $z^{m+1}\left(z(z-1)\ldots(z-k)\right)^{-1}$ $(|z|>k)$ or equivalently the coefficient of $z^{-2}$ in the Laurent expansion of

$$z^m\left(z(z-1)\ldots(z-k)\right)^{-1} \qquad (|z|>k).$$

Using (2) we see that for $k=2,\ldots m$ this comes from

$$\frac{S(m,k-1)}{(z-k+1)(z-k)} + \frac{S(m,k)}{z-k} = \frac{S(m,k-1)}{z^2}\left(1-\frac{k-1}{z}\right)^{-1}\left(1-\frac{k}{z}\right)^{-1} + \frac{S(m,k)}{z}\left(1-\frac{k}{z}\right)^{-1}$$

and is therefore $S(m,k-1)+kS(m,k)$. Thus we have

$$S(m+1,k) = S(m,k-1)+kS(m,k) \quad (k=2,\ldots,m;\ m=2,3,\ldots).$$

Apparently Stirling did not notice this fact which, in conjunction with (3), allows us to construct the table much more easily.

For a positive integer $m$ we can expand $z^{-m-1}$ uniquely in the form[16]

$$\frac{1}{z^{m+1}} = \frac{\sigma(m,m)}{z(z+1)\ldots(z+m)} + \frac{\sigma(m+1,m)}{z(z+1)\ldots(z+m+1)} + \frac{\sigma(m+2,m)}{z(z+1)\ldots(z+m+2)} + \ldots \qquad (\operatorname{Re} z>0). \tag{4}$$

The coefficients $\sigma(k,m)$ are given in the *Second Table* for $k=1,2,\ldots,9$. According to the rule given by Stirling just before the table the numbers in the rows are the coefficients on the right-hand sides of the identities

[16]See, for example, [78, 7.82], where it is shown that a function of the form $f(z) = \sum_{n=0}^{\infty} a_n z^{-n}$ $(|z|>r)$ can be expanded in the half-plane $\operatorname{Re} z > r$ in the form $f(z) = b_0 + \sum_{n=1}^{\infty} b_n\left((z+1)\ldots(z+n)\right)^{-1}$. The required development comes from taking $f(z)=z^{-m}$, for which $r=0$, and dividing by $z$. Using properties of inverse factorial series (see [40, Chapter X]) we see that the expansion is unique and that conversely we will recover the original series if we form the Laurent expansion of each of the functions $b_n\left((z+1)\ldots(z+n)\right)^{-1}$ $(|z|>n)$ and add together the finite number of terms of each degree. The case $m=1$ is applied in Example 6 of Proposition 2 (see its note).

$$
\begin{aligned}
n &= n,\\
n(1+n) &= n+n^2,\\
n(1+n)(2+n) &= 2n+3n^2+n^3,
\end{aligned}
$$

and so on.[17] Again we can explain this by considering Laurent expansions. Let $k$ be a positive integer and let $m \in \{1,2,\ldots,k\}$. Then from (4)

$$
\begin{aligned}
\frac{z(z+1)\ldots(z+k-1)}{z^{m+1}} &= \sigma(m,m)(z+m+1)\ldots(z+k-1)+\ldots\\
&\quad + \sigma(k-2,m)(z+k-1)+\sigma(k-1,m)\\
&\quad + \frac{\sigma(k,m)}{z+k} + \frac{\sigma(k+1,m)}{(z+k)(z+k+1)} + \ldots .
\end{aligned}
$$

Now the coefficient of $z^{-1}$ on the left-hand side is the coefficient of $z^m$ in the expansion of $z(z+1)\ldots(z+k-1)$, while on the right-hand side only

$$
\frac{\sigma(k,m)}{z+k} = \frac{\sigma(k,m)}{z}\left(1+\frac{k}{m}\right)^{-1}
$$

can contribute to the term in $z^{-1}$. The required coefficient must therefore be $\sigma(k,m)$. It follows that the entries in the $k$-th row of the table are the numbers $\sigma(k,m)$ $(m=1,2,\ldots,k)$.

It is clear from Stirling's rule that $\sigma(k,k) = 1$ and $\sigma(k,1) = (k-1)!$ $(k=1,2,\ldots)$. Again there is a simple recurrence relation:

$$
\begin{aligned}
&z(z+1)\ldots(z+k-1)(z+k)\\
&= \left(\sigma(k,1)z+\sigma(k,2)z^2+\ldots\sigma(k,k)z^k\right)(z+k)\\
&= k\sigma(k,1)z+\left(\sigma(k,1)+k\sigma(k,2)\right)z^2+\ldots+\left(\sigma(k,k-1)+k\sigma(k,k)\right)z^k\\
&\qquad + \sigma(k,k)z^{k+1}
\end{aligned}
$$

so that

$$
\sigma(k+1,m) = \sigma(k,m-1)+k\sigma(k,m) \quad (m=2,3,\ldots,k).
$$

What are called nowadays the *Stirling numbers of the first kind* are in fact the numbers $(-1)^{k-m}\sigma(k,m)$, that is to say, in each column of Stirling's *Second Table* every second entry is to be multiplied by $-1$. In combinatorial terms $\sigma(k,m)$ is the number of permutations of $k$ symbols which can be decomposed into exactly $m$ disjoint cycles.

---

[17] As an easy consequence we obtain the inverse of (1), which Stirling does not give:

$$
\begin{aligned}
z(z-1)\ldots(z-k+1) &= \sigma(k,k)z^k-\sigma(k,k-1)z^{k-1}+\sigma(k,k-2)z^{k-2}\\
&\quad -\ldots+(-1)^{k+1}\sigma(k,1)z.
\end{aligned}
$$

As an application Stirling obtains the following expansion, which he applies in Example 5 of Proposition 2 and Example 2 of Proposition 3 (see their notes and the note on Proposition 8):

$$\frac{1}{z^2+nz}=\sum_{k=0}^{\infty}\frac{(1-n)_k}{z(z+1)\dots(z+k+1)}.$$

This is done by first finding the Laurent expansion

$$\frac{1}{z^2+nz}=\sum_{k=0}^{\infty}\frac{(-1)^k n^k}{z^{k+2}}\qquad(|z|>|n|),$$

and then expressing each $1/z^{k+2}$ by its inverse factorial series. In fact his series is

$$\frac{1}{z(z+1)}F(1-n,1;z+2;1),$$

which, according to Gauss's formula (p. 5, (1)), converges to the required value provided $\mathrm{Re}(z+n)>0$ and $z\neq 0,-1,-2,\dots$.

We consider finally the rule which Stirling gives at the end of the Introduction for transforming a series of the form

$$\frac{\alpha_1}{z(z+1)}+\frac{\alpha_2}{z(z+1)(z+2)}+\frac{\alpha_3}{z(z+1)(z+2)(z+3)}+\dots$$

into one of the form

$$\frac{a_1}{z^2}+\frac{a_2}{z^3}+\frac{a_3}{z^4}+\frac{a_4}{z^5}+\dots$$

by means of the columns of the *First Table*. Let each term be developed in a series of negative integral powers of $z$. Then for $m=k+1,k+2,\dots$ the coefficient of $z^{-m}$ in the expansion of $(z(z+1)\dots(z+k))^{-1}$ for $|z|>k$ is the coefficient of $z^{-m+k+1}$ in the expansion of

$$\left(1+\frac{1}{z}\right)^{-1}\dots\left(1+\frac{k}{z}\right)^{-1}\qquad(|z|>k),$$

which is $(-1)^{m-k-1}$ times the coefficient of $z^{-m+k+1}$ in the expansion of

$$\left(1-\frac{1}{z}\right)^{-1}\dots\left(1-\frac{k}{z}\right)^{-1}\qquad(|z|>k),$$

or equivalently $(-1)^{m-k-1}$ times the coefficient of $z^{-m+1}$ in the expansion of

$$\frac{1}{(z-1)(z-2)\dots(z-k)}\qquad(|z|>k).$$

Thus from our previous discussion of the *First Table* we see that the required coefficient is $(-1)^{m-k-1}S(m-1,k)$ $(m=k+1,k+2,\dots)$. Precisely the first

$m-1$ terms of the given inverse factorial series will contribute to the term in $z^{-m}$, whose combined coefficient will therefore be

$$a_{m-1} = (-1)^m \sum_{k=1}^{m-1} (-1)^{k+1} \alpha_k S(m-1, k) \qquad (m = 2, 3, \ldots).$$

This is Stirling's final rule, which will apply whenever both series are suitably convergent; in general, the series of reciprocal powers may only be asymptotic.

There is no mention of the Bernoulli numbers in the *Methodus Differentialis*. However, Stirling appears to have been aware at a later date of a relationship between these and the numbers $S(m, k)$ (see [70, pp. 15–16]).

Stirling was certainly not the first to realise the importance of factorial expressions and their relevance for the calculus of finite differences and operations with series; indeed he refers to earlier work of the French mathematician François Nicole (1683–1758) in connection with Corollary 1 of Proposition 1. Nicole's contributions are contained in his *Traité du Calcul des Differences Finies* [49] (1717 and 1723) and its supplement [50] (1724). These have been discussed in [72] by Charles Tweedie, who has also given an interesting account of the Stirling numbers in [73].

**Stirling's Introductory Remarks in Part I (pp. 33–37).** In the section *On Simpler Series* Stirling quotes the series

$$\int_0^x t^{\theta-1}(e+ft^\eta)^{\lambda-1}dt = \frac{x^\theta}{\theta e}(e+fx^\eta)^\lambda - \frac{s(e+fx^\eta)^\lambda f x^{\theta+\eta}}{r\theta e^2} + \frac{s(s+1)(e+fx^\eta)^\lambda f^2 x^{\theta+2\eta}}{r(r+1)\theta e^3} - \cdots,$$

where $r = \dfrac{\theta+\eta}{\eta}$, $s = \dfrac{\theta+\lambda\eta}{\eta}$. Apparently he obtained this by using Proposition 7 from Newton's *De Quadratura Curvarum* [47, pp. 39–66], in which Newton discusses the areas of binomial curves (see the note on Proposition 16 and its scholion and [77, vol. VII, pp. 26–29]). The series may be developed by splitting up the integral and integrating by parts as follows:

$$\int_0^x t^{\theta-1}(e+ft^\eta)^{\lambda-1}dt$$

$$= \frac{1}{e}\int_0^x t^{\theta-1}(e+ft^\eta)^\lambda dt - \frac{f}{e}\int_0^x t^{\theta-1+\eta}(e+ft^\eta)^{\lambda-1}dt$$

$$= \frac{x^\theta(e+fx^\eta)^\lambda}{e\theta} - \frac{1}{e}\int_0^x \frac{t^\theta}{\theta}\lambda(e+ft^\eta)^{\lambda-1} f\eta t^{\eta-1} dt$$

$$- \frac{f}{e}\int_0^x t^{\theta-1+\eta}(e+ft^\eta)^{\lambda-1} dt$$

$$= \frac{x^\theta(e+fx^\eta)^\lambda}{e\theta} - \frac{f}{e}\left(\frac{\lambda\eta}{\theta}+1\right)\int_0^x t^{\theta-1+\eta}(e+ft^\eta)^{\lambda-1} dt$$

$$= \frac{x^\theta(e+fx^\eta)^\lambda}{e\theta} - \frac{f}{e}\left(\frac{\lambda\eta}{\theta}+1\right)\frac{x^{\theta+\eta}(e+fx^\eta)^\lambda}{e(\theta+\eta)}$$

$$+ \frac{f}{e}\left(\frac{\lambda\eta}{\theta}+1\right)\frac{f}{e}\left(\frac{\lambda\eta}{\theta+\eta}+1\right)\int_0^x t^{\theta-1+2\eta}(e+ft^\eta)^{\lambda-1} dt$$

and so on – this of course assumes that the parameters are such that all evaluations at $x = 0$ are zero and that the sequence of integrals tends to zero. In Example 3 of Proposition 21 Stirling uses interpolation methods along with a series transformation to obtain the above result (see the note on Proposition 21). Newton's Proposition 7 also finds application in Stirling's Propositions 24 and 25.

In the case $e = 1$, $f = -1$, $\theta = 1$, $\eta = 2$, $\lambda = \frac{1}{2}$ we obtain

$$\sin^{-1} x = \int_0^x \frac{1}{\sqrt{1-t^2}}\, dt$$

$$= x\sqrt{1-x^2}\left(1 + \sum_{n=1}^{\infty} \frac{2\times 4\times \ldots \times 2n}{3\times 5\times \ldots \times (2n+1)} x^{2n}\right)$$

$$= \sqrt{1-x^2}\sum_{n=0}^{\infty} \frac{2^{2n}(n!)^2 x^{2n+1}}{(2n+1)!},$$

which is valid for $-1 < x < 1$. Stirling also notes the MacLaurin series

$$\sin^{-1} x = x + \sum_{n=1}^{\infty} \frac{(1\times 3\times \ldots \times (2n-1))^2}{(2n+1)!} x^{2n+1}$$

$$= \sum_{n=0}^{\infty} \binom{2n}{n} \frac{x^{2n+1}}{(2n+1)2^{2n}},$$

which is valid for $-1 \le x \le 1$, and comments on the relative merits of the two series.

Stirling's remarks in the section *On Series Which Converge More Rapidly* seem rather obscure. Presumably his comments about the use of the sine or tangent relate to the problem of finding the angle (from which the area and the arc length follow) when its sine or tangent is given – the MacLaurin series for the inverse sine converges for all possible $x$, namely $-1 \leq x \leq 1$, whereas the inverse tangent series

$$\tan^{-1} x = \sum_{n=0}^{\infty} \frac{(-1)^n x^{2n+1}}{2n+1} \qquad (-1 \leq x \leq 1)$$

applies only over a bounded part of the range of the tangent. In the case of the hyperbola the angle is limited by that between the asymptotes.

Finally, in the section *On Successive Sums* we have Stirling's definitions which are fundamental for his subsequent discussions. For a convergent infinite series $\sum_{n=n_0}^{\infty} a_n$ the ***successive sums*** are the terms of the sequence

$$\sum_{n=n_0}^{\infty} a_n, \quad \sum_{n=n_0+1}^{\infty} a_n, \quad \sum_{n=n_0+2}^{\infty} a_n, \quad \ldots .$$

In the case of a finite sum $\sum_{n=n_0}^{N} b_n$, however, the successive sums are the terms of the finite sequence

$$\sum_{n=n_0}^{N} b_n, \quad \sum_{n=n_0}^{N-1} b_n, \quad \sum_{n=n_0}^{N-2} b_n, \quad \ldots, \quad \sum_{n=n_0}^{n_0} b_n = b_{n_0}.$$

We are usually concerned with the former. Just as Stirling uses $T$, $T'$, $T''$, $T'''$, $T^{\text{iv}}$, ... to denote the successive terms of a series, so he uses $S$, $S'$, $S''$, $S'''$, $S^{\text{iv}}$, ... to represent the successive sums.[18] He also labels the successive sums by means of an abcscissa in such a way that $S$, $S'$, $S''$, $S'''$, ... correspond to $z$, $z+1$, $z+2$, $z+3$, ... in the case of an infinite series and to $z$, $z-1$, $z-2$, $z-3$, ... in the case of a finite sum; in both cases $z$ is a conveniently chosen initial value, which is not necessarily an integer.

Stirling's diagram represents a convergent infinite series: $Sa$ is the sum of the series and the successive sums are $Sa$, $S^{\text{i}}b$, $S^{\text{ii}}c$, ... , which tend to zero since the curve approaches its asymptote; the differences of the successive sums, $S\alpha$, $S^{\text{i}}\beta$, $S^{\text{ii}}\gamma$, ... , represent the terms of the series.

**Proposition 1 and Its Scholion (pp. 37–40).** Here Stirling shows that if

$$a_r = r! \sum_{s=0}^{r} \frac{\alpha_s}{(r-s)!} \qquad (r = 1, 2, \ldots)$$

then

$$\sum_{r=1}^{z} a_r = \alpha_0 z + (z+1)! \sum_{r=1}^{z} \frac{\alpha_r}{(z-r)!\,(r+1)} \qquad (z = 1, 2, \ldots).$$

[18]Strictly $S'$, $S''$, $S'''$ should be $S^{\text{i}}$, $S^{\text{ii}}$, $S^{\text{iii}}$, but it is only in the associated diagram that this notation is used.

His proof is essentially an induction argument except that, instead of verifying directly that the result holds when $z = 1$, Stirling observes that when $z = 0$ (no terms) the right hand side is also zero.

In *Examples 1–5* he deduces that

$$\sum_{r=1}^{z} r = \frac{1}{2}z(z+1) \quad (\alpha_0 = 0, \alpha_1 = 1, \alpha_r = 0\ (r \geq 2)),$$

$$\sum_{r=1}^{z} (2r-1) = z^2 \quad (\alpha_0 = -1, \alpha_1 = 2, \alpha_r = 0\ (r \geq 2)),$$

$$\sum_{r=1}^{z} r^2 = \frac{1}{6}z(z+1)(2z+1) \quad (\alpha_0 = -1, \alpha_1 = \alpha_2 = 1, \alpha_r = 0\ (r \geq 3)),$$

$$\sum_{r=1}^{z} (2r-1)^2 = \frac{1}{3}z(4z^2-1) \quad (\alpha_0 = 1, \alpha_1 = 0, \alpha_2 = 4, \alpha_r = 0\ (r \geq 3)),$$

$$\sum_{r=1}^{z} r^3 = \frac{1}{4}z^2(z+1)^2 \quad (\alpha_0 = 0, \alpha_1 = 1, \alpha_2 = 3, \alpha_3 = 1, \alpha_r = 0\ (r \geq 4)).$$

The summation formula which Stirling states in the *scholion* results from applying Newton's forward difference formula to the sequence of partial sums

$$0, \quad A, \quad A + A_2, \quad A + A_2 + A_3, \quad A + A_2 + A_3 + A_4, \quad \ldots,$$

whose first differences are just the terms of the series and whose subsequent differences are therefore as given in Stirling's table (see Proposition 19 and its note).

**Proposition 2 (pp. 40–47).** Stirling's expression for the sum is of the form

$$\sum_{m=1}^{\infty} \frac{b_m}{mz(z+1)(z+2)\ldots(z+m-1)}.$$

Provided such a series does not diverge everywhere we know from properties of inverse factorial series (see pp. 6–7) that there is a real number $a$ such that its sum defines a function $f(z)$ on $(a, \infty)$ and moreover $f(z) \to 0$ as $z \to \infty$. Stirling's proof shows that

$$f(z) - f(z+1) = \sum_{m=1}^{\infty} \frac{b_m}{z(z+1)\ldots(z+m)}.$$

Thus if we give $z$ any value $z_0 \in (a, \infty)$ and define

$$a_n = \sum_{m=1}^{\infty} \frac{b_m}{(z_0+n)(z_0+n+1)\ldots(z_0+n+m)} \qquad (n = 0, 1, \ldots),$$

we have

$$a_n = f(z_0 + n) - f(z_0 + n + 1)$$

and consequently

$$\sum_{n=0}^{N} a_n = f(z_0) - f(z_0 + N + 1) \to f(z_0) \quad \text{as} \quad N \to \infty.$$

Thus

$$\sum_{n=0}^{\infty}\sum_{m=1}^{\infty} \frac{b_m}{(z_0+n)(z_0+n+1)\dots(z_0+n+m)}$$

$$= \sum_{m=1}^{\infty} \frac{b_m}{mz_0(z_0+1)\dots(z_0+m-1)}\,.$$

This is the essence of *Proposition 2* and its proof.

*Corollary 1* and *Example 1* deal with the simple case where only one $b_m$ is nonzero. *Corollary 2* and *Examples 2–4* are concerned with situations where other rational terms can be expressed as sums of a finite number of terms of the type dealt with in the proposition. Such expressions are obtained by means of the techniques developed in Stirling's Introduction. *Example 5* and *Example 6* are much more significant and provide striking illustrations of Stirling's stated aims: the transformation of slowly converging series into rapidly converging ones.

Viscount Brouncker, the first President of the Royal Society, gave the result

$$\sum_{n=1}^{\infty} \frac{1}{2n(2n-1)} = \sum_{n=1}^{\infty} \frac{(-1)^{n-1}}{n} = \ln 2$$

in an article in the *Philosophical Transactions* for 1668 [11]. Stirling deals with the series in its first form in *Example 5*. Its successive sums can be bounded by integrals in the usual way:

$$\int_{N+1}^{\infty} \frac{dx}{2x(2x-1)} < \sum_{n=N+1}^{\infty} \frac{1}{2n(2n-1)} < \int_{N+\frac{1}{2}}^{\infty} \frac{dx}{2x(2x-1)}\,,$$

so that

$$\frac{1}{2}\ln\frac{2N+2}{2N+1} < \sum_{n=N+1}^{\infty} \frac{1}{2n(2n-1)} < \frac{1}{2}\ln\frac{2N+1}{2N}\,.$$

We see from these inequalities that if, for example, we add up the first million terms of the series, the residual sum would be approximately $2.5 \times 10^{-7}$, producing an error in the seventh decimal place (cf. Stirling's list of partial sums at the end of *Example 5*).

Now the terms of Brouncker's series are given by

$$\frac{1}{4z(z+\frac{1}{2})} \qquad (z = 1/2\,, 3/2\,, 5/2\,, \dots)$$

and if we apply Stirling's expression for $\dfrac{1}{z^2+nz}$ (Introduction, p. 30) with $n = 1/2$ we can express the terms as

$$\frac{1}{4z(z+\frac{1}{2})} = \frac{1}{4z(z+1)} + \sum_{m=2}^{\infty} \frac{1 \times 3 \times \ldots \times (2m-3)}{2^{m+1} z(z+1)\ldots(z+m)}, \qquad (*)$$

which is of the form required in the proposition. In order to find the sum of the series Stirling adds up directly the first thirteen terms and transforms the residual sum. We have

$$\sum_{m=14}^{\infty} \frac{1}{2n(2n-1)} = \sum_{n=0}^{\infty} \frac{1}{4(13.5+n)(14+n)}$$

$$= \sum_{n=0}^{\infty} \Bigg( \frac{1}{4(13.5+n)(13.5+n+1)} + \sum_{m=2}^{\infty} \frac{1 \times 3 \times \ldots \times (2m-3)}{2^{m+1}(13.5+n)(13.5+n+1)\ldots(13.5+n+m)} \Bigg)$$

$$= \frac{1}{4 \times 13.5} + \sum_{m=2}^{\infty} \frac{1 \times 3 \times \ldots \times (2m-3)}{m 2^{m+1} \times 13.5(13.5+1)\ldots(13.5+m-1)}$$

by the proposition. Stirling adds directly the first nine terms of this series, presumably because he wants nine decimal places and the remaining terms all have zeros in their first nine decimal places. Correctly rounded to nine decimal places ln 2 is .693147181, so that Stirling's calculated value is out by 1 in the final decimal place.

We can obtain error bounds as follows. In general the expression given by the proposition for the sum of the terms (*) is

$$\frac{1}{4z} + \sum_{m=2}^{\infty} \frac{1 \times 3 \times \ldots (2m-3)}{m 2^{m+1} z(z+1)\ldots(z+m-1)}.$$

If we add up the first $k$ terms, the residual sum is

$$R_k = \frac{1 \times 3 \times \ldots \times (2k-1)}{2^{k+2} z(z+1)\ldots(z+k)} \times \left( \frac{1}{k+1} + \frac{2k+1}{2(k+2)(z+k+1)} + \frac{(2k+1)(2k+3)}{2^2(k+3)(z+k+1)(z+k+2)} + \ldots \right)$$

$$= \frac{1 \times 3 \times \ldots \times (2k-1)}{2^{k+2} z(z+1)\ldots(z+k)} \times \left( \frac{1}{k+1} + \frac{k+\frac{1}{2}}{(k+2)(z+k+1)} + \frac{(k+\frac{1}{2})(k+\frac{3}{2})}{(k+3)(z+k+1)(z+k+2)} + \ldots \right)$$

$$= \frac{1 \times 3 \times \ldots \times (2k-1)}{2^{k+2} z(z+1)\ldots(z+k)} \times$$

$$\left(\frac{1}{k+1} + \frac{1}{z+k+1}\left\{\frac{k+\frac{1}{2}}{k+2} + \frac{k+\frac{1}{2}}{z+k+2}\left(\frac{k+\frac{3}{2}}{k+3}\right)\right.\right.$$

$$\left.\left. + \frac{(k+\frac{1}{2})(k+\frac{3}{2})}{(z+k+2)(z+k+3)}\left(\frac{k+\frac{5}{2}}{k+4}\right) + \ldots\right\}\right) \quad (**)$$

An obvious inequality and an application of Gauss's formula for $F(a, b; c; 1)$ now lead to an upper bound:

$$R_k < \frac{1 \times 3 \times \ldots \times (2k-1)}{2^{k+2} z(z+1)\ldots(z+k)}$$

$$\times \left(\frac{1}{k+1} + \frac{1}{z+k+1}\left\{1 + \frac{k+\frac{1}{2}}{z+k+2} + \frac{(k+\frac{1}{2})(k+\frac{3}{2})}{(z+k+2)(z+k+3)} + \ldots\right\}\right)$$

$$= \frac{1 \times 3 \times \ldots \times (2k-1)}{2^{k+2} z(z+1)\ldots(z+k)}\left(\frac{1}{k+1} + \frac{1}{z+k+1} F(1, k+\tfrac{1}{2}; z+k+2; 1)\right)$$

$$= \frac{1 \times 3 \times \ldots \times (2k-1)}{2^{k+2} z(z+1)\ldots(z+k)}\left(\frac{1}{k+1} + \frac{1}{z+k+1} \times \frac{\Gamma(z+k+2)\Gamma(z+\frac{1}{2})}{\Gamma(z+k+1)\Gamma(z+\frac{3}{2})}\right)$$

$$= \frac{1 \times 3 \times \ldots \times (2k-1)}{2^{k+2} z(z+1)\ldots(z+k)}\left(\frac{1}{k+1} + \frac{1}{z+\frac{1}{2}}\right),$$

so that

$$R_k < \frac{1 \times 3 \times \ldots \times (2k-1)(z+k+\frac{3}{2})}{2^{k+2} z(z+1)\ldots(z+k)(k+1)(z+\frac{1}{2})}.$$

Putting $k = 9$, $z = 13.5$ we obtain from this an upper bound of $9.203 \times 10^{-10}$ for the error in Stirling's calculation due to taking just the first nine terms of the transformed series.

Since $\dfrac{x+a}{x+b}$ increases to 1 as $x \to \infty$ provided $a < b$, in (**) we may replace each of the ratios

$$\frac{k+\frac{3}{2}}{k+3}, \quad \frac{k+\frac{5}{2}}{k+4}, \quad \ldots \quad \text{by} \quad \frac{k+\frac{1}{2}}{k+2}$$

to obtain

$$R_k > \frac{1 \times 3 \times \ldots \times (2k-1)}{2^{k+2} z(z+1)\ldots(z+k)} \times \left( \frac{1}{k+1} + \frac{k+\frac{1}{2}}{(k+2)(z+k+1)} \right.$$

$$\left. \times \left\{ 1 + \frac{k+\frac{1}{2}}{z+k+2} + \frac{(k+\frac{1}{2})(k+\frac{3}{2})}{(z+k+2)(z+k+3)} + \ldots \right\} \right)$$

$$= \frac{1 \times 3 \times \ldots \times (2k-1)}{2^{k+2} z(z+1)\ldots(z+k)}$$

$$\times \left( \frac{1}{k+1} + \frac{k+\frac{1}{2}}{(k+2)(z+k+1)} F(1, k+\tfrac{1}{2}\,; z+k+2\,; 1) \right)$$

$$= \frac{1 \times 3 \times \ldots \times (2k-1)}{2^{k+2} z(z+1)\ldots(z+k)} \left( \frac{1}{k+1} + \frac{k+\frac{1}{2}}{(k+2)(z+\frac{1}{2})} \right).$$

With $z = 13.5$ and $k = 9$ this gives a lower bound of $8.68 \times 10^{-10}$ for the error.

Stirling had already applied a different transformation technique in [61] to evaluate $\sum_{n=1}^{\infty} \frac{1}{2n(2n-1)}$ (see [28, 33, 71]). He obtained the same number of correct decimal places in [61] but the transformation used there is much more complicated both in application and in analysis; it is not discussed in the present work and was presumably abandoned by Stirling in favour of the more straightforward method based on inverse factorial series. It is interesting that Stirling refers to the work of the French mathematician François Nicole (1683–1758) in his discussion of *Proposition 2*. Nicole's papers dealing with differences and inverse factorials ([49, 50]) are dated 1717, 1723 and 1724 but were each published two years later, so it is highly unlikely that Stirling knew any of Nicole's work in this area when he wrote his paper [61], receipt of which was noted at the meeting of the Royal Society on 18 June 1719. Another transformation of Brouncker's series is discussed in Example 2 of Proposition 8.

In *Example 6* Stirling gives a similar treatment of $\sum_{n=1}^{\infty} \frac{1}{n^2}$, whose successive sums satisfy the inequalities

$$\frac{1}{N+1} = \int_{N+1}^{\infty} \frac{1}{x^2}\,dx < \sum_{n=N+1}^{\infty} \frac{1}{n^2} < \int_{N+\frac{1}{2}}^{\infty} \frac{1}{x^2}\,dx = \frac{1}{N+\frac{1}{2}}.$$

The problem of determining $\sum_{n=1}^{\infty} 1/n^2$ exactly was at the time a celebrated problem and was first resolved by Euler in the early 1730s [16]. In a letter of 8 June 1736 Euler communicated to Stirling the values of $\sum_{n=1}^{\infty} 1/n^{2k}$ ($k = 1, 2, 3, 4$), in particular $\sum_{n=1}^{\infty} 1/n^2 = \pi^2/6$, and in his reply of 16 April 1738 Stirling said of Euler's results: "I acknowledge this to be quite ingenious and entirely new and I do not see that it has anything in common with the accepted methods, so that I readily believe that you have drawn it from a new source." (See [70, p. 144; 74, p. 179].)

Stirling employs the representation

$$\frac{1}{z^2} = \sum_{m=1}^{\infty} \frac{(m-1)!}{z(z+1)\dots(z+m)}$$

from p. 29 of the Introduction. Then, transforming the series from the $(N+1)$-th term onwards according to the proposition, we have

$$\begin{aligned}\sum_{n=1}^{\infty} \frac{1}{n^2} &= \sum_{n=1}^{N} \frac{1}{n^2} + \sum_{n=N+1}^{\infty} \sum_{m=1}^{\infty} \frac{(m-1)!}{n(n+1)\dots(n+m)} \\ &= \sum_{n=1}^{N} \frac{1}{n^2} + \sum_{n=0}^{\infty} \sum_{m=1}^{\infty} \frac{(m-1)!}{(N+1+n)(N+1+n+1)\dots(N+1+n+m)} \\ &= \sum_{n=1}^{N} \frac{1}{n^2} + \sum_{m=1}^{\infty} \frac{(m-1)!}{m(N+1)(N+2)\dots(N+m)}.\end{aligned}$$

Stirling takes $N = 12$, adds up $\sum_{n=1}^{12} 1/n^2$ directly and takes the first thirteen terms of the last series to obtain 1.644934065 as an approximation to $\sum_{n=1}^{\infty} 1/n^2$. This agrees with $\pi^2/6$ except in the last figure where the true value, correctly rounded, has 7.[19] In Example 1 of Proposition 11 Stirling applies a different transformation to obtain the sum correct to sixteen decimal places. (See also the related Example 2 of Proposition 12.)

An error analysis similar to that given for *Example 5* produces the following bounds for the error introduced by taking only the first $k$ terms of the transformed series:

$$\text{(upper bound)} \qquad \frac{k!(z+k+1)}{z^2(z+1)(z+2)\dots(z+k)(k+1)};$$

$$\text{(lower bound)} \qquad \frac{k!}{z(z+1)\dots(z+k)}\left(\frac{1}{k+1} + \frac{k+1}{z(k+2)}\right).$$

With $z = 13\,(= N+1)$ and $k = 13$ these show that the error lies between $1.1 \times 10^{-9}$ and $1.059 \times 10^{-9}$.

**Proposition 3 (pp. 47–51).** Stirling's demonstration of *Proposition 3* follows the same pattern as that of Proposition 2. Here $x$ and $n$ are constants and clearly we require that $x \neq 1$, otherwise the expression for the sum is not defined; however, we should also restrict $x$ to the interval $[-1, 1)$ in general,

[19] In fact, Stirling's calculated value, correctly rounded to nine decimal places, should have 6 in the last place.

since the terms must tend to zero as $z \to \infty$. Then, provided the series does not diverge everywhere, an expression of the form

$$f(z) = x^{z+n} \sum_{m=1}^{\infty} \frac{A_m}{z(z+1)\dots(z+m-1)}$$

defines a function on $(a, \infty)$ for some real number $a$ and $f(z) \to 0$ as $z \to \infty$. Using the previously established identity (see pp. 27–28)

$$\frac{1}{(z+1)(z+2)\dots(z+k)} = \frac{1}{z(z+1)\dots(z+k-1)} - \frac{k}{z(z+1)\dots(z+k)}$$

for $k = 1, 2, \dots$, Stirling effectively shows that

$$f(z) - f(z+1) = x^{z+n} \left( \frac{A_1(1-x)}{z} + \sum_{m=2}^{\infty} \frac{A_m(1-x) + (m-1)xA_{m-1}}{z(z+1)\dots(z+m-1)} \right).$$

Thus[20] the sum of a suitable series

$$\sum_{k=0}^{\infty} \left( x^{z_0+k+n} \sum_{m=1}^{\infty} \frac{a_m}{(z_0+k)(z_0+k+1)\dots(z_0+k+m-1)} \right)$$

is given by $f(z_0)$ if we define the coefficients $A_m$ by

$$A_1(1-x) = a_1 \quad \text{and} \quad A_m(1-x) + (m-1)xA_{m-1} = a_m \quad (m = 2, 3, \dots);$$

consequently we have

$$A_1 = \frac{a_1}{1-x} \quad \text{and} \quad A_m = \frac{a_m - (m-1)xA_{m-1}}{1-x} \qquad (m = 2, 3, \dots).$$

In *Example 1* Stirling considers the series

$$\sum_{k=0}^{\infty} \frac{t^{z_0+k-\frac{1}{2}}}{2(z_0+k)} \qquad (*)$$

for which

$$x = t, \quad n = -\tfrac{1}{2}, \quad a_1 = \tfrac{1}{2}, \quad a_m = 0 \qquad (m = 2, 3, \dots).$$

Then

$$A_1 = \frac{1}{2(1-t)} \quad \text{and} \quad A_m = -\frac{(m-1)tA_{m-1}}{1-t} \qquad (m = 2, 3, \dots),$$

from which it follows that

$$A_m = \frac{(-1)^{m-1}(m-1)!\, t^{m-1}}{2(1-t)^m} \qquad (m = 1, 2, \dots)$$

[20]Cf. the note on Proposition 2.

and the sum is

$$t^{z_0-\frac{1}{2}} \sum_{m=1}^{\infty} \frac{(-1)^{m-1}(m-1)!\,t^{m-1}}{2(1-t)^m z_0(z_0+1)\dots(z_0+m-1)}.$$

With $t=-1$ this becomes

$$(-1)^{z_0-\frac{1}{2}} \sum_{m=1}^{\infty} \frac{(m-1)!}{2^{m+1} z_0(z_0+1)\dots(z_0+m-1)}. \qquad (**)$$

Stirling proposes to determine numerically the sum of Leibniz's series[21]

$$\sum_{r=0}^{\infty} \frac{(-1)^r}{2r+1} = \tan^{-1} 1 = \frac{\pi}{4}.$$

For this he adds directly the first twelve terms and transforms the residual sum using (*) and (**) with $z_0 = 12.5$:

$$\begin{aligned} \sum_{r=12}^{\infty} \frac{(-1)^r}{2r+1} &= \sum_{k=0}^{\infty} \frac{(-1)^{12.5+k-\frac{1}{2}}}{2(12.5+k)} \\ &= \sum_{m=1}^{\infty} \frac{(m-1)!}{2^{m+1} 12.5(12.5+1)\dots(12.5+m-1)}; \end{aligned}$$

his approximation is obtained by adding the sum of the first ten terms of the last series to the initial sum of twelve terms. All the values given by Stirling are rounded and are correct to the number of places shown.

Again we can easily find bounds for the error $E$ introduced by neglecting subsequent terms in the transformed series. We have

[21] Stirling remarks that this had been greatly desired by Leibniz. His authority for this assertion may have been remarks in a letter from Leibniz to Oldenburg dated 26 October 1674 at Paris [36, I, Letter XXI, pp. 51–56]. Leibniz does not give a formal statement of his series there but he describes it in general terms and expresses his belief that he is the first person to have come up with a means by which the quadrature of the circle might be obtained exactly; he remarks: "It therefore remains only that the Doctrine of the sums of Series or numerical Progressions be perfected." The relevant extract from Leibniz's letter was published in the *Commercium Epistolicum* [12, (2e) Item XXXIII, pp. 115–116], which was certainly available to Stirling (see his Preface). Leibniz discussed the series in a paper of 1682 [35] (reproduced in [36, v5, 118–122]), but he had communicated it earlier to some friends: for example, Huygens refers to Leibniz's discovery in a letter to Leibniz dated 6 November 1674(?) [36, II, Letter III, pp. 16–17] and Oldenburg received a statement of the series from Leibniz in a letter of 27 August 1676 [36, I, Letter XXXVII, pp. 114–122; 12, (2e) Items LI–LIII, pp. 129–141].

$$E = \sum_{m=11}^{\infty} \frac{(m-1)!}{2^{m+1}12.5(12.5+1)\dots(12.5+m-1)}$$

$$= \frac{10!}{2^{12}12.5 \times 13.5 \times \dots \times 22.5}\left(1 + \frac{11}{2 \times 23.5} + \frac{11 \times 12}{2^2 \times 23.5 \times 24.5} + \dots\right)$$

$$< \frac{10!}{2^{12}12.5 \times 13.5 \times \dots \times 22.5}\left(1 + \frac{1}{2} + \frac{1}{2^2} + \dots\right)$$

$$= \frac{10!}{2^{12}12.5 \times 13.5 \times \dots \times 22.5} \times 2 < 4.523 \times 10^{-11}$$

and, since $\frac{x+a}{x+b}$ increases to 1 as $x \to \infty$ provided $a < b$,

$$E > \frac{10!}{2^{12}12.5 \times 13.5 \times \dots \times 22.5}\left(1 + \frac{11}{2 \times 23.5} + \left(\frac{11}{2 \times 23.5}\right)^2 + \dots\right)$$

$$= \frac{10!}{2^{12}12.5 \times 13.5 \times \dots \times 22.5}\left(1 - \frac{11}{2 \times 23.5}\right)^{-1} > 2.95 \times 10^{-11}.$$

The actual error in Stirling's calculated value is about $2.965 \times 10^{-11}$. In Example 3 of Proposition 12 Stirling applies a different transformation of Leibniz's series, there working to 17DP.

Finally in *Example 1* Stirling suggests the application of Proposition 3 to the series[22]

$$1 + \frac{1}{3} - \frac{1}{5} - \frac{1}{7} + \frac{1}{9} + \frac{1}{11} - \dots = \frac{\pi}{2\sqrt{2}},$$

$$1 + \frac{1}{2} - \frac{1}{4} - \frac{1}{5} + \frac{1}{7} + \frac{1}{8} - \dots = \frac{2\pi}{3\sqrt{3}}.$$

Each series has to be split into two parts, viz.

$$\left(1 - \frac{1}{5} + \frac{1}{9} - \frac{1}{13} + \dots\right) + \left(\frac{1}{3} - \frac{1}{7} + \frac{1}{11} - \frac{1}{15} + \dots\right)$$

and

$$\left(1 - \frac{1}{4} + \frac{1}{7} - \frac{1}{10} + \dots\right) + \left(\frac{1}{2} - \frac{1}{5} + \frac{1}{8} - \frac{1}{11} + \dots\right)$$

[22]These may be derived by evaluating the expansion

$$\ln(1-z) = -\sum_{n=1}^{\infty} \frac{z^n}{n} \quad (|z| \le 1,\ z \ne 1)$$

at $z = e^{i\pi/4}$, $-e^{i\pi/4}$, $e^{i\pi/3}$ and considering the imaginary parts.

respectively, and each part is to be transformed in a similar way to that applied to the first series of *Example 1*: in the respective cases we consider with $t = -1$

$$\sum_{k=0}^{\infty} \frac{t^{z_0+k-\frac{1}{4}}}{4(z_0+k)} \quad (z_0 = \tfrac{1}{4}, \tfrac{5}{4}, \tfrac{9}{4}, \ldots), \quad \sum_{k=0}^{\infty} \frac{t^{z_0+k-\frac{3}{4}}}{4(z_0+k)} \quad (z_0 = \tfrac{3}{4}, \tfrac{7}{4}, \tfrac{11}{4}, \ldots),$$

$$\sum_{k=0}^{\infty} \frac{t^{z_0+k-\frac{1}{3}}}{3(z_0+k)} \quad (z_0 = \tfrac{1}{3}, \tfrac{4}{3}, \tfrac{7}{3}, \ldots), \quad \sum_{k=0}^{\infty} \frac{t^{z_0+k-\frac{2}{3}}}{3(z_0+k)} \quad (z_0 = \tfrac{2}{3}, \tfrac{5}{3}, \tfrac{8}{3}, \ldots).$$

In *Example 2* Stirling applies the proposition to the series $\sum_{k=1}^{\infty} \frac{x^k}{2k(2k-1)}$, whose sum $S(x)$ may be shown to be

$$S(x) = \begin{cases} \frac{1}{2}\ln(1-x) - \sqrt{-x}\tan^{-1}\sqrt{-x} & (-1 \le x \le 0), \\ \frac{1}{2}\left((1+\sqrt{x})\ln(1+\sqrt{x}) + (1-\sqrt{x})\ln(1-\sqrt{x})\right) & (0 \le x < 1), \\ \ln 2 & (x = 1). \end{cases}$$

As in the case of Brouncker's series (Example 5 of Proposition 2 – see its note) Stirling employs his development of $(z^2 + nz)^{-1}$ as an inverse factorial series (Introduction, p. 30) to express the term in the required form:

$$\begin{aligned} \frac{x^z}{2z(2z-1)} &= \frac{x^z}{4z(z-\frac{1}{2})} \\ &= \frac{x^z}{4}\left(\frac{1}{z(z+1)} + \frac{\frac{3}{2}}{z(z+1)(z+2)} + \frac{\frac{3}{2}\times\frac{5}{2}}{z(z+1)(z+2)(z+3)} + \ldots\right) \\ &= \frac{x^z}{4z(z+1)} F(1, \tfrac{3}{2}\,; z+2\,; 1). \end{aligned}$$

However, he makes an error in his identification of this with the form in the proposition by ignoring the fact that there is no term in $z^{-1}$. Certainly $n = 0$, but the coefficients should be

$$a = 0, \quad b = \tfrac{1}{4}, \quad c = \tfrac{3}{8}, \quad d = \tfrac{15}{16}, \quad e = \tfrac{105}{32}, \quad \ldots,$$

and then the corresponding quantities $A$, $B$, $C$, $D$, $E$ are

$$A = 0, \quad B = \frac{1}{4(1-x)}, \quad C = \frac{3 - 16Bx}{8(1-x)},$$

$$D = \frac{15 - 48Cx}{16(1-x)}, \quad E = \frac{105 - 128Dx}{32(1-x)}.$$

The expression for the sum should be

$$x^z\left(\frac{1}{4(1-x)z(z+1)}+\frac{3-16Bx}{8(1-x)z(z+1)(z+2)}\right.$$
$$\left.+\frac{15-48Cx}{16(1-x)z(z+1)(z+2)(z+3)}+\ldots\right).$$

**Proposition 4 (pp. 51–52).** Stirling appears to be concerned here with the situation where there is a linear recurrence relation

$$\sum_{k=0}^{p}\phi_k(m)s_{m+k}=0 \tag{1}$$

for the successive sums $s_m=\sum_{n=m}^{\infty}a_n$ of a series $\sum a_n$. Since

$$s_{m+k}=s_m-\sum_{n=m}^{m+k-1}a_n \qquad (k=1,2,\ldots),$$

we can rewrite (1) in the form

$$\psi(m)s_m+\sum_{k=0}^{p-1}\psi_k(m)a_{m+k}=0. \tag{2}$$

Then with $m$ replaced by $m+1$ we have

$$\psi(m+1)(s_m-a_m)+\sum_{k=0}^{p-1}\psi_k(m+1)a_{m+k+1}=0. \tag{3}$$

Eliminating $s_m$ from (2) and (3), we obtain a recurrence relation for the terms, which involves $a_m\,, a_{m+1}\,,\ldots\,, a_{m+p}$. The illustrations in *Example 1* and *Example 2* are quite clear, although some of the details have been suppressed in *Example 2*.

**Proposition 5 and Its Scholion (pp. 52–54).** This proposition continues the theme of Proposition 4 and uses the same two examples. In general, the series whose terms are defined by Stirling's relation $(z-n)T = zT'$ $(z=m,m+1,\ldots)$ of *Example 1* is $AF(m-n,1\,;m\,;1)$, where $A$ is a constant, and by Gauss's formula its sum is

$$A\,\frac{\Gamma(m)\Gamma(n-1)}{\Gamma(n)\Gamma(m-1)}=A\,\frac{m-1}{n-1}$$

provided $n-1>0$ (for real $n$) and $m\neq 0,-1,-2,\ldots$.

In *Example 2* the relation $(z+2)T+3zT'=0$ $(z=1,2,\ldots)$ with first term 1 produces the series $F(1,3\,;1\,;-\frac{1}{3})$, whose general term is

$$\frac{(-1)^k \frac{1}{2}(k+1)(k+2)}{3^k} \qquad (k=0,1,2,\ldots).$$

As Stirling notes, the numerators contain the triangular numbers, viz.

$$\sum_{r=1}^{k+1} r = \tfrac{1}{2}(k+1)(k+2) \qquad (k=0,1,2,\ldots).$$

What Stirling says in the *scholion* about continuing the series of the first example backwards does not seem to be completely correct. Presumably we now determine $T$ from $T'$ with $z = m-1, m-2, \ldots$ and $A$ as initial value of $T'$. This is certainly consistent with his treatment of the geometric series in the *scholion* and produces

$$A\left(\frac{m-1}{m-n-1} + \frac{(m-1)(m-2)}{(m-n-1)(m-n-2)} + \frac{(m-1)(m-2)(m-3)}{(m-n-1)(m-n-2)(m-n-3)} + \cdots\right)$$

$$= A\,\frac{m-1}{m-n-1}\left(1 + \frac{2-m}{2-m+n} + \frac{(2-m)(3-m)}{(2-m+n)(3-m+n)} + \cdots\right)$$

$$= A\,\frac{m-1}{m-n-1}\,F(2-m,1;2-m+n;1)$$

$$= A\,\frac{(m-1)\,\Gamma(2-m+n)\,\Gamma(n-1)}{(m-n-1)\,\Gamma(n)\,\Gamma(1-m+n)} = -A\,\frac{m-1}{n-1}$$

provided $n-1>0$ (for real $n$) and $2-m+n \neq 0,-1,-2,\ldots$. Thus, unless $m$ or $2-m+n$ is a negative integer or zero and provided $n>1$, both the forward and backward series converge and the sum of one is equal to the sum of the other times $-1$.

Stirling's remarks on quadratures correspond to

$$\int_0^z t^{-n}\,dt = \frac{z^{1-n}}{1-n} \quad (n<1), \qquad \int_z^\infty t^{-n}\,dt = -\frac{z^{1-n}}{1-n} \quad (n>1).$$

Neither integral converges in the case $n=1$ corresponding to the *hyperbola of Apollonius*, the name applied to the simplest case of Fermat's general hyperbola: $x^m y^n = a$. His final remark concerning the correction of areas and sums presumably refers to the correct choice of range of integration and summation.

**Proposition 6 (pp. 54–57).** Here $S$ and $S'$ may be either *successive* sums or *partial* sums. What Stirling intends may be clearer if we ignore first of all the connection with series and consider an arbitrary *sequence* $\{u_n\}$ of positive terms which is defined recursively by a relation of the form

$$u_n p(n) = m u_{n+1} q(n) \qquad (n = n_0, n_0 + 1, \ldots),$$

where $m$ is a positive constant and $p$, $q$ are monic polynomials of the same degree:

$$p(n) = n^\theta + an^{\theta-1} + \ldots\ ; \quad q(n) = n^\theta + cn^{\theta-1} + \ldots .$$

Then

$$\begin{aligned}\frac{u_{n+1}}{u_n} = \frac{p(n)}{m\,q(n)} &= \frac{1}{m}\left(1 + \frac{(a-c)n^{\theta-1} + \ldots}{n^\theta + cn^{\theta-1} + \ldots}\right) \\ &= \frac{1}{m}\left(1 + \frac{a-c}{n} + O\left(\frac{1}{n^2}\right)\right).\end{aligned}$$

If $m > 1$ we can find $N \geq n_0$ such that

$$1 + \frac{(a-c)n^{\theta-1} + \ldots}{n^\theta + cn^{\theta-1} + \ldots} < \sqrt{m} \quad \text{if} \quad n \geq N.$$

Consequently

$$\frac{u_{n+1}}{u_n} < \frac{1}{\sqrt{m}} < 1 \quad \text{if} \quad n \geq N,$$

from which it follows that $\{u_n\}$ is eventually strictly decreasing to 0. We see in a similar way that if $m < 1$ then $\{u_n\}$ is eventually strictly increasing and unbounded.

The situation in which $m = 1$ is more complicated. Clearly, for all sufficiently large $n$,

$$1 + \frac{(a-c)n^{\theta-1} + \ldots}{n^\theta + cn^{\theta-1} + \ldots} \begin{cases} < 1 & \text{if } a < c, \\ > 1 & \text{if } a > c, \end{cases}$$

so that the sequence $\{u_n\}$ is eventually strictly decreasing if $a < c$ and eventually strictly increasing if $a > c$. We can see in the former case that the limit must be 0 by considering what would be its logarithm if the limit were positive, namely

$$\ln u_{n_0} + \sum_{n=n_0}^{\infty} \ln\left(1 + \frac{(a-c)n^{\theta-1} + \ldots}{n^\theta + cn^{\theta-1} + \ldots}\right);$$

if $a < c$ we can find $N \geq n_0$ such that

$$0 < -\frac{(a-c)n^{\theta-1} + \ldots}{n^\theta + cn^{\theta-1} + \ldots} < 1 \quad \text{if} \quad n \geq N,$$

and consequently[23]

[23]Note that $\ln(1-x) < -x$ on $(0,1)$.

$$\sum_{n=N}^{N+m} \ln\left(1+\frac{(a-c)n^{\theta-1}+\dots}{n^\theta+cn^{\theta-1}+\dots}\right) < \sum_{n=N}^{N+m} \frac{(a-c)n^{\theta-1}+\dots}{n^\theta+cn^{\theta-1}+\dots} \to -\infty$$

as $m \to \infty$. We therefore have that, when $m = 1$, the sequence $\{u_n\}$ converges to 0 if $a < c$ but is unbounded if $a > c$.

If the $u_n$ represent the *successive* sums of a series then convergence of the series is equivalent to the limit of the successive sums being 0. Thus by the above discussion the series will converge provided $m > 1$ or $m = 1$ and $a < c$ and will diverge to $\infty$ if $m < 1$ or $m = 1$ and $a > c$. This is the content of the *corollary* to *Proposition 6.*

However, if the $u_n$ represent the *partial* sums of a series, the sum can be finite and nonzero only if $m = 1$ and $a = c$. This is the situation illustrated in the *example*, where Stirling proposes a series whose partial sums $S_k$ satisfy

$$S_1 = 1, \qquad S_k(k^2+k+\tfrac{1}{4}) = S_{k+1}(k^2+k) \quad (k = 1, 2, \dots).$$

We have equivalently

$$(2k+1)^2 S_k = 2k(2k+2)S_{k+1}$$

and

$$S_k = \frac{1 \times 3^2 \times \dots \times (2k-1)^2}{2 \times 4^2 \times \dots \times (2k-2)^2 \times 2k} \qquad (k = 2, 3, \dots).$$

It follows from Wallis's product that the sum of the series must be $4/\pi$. Its terms may be determined by replacing $S_{k+1}$ by $S_k + a_{k+1}$:

$$a_1 = 1, \qquad 4k(k+1)a_{k+1} = S_k \qquad (k = 1, 2, \dots).$$

Stirling considers finally the two relations

$$Sz^2 = S'(z^2-1) \quad \text{and} \quad Sz = S'(z+1).$$

In the first of these $m = 1$ and $a = c = 0$, so that we are dealing with partial sums $S_k = \sum_{n=2}^{k} a_n$ which satisfy the relation

$$k^2 S_k = (k^2-1)S_{k+1} \qquad (k = 2, 3, \dots).$$

Replacing $S_{k+1}$ by $S_k + a_{k+1}$ we obtain

$$S_k = (k^2-1)a_{k+1}, \tag{1}$$

then

$$S_k + a_{k+1} = S_{k+1} = ((k+1)^2-1)a_{k+2},$$

and on subtraction

$$k^2 a_{k+1} = ((k+1)^2-1)a_{k+2},$$

or

$$ka_{k+1} = (k+2)a_{k+2} \qquad (k = 2, 3, \dots). \tag{2}$$

Stirling takes $\frac{1}{2}$ for the first term $a_2$. Then from (1) and (2) the terms come out as

$$a_3 = \frac{1}{2\times 3}, \quad a_4 = \frac{1}{3\times 4}, \quad a_5 = \frac{1}{4\times 5}, \quad \ldots .$$

In the second relationship $Sz = S'(z+1)$ we have $m = 1$, $a = 0$, $c = 1$, so that we are dealing with successive sums $S_k = \sum_{n=k}^{\infty} b_n$ whose limit is 0. From

$$kS_k = (k+1)S_{k+1} \quad (k = 1, 2, \ldots)$$

we obtain on putting $S_{k+1} = S_k - b_k$

$$S_k = (k+1)b_k; \tag{3}$$

thus

$$S_k - b_k = S_{k+1} = (k+2)b_{k+1},$$

and we obtain on subtraction

$$kb_k = (k+2)b_{k+1} \qquad (k = 1, 2, \ldots). \tag{4}$$

Taking $b_1 = \frac{1}{2}$ in (4) produces the same terms as before ($b_k = a_{k+1}$). According to (1)

$$\sum_{n=1}^{k-1} \frac{1}{n(n+1)} = \frac{k^2-1}{k(k+1)} = \frac{k-1}{k} \qquad (k = 2, 3, \ldots),$$

while (3) gives

$$\sum_{n=k}^{\infty} \frac{1}{n(n+1)} = \frac{k+1}{k(k+1)} = \frac{1}{k} \qquad (k = 1, 2, \ldots).$$

**Proposition 7 and Its Scholion (pp. 57–60).** The relation $(z-n)T + (m-1)zT' = 0$, where $n$ and $m$ are constants with $m \neq 1$, generates the series

$$T_0\left(1 + \frac{z_0 - n}{z_0(1-m)} + \frac{(z_0-n)(z_0-n+1)}{z_0(z_0+1)(1-m)^2} + \ldots\right), \tag{1}$$

where $T_0$ is the first term, which corresponds to the initial value $z_0$ ($\neq 0, -1, -2, \ldots$) of $z$. This series is

$$T_0 F\left(1, z_0 - n\,; z_0\,; \frac{1}{1-m}\right).$$

According to the proposition its sum is

$$\frac{m-1}{m} T_0 \left(1 + \frac{n}{z_0 m} + \frac{n(n+1)}{z_0(z_0+1)m^2} + \ldots\right) = \frac{m-1}{m} T_0 F\left(1, n\,; z_0\,; \frac{1}{m}\right). \tag{2}$$

This is a special case of the general linear transformation formula

$$F(a,b;c;z) = (1-z)^{-a} F\left(a, c-b; c; \frac{z}{z-1}\right).$$

In general both series will converge provided $|1-m|^{-1} < 1$ and $|m|^{-1} < 1$; for real $m$ these inequalities mean that $m < -1$ or $m > 2$. Stirling's remark in the *corollary* about the series being infinitely large when $m$ is negative does not apply when $m < -1$. The case $m = -1$ does in fact occur in Stirling's examples, but there the other parameters $n$ and $z_0$ are such as to ensure convergence of both series.[24] Stirling's proof consists of setting up a recurrence relation for the sum divided by the first term, which he solves by fitting an inverse factorial series. The argument is valid subject to the parameters being such as to produce suitably convergent series.

The series of *Example 1* is

$$\tfrac{1}{2} F\left(1,1;\tfrac{3}{2};\tfrac{1}{2}\right) \qquad \left(T_0 = \tfrac{1}{2}, z_0 = \tfrac{3}{2}, m = -1, n = \tfrac{1}{2}\right).$$

Now in general

$$(1-z^2)^{1/2} F\left(1,1;\tfrac{3}{2};z^2\right) = z^{-1}\sin^{-1} z,$$

so that the sum of Stirling's series is $\sin^{-1}(1/\sqrt{2}) = \pi/4$. To sum the series Stirling adds the first twelve terms directly and then employs the proposition to approximate to the residual sum, which is

$$\frac{12!}{2\left(\frac{3}{2}\times\frac{5}{2}\ldots\times\frac{25}{2}\right)2^{12}}\left(1+\frac{13}{\frac{27}{2}\times 2}+\frac{13\times 14}{\frac{27}{2}\times\frac{29}{2}\times 2^2}+\ldots\right)$$

$$= \frac{12!}{2\times 3\times 5\times\ldots\times 25} F\left(1,13;\tfrac{27}{2};\tfrac{1}{2}\right) \qquad \left(z_0 = \tfrac{27}{2}, m = -1, n = \tfrac{1}{2}\right)$$

$$= \frac{12!}{3\times 5\times\ldots\times 25} F\left(1,\tfrac{1}{2};\tfrac{27}{2};-1\right).$$

The first 12 terms of the last series, which is absolutely convergent, are now used to provide an approximation to the residual sum. Since the terms of the last series alternate in sign and their moduli decrease to 0, the error involved in neglecting the remaining terms lies strictly between the thirteenth term

$$t_{13} = \frac{12!}{3\times 5\times\ldots\times 25}\times\frac{\frac{1}{2}\times\frac{3}{2}\times\ldots\times\frac{23}{2}}{\frac{27}{2}\times\frac{29}{2}\times\ldots\times\frac{49}{2}} = \frac{12!}{25\times 27\times\ldots\times 49}$$
$$< 2.6\times 10^{-12},$$

and the sum of the thirteenth and fourteenth terms

[24]A non-terminating series $F(a,b;c;-1)$ converges if $\mathrm{Re}\,(c-a-b) > -1$, the convergence being absolute when $\mathrm{Re}\,(c-a-b) > 0$.

$$t_{13}\left(1-\tfrac{25}{51}\right) > 1.3\times 10^{-12}.$$

Working with double precision arithmetic produces 0.7853 9816 3395 7013 for the calculated value, which is less than $\pi/4$ by about $1.75\times 10^{-12}$ and is consistent with these error bounds.

In *Example 2* Stirling considers the conditionally convergent series

$$1+\frac{1}{2}(-1)+\frac{\frac{1}{2}\times\frac{3}{2}}{2!}(-1)^2+\frac{\frac{1}{2}\times\frac{3}{2}\times\frac{5}{2}}{3!}(-1)^3+\dots$$

$$= F\left(1,\tfrac{1}{2};1;-1\right) = (1-(-1))^{-1/2} = \frac{1}{\sqrt{2}}.$$

Here he adds directly the first 10 terms and transforms the residual sum:

$$\frac{\frac{1}{2}\times\frac{3}{2}\times\dots\times\frac{19}{2}}{10!}\left(1-\frac{\frac{21}{2}}{11}+\frac{\frac{21}{2}\times\frac{23}{2}}{11\times 12}-\dots\right)$$

$$=\frac{1\times 3\times\dots\times 19}{10!\times 2^{10}}F\left(1,\tfrac{21}{2};11;-1\right) \qquad \left(m=2,\, n=\tfrac{1}{2},\, z_0=11\right)$$

$$=\frac{1\times 3\times\dots\times 19}{10!\times 2^{11}}F\left(1,\tfrac{1}{2};11;\tfrac{1}{2}\right).$$

Ten terms of the last series are used to complete the approximation. The error is therefore

$$\frac{1\times 3\times\dots\times 19}{10!\times 2^{11}}\times\frac{1\times 3\times\dots\times 19}{11\times 12\times\dots\times 20\times 2^{20}}$$
$$\times\left(1+\frac{\frac{21}{2}}{21\times 2}+\frac{\frac{21}{2}\times\frac{23}{2}}{21\times 22\times 2^2}+\dots\right).$$

This is less than

$$\frac{(1\times 3\times\dots\times 19)^2}{20!\times 2^{31}}\left(1+\frac{1}{2}+\frac{1}{2^2}+\dots\right)=\frac{(1\times 3\times\dots\times 19)^2}{20!\times 2^{30}}$$
$$< 1.641\times 10^{-10},$$

and, since $(\frac{21}{2}+k)/(21+k) > \frac{1}{2}$ $(k>0)$, it is greater than

$$\frac{(1\times 3\times\dots\times 19)^2}{20!\times 2^{31}}\left(1+\frac{1}{2^2}+\frac{1}{2^4}+\dots\right)=\frac{(1\times 3\times\dots\times 19)^2}{20!\times 2^{29}\times 3}$$
$$> 1.093\times 10^{-10}.$$

Stirling finds 0.7071 0678 10 for the value of $1/\sqrt{2}$ and again the final digit is incorrect. This time it is not just due to the cumulative effects of rounding – the error bounds show that the calculation cannot give the tenth decimal place. Double precision arithmetic produces 0.7071 0678 1076 5897 for the calculated value, which is less than $1/\sqrt{2}$ by about $1.0996 \times 10^{-10}$.

As Stirling notes in the *scholion*, a second application of the transformation brings us back to the original series (cf. (1) and (2) above):

$$\frac{m-1}{m} T_0 F\left(1, n\,; z_0\,; \frac{1}{m}\right) = \frac{m-1}{m} T_0 F\left(1, z_0 - (z_0 - n)\,; z_0\,; \frac{1}{1-(1-m)}\right)$$

$$\longrightarrow \frac{(1-m)-1}{1-m} \times \frac{m-1}{m} T_0 F\left(1, z_0 - n\,; z_0\,; \frac{1}{1-m}\right)$$

$$= T_0 F\left(1, z_0 - n\,; z_0\,; \frac{1}{1-m}\right).$$

The coefficients in any non-terminating $F(a,b\,;c\,;z)$ are eventually positive and $1/m$ and $1/(1-m)$ have opposite signs except where $0 \le m \le 1$; this justifies Stirling's remark about a series whose terms have constant sign changing into a series whose terms have alternating signs and conversely. As noted above, in the alternating case the series may be only conditionally convergent, which generally produces slow convergence. The example in the *scholion* is Leibniz's series

$$\frac{\pi}{4} = \tan^{-1} 1 = 1 - \frac{1}{3} + \frac{1}{5} - \frac{1}{7} + \frac{1}{9} - \ldots = F\left(1, \tfrac{1}{2}\,; \tfrac{3}{2}\,; -1\right),$$

which transforms to the series of *Example 1*

$$\frac{1}{2} F\left(1, 1\,; \tfrac{3}{2}\,; \tfrac{1}{2}\right) = \sin^{-1} \frac{1}{\sqrt{2}} = \frac{\pi}{4}.$$

**Proposition 8 (pp. 60–64).** The given series is

$$T_0 \left(1 + \frac{(z_0 - m)(z_0 - n)}{z_0(z_0 - n + 1)} + \frac{(z_0 - m)(z_0 - m + 1)(z_0 - n)(z_0 - n + 1)}{z_0(z_0 + 1)(z_0 - n + 1)(z_0 - n + 2)} + \ldots\right), \qquad (1)$$

where $T_0$, $z_0$, $m$, $n$ are constants such that neither $z_0$ nor $z_0 - n + 1$ is zero or a negative integer. It may be written as (see p. 6)

$$T_0\, {}_3F_2 \left[\begin{matrix} z_0 - m, & z_0 - n, & 1 \\ z_0, & z_0 - n + 1 & \end{matrix}\right].$$

According to the proposition its sum is given by

$$(z_0 - n)T_0\left(\frac{1}{m} + \frac{n}{z_0(m+1)} + \frac{n(n+1)}{z_0(z_0+1)(m+2)} + \dots\right)$$

$$= \frac{(z_0-n)}{m} T_0\left(1 + \frac{nm}{z_0(m+1)} + \frac{n(n+1)m(m+1)}{z_0(z_0+1)(m+1)(m+2)} + \dots\right)$$

$$= \frac{(z_0-n)}{m} T_0\, {}_3F_2\left[\begin{matrix} n, & m, & 1 \\ z_0, & m+1 & \end{matrix}\right], \tag{2}$$

where of course neither $z_0$ nor $m$ may be zero or a negative integer.

For (1) the ratio of the $(k+1)$-th term to the $k$-th term $(k = 0, 1, 2, \dots)$ is

$$\frac{(z_0 - m + k)(z_0 - n + k)}{(z_0 + k)(z_0 - n + 1 + k)} = 1 - \frac{m+1}{k} + O\left(\frac{1}{k^2}\right).$$

For (2) this ratio is

$$\frac{(n+k)(m+k)}{(z_0+k)(m+1+k)} = 1 - \frac{z_0 - n + 1}{k} + O\left(\frac{1}{k^2}\right).$$

Thus by Gauss's test (1) converges if $m > 0$ and diverges if $m \leq 0$, while (2) converges if $z_0 - n > 0$ and diverges if $z_0 - n \leq 0$. In particular, as Stirling observes in the *corollary*, (1) diverges when $m$ is zero or a negative integer.[25]

Defining $y_k$ $(k = 0, 1, 2, \dots)$ by

$$S_k = T_k(z_0 + k - n)y_k, \tag{3}$$

where $T_k$ denotes the $k$-th term of (1) and $S_k = \sum_{n=k}^{\infty} T_n$, we find as in Stirling's proof that the $y_k$ satisfy the recurrence relation

$$y_k - y_{k+1} + \frac{m}{z_0+k} y_{k+1} - \frac{1}{z_0+k-n} = 0. \tag{4}$$

The term $1/(z_0 + k - n)$ (Stirling's $1/(z-n)$) is expanded as

$$\frac{1}{z_0+k-n} = \frac{1}{z_0+k} + \frac{n}{(z_0+k)(z_0+k+1)} + \frac{n(n+1)}{(z_0+k)(z_0+k+1)(z_0+k+2)} + \dots.$$

Stirling then solves the recurrence relation by assuming an inverse factorial series representation for $y_k$:

$$y_k = \frac{\alpha_1}{m} + \frac{\alpha_2}{(m+1)(z_0+k)} + \frac{\alpha_3}{(m+2)(z_0+k)(z_0+k+1)} + \dots. \tag{5}$$

---

[25] Of course he probably based this observation on the fact that (2) is not defined for such $m$.

Substitution of this in (4) determines the coefficients. The series (2) then results if we put $k = 0$ and substitute for $y_0$ in (3).

Curiously, Stirling states the expansion of $1/(z-n)$ without comment. Presumably its derivation is taken as an obvious application of the principles laid down in the Introduction and is similar to his development of $1/(z^2+nz)$ (Introduction, p. 30): for example,

$$\begin{aligned}\frac{1}{z-n} &= \frac{1}{z}\left(1-\frac{n}{z}\right)^{-1} = \frac{1}{z}+\frac{n}{z^2}+\frac{n^2}{z^3}+\frac{n^3}{z^4}+\dots\\ &= \frac{1}{z}+\frac{a_1}{z(z+1)}+\frac{a_2}{z(z+1)(z+2)}+\frac{a_3}{z(z+1)(z+2)(z+3)}+\dots,\end{aligned}$$

where, according to the rule given at the top of p. 30,

$$\begin{aligned}a_1 &= n,\\ a_2 &= n+n^2 = n(n+1),\\ a_3 &= 2n+3n^2+n^3 = n(n+1)(n+2),\end{aligned}$$

and so on. The resulting series, which is valid for $\mathrm{Re}\,(z) > \mathrm{Re}\,(n)$, is referred to as *Waring's formula* in [40, 10.2]. It is noted in [72] that the series is a special case of a result given by Nicole in 1724 [50] and that it is also contained in an article of 1717 by De Montmort [45]. Alternatively the series may be deduced from Stirling's result

$$\frac{1}{z^2+nz} = \frac{1}{z(z+1)}+\frac{1-n}{z(z+1)(z+2)}+\frac{(1-n)(2-n)}{z(z+1)(z+2)(z+3)}+\dots$$

by multiplying by $z$ and replacing $z$ by $z-1$ and $n$ by $-n+1$.

The recurrence relation (difference equation) (4) is discussed in detail in [52], where its general solution is obtained. Stirling notes in the *corollary* that in the case where $n$ is a negative integer or zero the series is exactly summable (because then (2) has only finitely many nonzero terms). It is pointed out in [52, §10] that we can also sum exactly in terms of the hypergeometric function when $z_0$ is a positive integer. For example, referring to (2), if $z_0 = 1$ this is just

$$\frac{1-n}{m}T_0\,F(n,m;m+1;1),$$

and for $z_0 = 2, 3, \dots$ we have, provided no zero terms are introduced in the denominator,

$$\begin{aligned}&\frac{nm}{z_0(m+1)}+\frac{n(n+1)m(m+1)}{z_0(z_0+1)(m+1)(m+2)}+\dots\\ &= \frac{(z_0-1)!\,m}{(n-z_0+1)(n-z_0+2)\dots(n-1)(m-z_0+1)}\\ &\quad\times\left(F(n-z_0+1,m-z_0+1;m-z_0+2;1)-P_{z_0}\right),\end{aligned}$$

where $P_{z_0}$ denotes the sum of the first $z_0$ terms of the preceding hypergeometric series. Using Gauss's formula, we may express this hypergeometric series in terms of the Gamma function:

$$F(n-z_0+1, m-z_0+1\,; m-z_0+2\,; 1) = \frac{\Gamma(m-z_0+2)\Gamma(z_0-n)}{\Gamma(m-n+1)},$$

provided $\operatorname{Re}(z_0-n) > 0$.

Stirling also refers in the *corollary* to the application of Propositions 7 and 8 to the quadrature of binomial curves. This relates to the following series which he has already stated on p. 33:[26]

$$\int_0^x t^{\theta-1}(e+ft^\eta)^{\lambda-1}dt = \frac{x^\theta}{\theta e}(e+fx^\eta)^\lambda - \frac{s(e+fx^\eta)^\lambda f x^{\theta+\eta}}{r\theta e^2} + \frac{s(s+1)(e+fx^\eta)^\lambda f^2 x^{\theta+2\eta}}{r(r+1)\theta e^3} - \dots,$$

where $r = \dfrac{\theta+\eta}{\eta}$, $s = \dfrac{\theta+\lambda\eta}{\eta}$. The terms of the series satisfy the relation

$$\begin{aligned} \frac{T'}{T} &= \frac{s+k}{r+k}\left(-\frac{fx^\eta}{e}\right) \qquad (k=0,1,2,\dots) \\ &= \frac{z-(r-s)}{z}\left(-\frac{fx^\eta}{e}\right) \qquad (z=r, r+1, r+2, \dots), \end{aligned}$$

that is,

$$(z-(r-s))T + \frac{e}{fx^\eta}zT' = 0.$$

Thus we may apply Proposition 7 with

$$n = r-s \quad \text{and} \quad m = 1+\frac{e}{fx^\eta}.$$

If, however, $x$ is such that $e+fx^\eta = 0$, then $\dfrac{fx^\eta}{e} = -1$ and we have

$$\begin{aligned} \int_0^x t^{\theta-1}(e+ft^\eta)^{\lambda-1}dt &= e^{\lambda-1}\int_0^x t^{\theta-1}\left(1+\frac{f}{e}t^\eta\right)^{\lambda-1}dt \\ &= e^{\lambda-1}\int_0^x \sum_{k=0}^{\infty}\binom{\lambda-1}{k}\left(\frac{f}{e}\right)^k t^{k\eta+\theta-1}dt \\ &= e^{\lambda-1}\sum_{k=0}^{\infty}\binom{\lambda-1}{k}\left(\frac{f}{e}\right)^k \frac{x^{k\eta+\theta}}{k\eta+\theta} \\ &= x^\theta e^{\lambda-1}\sum_{k=0}^{\infty}\frac{(-1)^k}{k\eta+\theta}\binom{\lambda-1}{k}. \end{aligned}$$

[26] See also the note on Stirling's introductory remarks in Part I (p. 174), and Propositions 24 and 25 as well as their notes.

Now the terms of this series satisfy the relation

$$\frac{T'}{T} = -\frac{(\lambda - 1 - k)(k\eta + \theta)}{(k+1)((k+1)\eta + \theta)} \qquad (k = 0, 1, 2, \ldots)$$

$$= \frac{(z - \lambda)\left(z - \left(1 - \frac{\theta}{\eta}\right)\right)}{z\left(z - \left(1 - \frac{\theta}{\eta}\right) + 1\right)} \qquad (z = 1, 2, \ldots),$$

which corresponds to the relation of Proposition 8 with $m = \lambda$ and $n = 1 - \dfrac{\theta}{\eta}$.

In *Example 1* Stirling considers

$$\frac{\pi}{2} = \sin^{-1} 1 = 1 + \sum_{k=1}^{\infty} \frac{\prod_{r=1}^{k}(2r-1)^2}{(2k+1)!}$$

$$= 1 + \sum_{k=1}^{\infty} \frac{\left(\frac{1}{2} \times \frac{3}{2} \times \ldots \times \left(\frac{1}{2} + k - 1\right)\right)^2}{\frac{3}{2} \times \frac{5}{2} \times \ldots \times \left(\frac{3}{2} + k - 1\right) \times k!}$$

$$= {}_3F_2\left[\begin{matrix} \frac{1}{2}, & \frac{1}{2}, & 1 \\ 1, & \frac{3}{2} & \end{matrix}\right].$$

He adds up the first 12 terms directly and transforms the remaining part of the series:

$$\frac{\left(\frac{1}{2} \times \frac{3}{2} \times \ldots \times \frac{23}{2}\right)^2}{\frac{3}{2} \times \frac{5}{2} \times \ldots \times \frac{25}{2} \times 12!}\left(1 + \frac{\left(\frac{25}{2}\right)^2}{\frac{27}{2} \times 13} + \frac{\left(\frac{25}{2} \times \frac{27}{2}\right)^2}{\frac{27}{2} \times \frac{29}{2} \times 13 \times 14} + \ldots\right)$$

$$= \frac{\left(\frac{1}{2} \times \frac{3}{2} \times \ldots \times \frac{23}{2}\right)^2}{\frac{3}{2} \times \frac{5}{2} \times \ldots \times \frac{25}{2} \times 12!}\,{}_3F_2\left[\begin{matrix} \frac{25}{2}, & \frac{25}{2}, & 1 \\ 13, & \frac{27}{2} & \end{matrix}\right] \qquad \left(z_0 = 13,\ m = n = \tfrac{1}{2}\right)$$

$$= 25 \times \frac{(1 \times 3 \times \ldots \times 23)^2}{25!}\,{}_3F_2\left[\begin{matrix} \frac{1}{2}, & \frac{1}{2}, & 1 \\ 13, & \frac{3}{2} & \end{matrix}\right]$$

$$= \frac{(1 \times 3 \times \ldots \times 23)^2}{24!}\left(1 + \frac{\left(\frac{1}{2}\right)^2}{13 \times \frac{3}{2}} + \frac{\left(\frac{1}{2} \times \frac{3}{2}\right)^2}{13 \times 14 \times \frac{3}{2} \times \frac{5}{2}} + \ldots\right).$$

The first 12 terms of this series are now used to complete the calculation – they are given in the second column of the table on p. 63, all correctly rounded to the number of places shown.[27] The final approximation suffers from the cumulative effect of rounding – repeating Stirling's calculation in double precision arithmetic produces 1.5707 9632 6052 944, which is less than

[27]However, the entry for *I* should have 251 in place of 250; also in the original the entry for *E* is given as 24261 (the last two digits have been transposed).

$\pi/2$ by about $7.42\times 10^{-10}$. I have not been able to find simple, effective error bounds for this calculation.

The rule stated at the end of *Example 1* is an iterative method for calculating the terms of the given series and may be established easily by induction:

$$\frac{1}{2k+3}\left(\frac{(1\times 3\times\ldots\times(2k-1))^2}{(2k)!}-\frac{1}{2k+2}\times\frac{(1\times 3\times\ldots\times(2k-1))^2}{(2k)!}\right)$$

$$=\frac{(1\times 3\times\ldots\times(2k-1))^2(2k+1)}{(2k+3)(2k+2)\times(2k)!}$$

$$=\frac{(1\times 3\times\ldots\times(2k+1))^2}{(2k+3)!}.$$

In *Example 2* Stirling returns to Brouncker's series.[28] We have

$$\frac{1}{1\times 2}+\frac{1}{3\times 4}+\frac{1}{5\times 6}+\ldots=\frac{1}{2}\,{}_3F_2\begin{bmatrix}\frac{1}{2}, & 1, & 1\\ \frac{3}{2}, & 2 & \end{bmatrix}.$$

In the above notation we may take, as Stirling notes,

$$z_0=\tfrac{3}{2},\quad n=z_0+1-2=\tfrac{1}{2},\quad m=z_0-\tfrac{1}{2}=1, \tag{5}$$

or, interchanging 3/2 and 2,

$$z_0=2,\quad n=z_0+1-\tfrac{3}{2}=\tfrac{3}{2},\quad m=1. \tag{6}$$

If we transform the series from the $k$-th term ($k=0,1,2,\ldots$) we obtain for (5)

$$\frac{\frac{3}{2}+k-\frac{1}{2}}{(2k+1)(2k+2)}\left(1+\frac{\frac{1}{2}}{2(\frac{3}{2}+k)}+\frac{\frac{1}{2}\times\frac{3}{2}}{3(\frac{3}{2}+k)(\frac{5}{2}+k)}+\ldots\right)$$

$$=\frac{1}{2(2k+1)}\,{}_3F_2\begin{bmatrix}\frac{1}{2}, & 1, & 1\\ \frac{3}{2}+k, & 2 & \end{bmatrix},$$

and for (6)

$$\frac{2+k-\frac{3}{2}}{(2k+1)(2k+2)}\left(1+\frac{\frac{3}{2}}{2(2+k)}+\frac{\frac{3}{2}\times\frac{5}{2}}{3(2+k)(3+k)}+\ldots\right)$$

$$=\frac{1}{2(2k+2)}\,{}_3F_2\begin{bmatrix}\frac{3}{2}, & 1, & 1\\ 2+k, & 2 & \end{bmatrix}.$$

These correspond to the formulae given in *Example 2*. Stirling does not provide a calculation here, but, ignoring the fact that he has already dealt with

[28]See Example 4 of Proposition 2 and the note on Proposition 2.

Brouncker's series in Example 4 of Proposition 2, he refers to Proposition 3 or Proposition 7 for a means of summing the series when it is expressed in the form

$$1-\frac{1}{2}+\frac{1}{3}-\frac{1}{4}+\frac{1}{5}-\dots.$$

For the application of Proposition 3 we have $z=1,2,3,\dots$ with

$$x=-1,\quad n=-1,\quad a=1,\quad b=c=\dots=0.$$

Transforming from the term corresponding to $z=k$ gives the sum from the $k$-th term as

$$\begin{aligned}(-1)^{k-1}\biggl(\frac{1}{2k}&+\frac{1!}{2^2k(k+1)}+\frac{2!}{2^3k(k+1)(k+2)}\\&+\frac{3!}{2^4k(k+1)(k+2)(k+3)}+\dots\biggr)\\&=\frac{(-1)^{k-1}}{2k}F\left(1,1;k+1;\tfrac{1}{2}\right)\qquad(k=1,2,\dots).\end{aligned}$$

Proposition 7 yields an equivalent expression: the defining relation for the terms is $(z-1)T+zT'=0$ $(z=2,3,\dots)$, so that $n=1$ and $m=2$; the sum from the term corresponding to $z=k$ is then given by

$$\begin{aligned}&\frac{(-1)^k}{2(k-1)}\left(1+\frac{1!}{2k}+\frac{2!}{2^2k(k+1)}+\frac{3!}{2^3k(k+1)(k+2)}+\dots\right)\\&=\frac{(-1)^k}{2(k-1)}F\left(1,1;k;\tfrac{1}{2}\right)\qquad(k=2,3,\dots).\end{aligned}$$

**Proposition 9 and Its Scholion (pp. 65–68).** The aim of the proposition is clearly illustrated by Stirling in *Example 1* and *Example 2*, so we restrict attention to the examples outlined in the *scholion*. In the first of these we have

$$\text{(i)}\quad S=\frac{2z-1}{2}T+S_2\qquad\text{and}\qquad\text{(ii)}\quad T'=\left(\frac{z-1}{z}\right)^2T.$$

Proceeding as in *Example 1* and *Example 2*, we obtain from (i)

$$S-T=\frac{2(z+1)-1}{2}T'+S_2-T_2.$$

Subtraction of this from (i) leads to

$$T=\frac{2z-1}{2}T-\frac{2z+1}{2}T'+T_2,$$

and then by (ii)

$$\left(1-\frac{2z-1}{2}+\frac{2z+1}{2}\left(\frac{z-1}{z}\right)^2\right)T=T_2,$$

which simplifies to the relation stated by Stirling, viz.

$$T_2=\frac{1}{2z^2}T. \qquad (*)$$

Finally, from (*) and (ii)

$$\begin{aligned} T_2' &= \frac{1}{2(z+1)^2}T' = \frac{1}{2}\left(\frac{z-1}{z(z+1)}\right)^2 T \\ &= \left(\frac{z-1}{z+1}\right)^2 T_2. \qquad (**) \end{aligned}$$

When $z = 2, 3, 4, \ldots$ and the initial term is 1, the first series is easily found from (ii) to be

$$\sum_{n=2}^{\infty}\frac{1}{(n-1)^2}.$$

Then from (*) the initial term of the second series is 1/8 and we can prove by induction using (**) that the second series is

$$\frac{1}{2}\sum_{n=2}^{\infty}\frac{1}{n^2(n-1)^2}=\frac{1}{8}\sum_{n=2}^{\infty}\frac{1}{(\frac{1}{2}n(n-1))^2}.$$

In the latter form the denominators of the terms inside the summation are the squares of the triangular numbers (see the note on Proposition 5).

Putting $z = 2$ and $T = 1$ in (i), we obtain

$$\sum_{n=2}^{\infty}\frac{1}{(n-1)^2}=\frac{3}{2}+\frac{1}{2}\sum_{n=2}^{\infty}\frac{1}{n^2(n-1)^2},$$

which is equivalent to Stirling's relation $8S - 12 = 8S_2$. The sum of the first series is of course $\pi^2/6$.

In the case of the pyramidal numbers, viz.

$$\sum_{r=1}^{k} r^2 = \tfrac{1}{6}k(k+1)(2k+1),$$

we can show by means of partial fractions that for $k_0 = 1, 2, \ldots$

$$\sum_{k=k_0}^{\infty}\frac{36}{k^2(k+1)^2(2k+1)^2}=72\sum_{k=k_0}^{\infty}\frac{1}{k^2}+576\sum_{k=k_0}^{\infty}\frac{1}{(2k+1)^2}-\frac{216}{k_0}-\frac{36}{k_0^2}.$$

Then, since

$$\sum_{k=k_0}^{\infty} \frac{1}{(2k+1)^2} = \sum_{k=2k_0}^{\infty} \frac{1}{k^2} - \frac{1}{4}\sum_{k=k_0}^{\infty} \frac{1}{k^2},$$

we have

$$\sum_{k=k_0}^{\infty} \frac{36}{k^2(k+1)^2(2k+1)^2} = 576 \sum_{k=2k_0}^{\infty} \frac{1}{k^2} - 72 \sum_{k=k_0}^{\infty} \frac{1}{k^2} - \frac{216}{k_0} - \frac{36}{k_0^2},$$

so that, as Stirling remarks, the sums may be obtained from those of the first series.

Stirling also mentions the tetrahedral numbers,[29] viz.

$$\sum_{r=1}^{k} \tfrac{1}{2}r(r+1) = \tfrac{1}{6}k(k+1)(k+2).$$

In this case we have the corresponding expression

$$\sum_{k=k_0}^{\infty} \frac{36}{k^2(k+1)^2(k+2)^2} = 54 \sum_{k=k_0}^{\infty} \frac{1}{k^2} - \frac{45}{k_0^2} - \frac{9}{(k_0+1)^2} - \frac{27}{k_0} - \frac{27}{k_0+1}.$$

Concerning the series introduced at the end of the *scholion* we note that by Gauss's test they converge if and only if $b + c > 1$.

**Proposition 10 (pp. 68–70).** *Case 1* is by way of being a lemma for Proposition 11, although in fact a more general version is required there. We sketch the details following Stirling's procedures. Suppose that the relation of the terms is

$$T' = \left(\frac{z-m}{z+k}\right)\left(\frac{z-n}{z-n+k+1}\right)T \qquad (z = z_0, z_1, z_2, \ldots),$$

or in modern terminology,

$$a_{i+1} = \left(\frac{z_0+i-m}{z_0+i+k}\right)\left(\frac{z_0+i-n}{z_0+i-n+k+1}\right)a_i \qquad (i = 0, 1, 2, \ldots), \qquad (1)$$

where $m$, $k$, $n$ are constants ($k = 0$ in Stirling's version). By Gauss's test we require $m + 2k > 0$ for convergence; in the applications $k$ is an arbitrary non-negative integer, so the condition then reduces to $m > 0$. The series defined by (1) is

---

[29]The geometrical significance of the triangular, tetrahedral and pyramidal numbers is nicely illustrated in [29, pp. 58–59].

$$a_0\left(1+\frac{(z_0-m)(z_0-n)}{(z_0+k)(z_0-n+k+1)}\right.$$

$$\left.+\frac{(z_0-m)(z_0-m+1)(z_0-n)(z_0-n+1)}{(z_0+k)(z_0+k+1)(z_0-n+k+1)(z_0-n+k+2)}+\dots\right)$$

$$=a_0\sum_{i=0}^{\infty}\frac{(z_0-m)_i(z_0-n)_i}{(z_0+k)_i(z_0-n+k+1)_i}\,. \tag{2}$$

In particular, when $k=0$ and $z=1,2,3,\dots$ the series is

$$a\,F(1-m,1-n\,;2-n\,;1), \tag{3}$$

whose sum according to Gauss's formula is

$$a\,\frac{\Gamma(2-n)\,\Gamma(m)}{\Gamma(1-n+m)}=a(1-n)\,B(1-n,m), \tag{4}$$

provided $2-n$ is not a negative integer or zero and $m>0$.

We wish to determine constants $p$, $q$ and $a_i^{(2)}$ $(i=0,1,2,\dots)$ such that

$$\sum_{i=j}^{\infty}a_i=\frac{z_0+j+p}{q}\,a_j+\sum_{i=j}^{\infty}a_i^{(2)} \qquad (j=0,1,2,\dots) \tag{5}$$

and $(z_0+j+p)a_j/q$ is a good approximation to $\sum_{i=j}^{\infty}a_i$ at least for large $j$. Subtracting the expressions given by (5) for $j$ and $j+1$ we obtain

$$a_j=\frac{z_0+j+p}{q}\,a_j-\frac{z_0+j+1+p}{q}\,a_{j+1}+a_j^{(2)},$$

so that by (1)

$$a_j^{(2)}=\left(1-\frac{z_0+j+p}{q}+\frac{z_0+j+1+p}{q}\cdot\frac{z_0+j-m}{z_0+j+k}\cdot\frac{z_0+j-n}{z_0+j-n+k+1}\right)a_j$$

$$=\frac{h(j)a_j}{q(z_0+j+k)(z_0+j-n+k+1)}\,, \tag{6}$$

where

$$h(j)=(q-m-2k)(z_0+j)^2$$
$$+(mn-pm-m-n-k^2+kn-k-2pk-p+2kq-qn+q)(z_0+j)$$
$$+pmn+mn-pk^2+pkn-pk+qk^2-qkn+qk.$$

To optimise the proposed approximation we want the terms $a_j^{(2)}$ to be as small as possible; at least for large $j$ this means that the coefficients of $(z_0+j)^2$ and $z_0+j$ in $h(j)$, which are independent of $j$, should be zero.[30] Clearly, then,

[30]Stirling describes a method of transforming series in [61], which may also be explained in terms of a similar optimisation procedure (see [71]).

$$q = m + 2k,$$

and it follows from the coefficient of $z_0 + j$ that

$$p = \frac{3k^2 - kn + k + 2km - n}{m + 2k + 1}.$$

Note that when $k = 0$ these values for $p$ and $q$ reduce to those obtained by Stirling.

The identity (5) now becomes

$$\sum_{i=j}^{\infty} a_i = \frac{(m + 2k + 1)(z_0 + j + 2k) - (k + 1)(k + n)}{(m + 2k)(m + 2k + 1)} a_j + \sum_{i=j}^{\infty} a_i^{(2)}. \quad (7)$$

From (6) and (1) we obtain

$$\begin{aligned} \frac{a_{i+1}^{(2)}}{a_i^{(2)}} &= \frac{(z_0 + i + k)(z_0 + i - n + k + 1)}{(z_0 + i + k + 1)(z_0 + i - n + k + 2)} \times \frac{a_{i+1}}{a_i} \\ &= \frac{(z_0 + i - m)(z_0 + i - n)}{(z_0 + i + k + 1)(z_0 + i - n + k + 2)}, \end{aligned} \quad (8)$$

which is just (1) with $k$ replaced by $k + 1$. From (6) and (2), after some manipulation, we find for $0 \le j \le i$

$$\begin{aligned} a_i^{(2)} &= \frac{(k + 1)(m + k)(n + k)(m - n + k + 1)a_i}{(m + 2k)(m + 2k + 1)(z_0 + i + k)(z_0 + i - n + k + 1)} \\ &= a_j \frac{(k + 1)(m + k)(n + k)(m - n + k + 1)(z_0 + j - m)_{i-j}(z_0 + j - n)_{i-j}}{(m + 2k)(m + 2k + 1)(z_0 + j + k)_{i-j+1}(z_0 + j - n + k + 1)_{i-j+1}}. \end{aligned} \quad (9)$$

We will make use of these identities in our discussion of Proposition 11 below.

In *Case 2* Stirling causes a little confusion by reformulating the equation for the terms: $T$ and $T'$ are now on the same side and $r$ has been replaced by $-r$. In the original he asserts that $r \neq 1$, as one would expect from $r = 1$ in *Case 1*; however, for the reformulated equation the restriction should be $r \neq -1$. He proposes the representation

$$S = p\left(\frac{z + m}{z + n}\right) T + S_2,$$

where $m$, $n$, $p$ are constants to be determined following procedures and criteria similar to those employed in *Case 1*. We expand a little on Stirling's description of how the constants are determined. Since he does not determine the terms of $S_2$ at this stage, it is convenient for simplicity to retain his notation. The stated relation between $T_2$ and $T$ leads to

$$T_2 = \frac{h(z)}{r(z+n)(z+n+1)(z^2+cz+d)} T,$$

where

$$\begin{aligned} h(z) = &(r-(r+1)p)z^4 \\ &+ (r(1-p)(n+1+c) + rn - rmp - p(n+m+1) - ap)z^3 \\ &+ (r(n-mp)(n+1) + rc((1-p)(n+1)+n-mp) + rd(1-p) \\ &\quad - p(n(m+1) + a(n+m+1) + b))z^2 \\ &+ (rc(n-mp)(n+1) + rd((1-p)(n+1)+n-mp) \\ &\quad - p(an(m+1) + b(n+m+1)))z \\ &+ rd(n-mp)(n+1) - bpn(m+1). \end{aligned}$$

In order to minimise $T_2$ for large $z$ we should choose the three constants $m$, $n$, $p$ so as to make the coefficients of the three highest powers in $h(z)$ zero. From the coefficient of $z^4$ we obtain

$$p = \frac{r}{r+1},$$

and substitution of this into the coefficient of $z^3$ produces

$$n - m = \frac{a-c}{r+1};$$

using these values for $p$ and $n-m$, we obtain from the coefficient of $z^2$ after some manipulation

$$n = \frac{1}{r+1}(cr+a-1) - \frac{b-d}{a-c},$$

and consequently

$$m = c - \frac{1}{r+1} - \frac{b-d}{a-c}.$$

For later application it is useful to have a statement of the transformation in the notation used in the first part of this note: we have

$$\left((z_0+i)^2 + a(z_0+i) + b\right) a_i + r\left((z_0+i)^2 + c(z_0+i) + d\right) a_{i+1} = 0,$$

and

$$\sum_{i=j}^{\infty} a_i = p\left(\frac{z_0+j+m}{z_0+j+n}\right) a_j + \sum_{i=j}^{\infty} a_i^{(2)}, \tag{10}$$

where

$$p = \frac{r}{r+1},$$

$$m = c - \frac{1}{r+1} - \frac{b-d}{a-c} = n - \frac{a-c}{r+1},$$

$$n = \frac{1}{r+1}(cr+a-1) - \frac{b-d}{a-c} = m + \frac{a-c}{r+1}.$$

This transformation is applied in Proposition 12.

As Stirling notes, the above procedures may be carried through in the general case where

$$a_{i+1} = \frac{1}{r}\frac{f(z_0+i)}{g(z_0+i)}\, a_i$$

and $f$, $g$ are polynomials of the form

$$f(z) = z^\theta + az^{\theta-1} + bz^{\theta-2} + \dots$$
$$g(z) = z^\theta + cz^{\theta-1} + dz^{\theta-2} + \dots .$$

The same expressions result for $p$, $m$ and $n$. This situation arises in Examples 2 and 3 of Proposition 12.

**Proposition 11 (pp. 70–75).** Here we have Stirling's series of Proposition 10, Case 1, and the result follows by repeated application of the general process described in the previous note. The stated relation of the terms corresponds to $k = 0$ in (1), and (7), (8) and (9) then give

$$\sum_{i=j}^{\infty} a_i = \frac{(m+1)(z_0+j)-n}{m(m+1)}\, a_j + \sum_{i=j}^{\infty} a_i^{(2)},$$

$$\frac{a_{i+1}^{(2)}}{a_i^{(2)}} = \frac{(z_0+i-m)(z_0+i-n)}{(z_0+i+1)(z_0+i-n+2)},$$

$$a_i^{(2)} = \frac{mn(m-n+1)}{m(m+1)(z_0+i)(z_0+i-n+1)}\, a_i.$$

Then from (7), (8) and (9) with $k = 1$ we obtain

$$\sum_{i=j}^{\infty} a_i = \frac{(m+1)(z_0+j)-n}{m(m+1)}\, a_j + \frac{(m+3)(z_0+j+2)-2(n+1)}{(m+2)(m+3)}\, a_j^{(2)} + \sum_{i=j}^{\infty} a_i^{(3)},$$

$$\frac{a_{i+1}^{(3)}}{a_i^{(3)}} = \frac{(z_0+i-m)(z_0+i-n)}{(z_0+i+2)(z_0+i-n+3)},$$

$$a_i^{(3)} = \frac{2(m+1)(n+1)(m-n+2)}{(m+2)(m+3)(z_0+i+1)(z_0+i-n+2)}\, a_i^{(2)}.$$

Continuing with $k = 2$ leads to

$$\sum_{i=j}^{\infty} a_i = \frac{(m+1)(z_0+j)-n}{m(m+1)} a_j + \frac{(m+3)(z_0+j+2)-2(n+1)}{(m+2)(m+3)} a_j^{(2)}$$

$$+ \frac{(m+5)(z_0+j+4)-3(n+2)}{(m+4)(m+5)} a_j^{(3)} + \sum_{i=j}^{\infty} a_i^{(4)},$$

$$\frac{a_{i+1}^{(4)}}{a_i^{(4)}} = \frac{(z_0+i-m)(z_0+i-n)}{(z_0+i+3)(z_0+i-n+4)},$$

$$a_i^{(4)} = \frac{3(m+2)(n+2)(m-n+3)}{(m+4)(m+5)(z_0+i+2)(z_0+i-n+3)} a_i^{(3)}.$$

The process continues in an obvious way. Stirling's $T$, $T_2$, $T_3$, $T_4$, ... are our $a_j$, $a_j^{(2)}$, $a_j^{(3)}$, $a_j^{(4)}$, ... and the successive iterates produce the partial sums of Stirling's series with remainders.

The factorial nature of the $a_i^{(s)}$ is clear from (9) and (2) above: for $s = 0, 1, 2, \ldots$ with $a_i^{(1)} = a_i$,

$$a_i^{(s+1)} = \frac{s!\,(m)_s(n)_s(m-n+1)_s}{(m)_{2s}(z_0+i)_s(z_0+i-n+1)_s} a_i$$

$$= \frac{s!\,(m)_s(n)_s(m-n+1)_s(z_0-m)_i(z_0-n)_i}{(m)_{2s}(z_0)_{s+i}(z_0-n+1)_{s+i}} a_0 \qquad (i = 0, 1, 2, \ldots)$$

$$= \frac{s!\,(m)_s(n)_s(m-n+1)_s(z_0-m)_i(z_0-n)}{(m)_{2s}(z_0)_{s+i}(z_0-n+i)_{s+1}} a_0 \qquad (i = 1, 2, \ldots).$$

The terms of the new series are for $s = 0, 1, 2, \ldots$, $j = 0, 1, 2, \ldots$

$$\frac{s!\,(m)_s(n)_s(m-n+1)_s((m+2s+1)(z_0+j+2s)-(s+1)(n+s))}{(m)_{2s+2}(z_0+j)_s(z_0+j-n+1)_s} a_j$$

$$= \frac{s!\,(m)_s(n)_s(m-n+1)_s(z_0-m)_j(z_0-n)_j}{(m)_{2s+2}(z_0)_{s+j}(z_0-n+1)_{s+j}}$$

$$\times ((m+2s+1)(z_0+j+2s)-(s+1)(n+s))a_0. \qquad (\alpha)$$

If the sum of the first $s$ terms ($s > 0$) of the new series is used to approximate the sum of the given series $\sum_{i=j}^{\infty} a_i$ then the error is

$$\sum_{i=j}^{\infty} a_i^{(s+1)} = a_0 \frac{s!\,(m)_s(n)_s(m-n+1)_s(z_0-n)}{(m)_{2s}}$$

$$\times \sum_{i=j}^{\infty} \frac{(z_0-m)_i}{(z_0)_{s+i}(z_0-n+i)_{s+1}}. \qquad (\beta)$$

Stirling applies the proposition most effectively in four examples. The calculations are carried to an impressive seventeen decimal places, which we can justify by examining the error involved. In each case Stirling tabulates on the left the terms $T$, $T_2$, $T_3$, ... and on the right the corresponding terms in his series for $S$. We discuss these examples using the notation developed above and in the previous note.

*Example 1.* Here we have

$$\sum_{i=0}^{\infty} \frac{1}{(1+i)^2} = \frac{\pi^2}{6},$$

for which (see (1) of previous note with $k = 0$)

$$a_0 = 1, \quad a_{i+1} = \left(\frac{1+i}{2+i}\right)^2 a_i = \left(\frac{2+i-1}{2+i}\right)\left(\frac{2+i-1}{2+i-1+1}\right) a_i,$$

where $i = 0, 1, 2, \ldots$, and $z_0 = 2$, $m = n = 1$. Stirling finds directly $\sum_{i=0}^{9} a_i$. He then applies the proposition to transform $\sum_{i=10}^{\infty} a_i$, taking nine terms of the new series, viz. (from $(\alpha)$)

$$(10!)^2 \sum_{s=0}^{8} \frac{(s!)^4 \left(4(s+1)(s+6) - (s+1)^2\right)}{(2s+2)!\,\left((s+11)!\right)^2}.$$

By $(\beta)$ the error is

$$E = \sum_{i=10}^{\infty} a_i^{(10)} = \frac{(9!)^4}{18!} \sum_{i=10}^{\infty} \frac{i!}{(10+i)!\,(1+i)_{10}} = \frac{(9!)^4}{18!} \sum_{i=10}^{\infty} \frac{1}{\left((1+i)_{10}\right)^2},$$

from which it follows that $18!\,(9!)^{-4}E$ lies between the values of the following integrals:[31]

$$\int_{10}^{\infty} \frac{dx}{(x+1)^2(x+2)^2 \ldots (x+10)^2};$$

$$\int_{9.5}^{\infty} \frac{dx}{(x+1)^2(x+2)^2 \ldots (x+10)^2}.$$

With the aid of Maple (see p. 10) I find from these bounds that

$$4.8 \times 10^{-18} < E < 9.1 \times 10^{-18}.$$

Repeating Stirling's calculation using quadruple precision arithmetic produces 2, 8 for the seventeenth and eighteenth digits after the decimal point,

[31]The range of integration in the second integral, the upper bound, follows from the midpoint rule on noting that the second derivative of the integrand is positive if $x > -1$.

the corresponding digits in $\pi^2/6$ being 3, 6, so that the error is about $8\times10^{-18}$. (Cf. Proposition 2, Example 6; also Proposition 12, Example 2.)

*Example 2.* The series is

$$\sum_{i=0}^{\infty} \frac{\left(\left(\frac{1}{2}\right)_i\right)^2}{i!\left(\frac{3}{2}\right)_i} = F\left(\tfrac{1}{2}, \tfrac{1}{2}; \tfrac{3}{2}; 1\right) = \frac{\Gamma(\frac{3}{2})\,\Gamma(\frac{1}{2})}{(\Gamma(1))^2} = \tfrac{1}{2}\left(\Gamma\left(\tfrac{1}{2}\right)\right)^2 = \frac{\pi}{2},$$

for which

$$a_0 = 1, \quad a_{i+1} = \frac{\left(\frac{1}{2}+i\right)^2}{(1+i)(\frac{3}{2}+i)}\, a_i = \left(\frac{1+i-\frac{1}{2}}{1+i}\right)\left(\frac{1+i-\frac{1}{2}}{1+i-\frac{1}{2}+1}\right) a_i$$

for $i = 0, 1, 2, \ldots$, and so $z_0 = 1$, $m = n = \frac{1}{2}$. Again Stirling finds $\sum_{i=0}^{9} a_i$ by direct addition and takes the first nine terms of the transformed series $(\alpha)$, viz.

$$\left(\tfrac{1}{2}\right)_{10}^2 \sum_{s=0}^{8} \frac{(s!)^2\left(\left(\frac{1}{2}\right)_s\right)^2\left((2s+\frac{3}{2})(2s+11)-(s+1)(s+\frac{1}{2})\right)}{\left(\frac{1}{2}\right)_{2s+2}(s+10)!\left(\frac{3}{2}\right)_{s+10}};$$

from $(\beta)$ the error is

$$E = \sum_{i=10}^{\infty} a_i^{(10)} = \frac{(9!)^2\left(\left(\frac{1}{2}\right)_9\right)^2}{2\left(\frac{1}{2}\right)_{18}} \sum_{i=10}^{\infty} \frac{\left(\frac{1}{2}\right)_i}{(9+i)!\,(\frac{1}{2}+i)_{10}}.$$

Now[32]

$$\frac{\left(\frac{1}{2}\right)_i}{i!} = 2^{-2i}\binom{2i}{i} \qquad \text{and} \qquad \frac{1}{\sqrt{\left(i+\frac{1}{2}\right)\pi}} < 2^{-2i}\binom{2i}{i} < \frac{1}{\sqrt{i\pi}}.$$

Thus, as in the previous example, we can deduce that

$$\frac{2\left(\frac{1}{2}\right)_{18}\sqrt{\pi}}{(9!)^2\left(\left(\frac{1}{2}\right)_9\right)^2}\,E$$

lies between the values of the following two integrals:

$$\int_{10}^{\infty} \frac{dx}{\sqrt{x+.5}\,(x+1)(x+2)\ldots(x+9)(x+.5)(x+1.5)\ldots(x+9.5)}$$

$$\int_{9.5}^{\infty} \frac{dx}{\sqrt{x}\,(x+1)(x+2)\ldots(x+9)(x+.5)(x+1.5)\ldots(x+9.5)}.$$

Evaluating these expressions with the aid of Maple produces the bounds

[32] We see in the note on Proposition 23 that the inequalities follow from other results of Stirling.

$$2.9 \times 10^{-18} < E < 5.9 \times 10^{-18}.$$

All the digits in Stirling's calculated value are in fact correct; however the eighteenth digit after the decimal point in $\pi/2$ is 9 and if we repeat Stirling's calculation using quadruple precision arithmetic, we find 3, 8 as the next two digits, so there is an error of about $5 \times 10^{-18}$.

*Example 3.* Here we have another hypergeometric series

$$\frac{1}{3}\sum_{i=0}^{\infty}\frac{\left(\frac{1}{2}\right)_i\left(\frac{3}{4}\right)_i}{i!\,\left(\frac{7}{4}\right)_i} = \frac{1}{3}F\left(\tfrac{1}{2},\tfrac{3}{4};\tfrac{7}{4};1\right) = \frac{\Gamma(\frac{7}{4})\,\Gamma(\frac{1}{2})}{3\,\Gamma(\frac{5}{4})} = \frac{\Gamma(\frac{3}{4})\sqrt{\pi}}{\Gamma(\frac{1}{4})}$$
$$= \frac{\pi^{3/2}}{\left(\Gamma(\frac{1}{4})\right)^2\sin\frac{\pi}{4}} = \frac{\sqrt{2}\pi^{3/2}}{\left(\Gamma(\frac{1}{4})\right)^2},$$

for which

$$a_0 = \tfrac{1}{3}, \quad a_{i+1} = \frac{\left(\frac{1}{2}+i\right)\left(\frac{3}{4}+i\right)}{(i+1)(\frac{7}{4}+i)}\,a_i = \left(\frac{1+i-\frac{1}{2}}{1+i}\right)\left(\frac{1+i-\frac{1}{4}}{1+i-\frac{1}{4}+1}\right)a_i$$

for $i = 0, 1, 2, \ldots$, and $z_0 = 1$, $m = \frac{1}{2}$, $n = \frac{1}{4}$. This time Stirling takes nine initial terms. He then transforms the series from its tenth term and takes the first nine terms of the transformed series, viz.

$$\frac{1}{3}\left(\tfrac{1}{2}\right)_9\left(\tfrac{3}{4}\right)_9\sum_{s=0}^{8}\frac{s!\,\left(\frac{1}{2}\right)_s\left(\frac{1}{4}\right)_s\left(\frac{5}{4}\right)_s\left((2s+\frac{3}{4})(2s+10)-(s+1)(s+\frac{1}{4})\right)}{\left(\frac{1}{2}\right)_{2s+2}\,(s+9)!\,\left(\frac{7}{4}\right)_{s+9}}.$$

In this case the error is

$$E = \sum_{i=9}^{\infty}a_i^{(10)} = \frac{9!\,\left(\frac{1}{2}\right)_9\left(\frac{1}{4}\right)_9\left(\frac{5}{4}\right)_9}{4\left(\frac{1}{2}\right)_{18}}\sum_{i=9}^{\infty}\frac{\left(\frac{1}{2}\right)_i}{(9+i)!\,(\frac{3}{4}+i)_{10}}.$$

Proceeding as in the previous example, we find that

$$\frac{4\left(\frac{1}{2}\right)_{18}\sqrt{\pi}}{9!\,\left(\frac{1}{2}\right)_9\left(\frac{1}{4}\right)_9\left(\frac{5}{4}\right)_9}\,E$$

lies between the values of the following two integrals:

$$\int_9^{\infty}\frac{dx}{\sqrt{x+.5}\,(x+1)(x+2)\ldots(x+9)(x+.75)(x+1.75)\ldots(x+9.75)};$$

$$\int_{8.5}^{\infty}\frac{dx}{\sqrt{x}\,(x+1)(x+2)\ldots(x+9)(x+.75)(x+1.75)\ldots(x+9.75)}.$$

According to Maple, we then have the error bounds

$$2.6 \times 10^{-18} < E < 5.5 \times 10^{-18}.$$

Repeating Stirling's calculation in quadruple precision arithmetic, I find that the sixteenth to nineteenth digits after the decimal point are 0, 9, 8, 8, so that, correctly rounded, Stirling's final two digits should be 1, 0 and not 1, 1. The sixteenth to nineteenth digits of the true value are 1, 0, 3, 7, so the error is about $5 \times 10^{-18}$.

*Example 4.* Continuing with hypergeometric series, we have

$$\sum_{i=0}^{\infty} \frac{\left(\frac{1}{2}\right)_i \left(\frac{1}{4}\right)_i}{i!\,\left(\frac{5}{4}\right)_i} = F(\tfrac{1}{2}, \tfrac{1}{4}; \tfrac{5}{4}; 1) = \frac{\Gamma(\frac{5}{4})\Gamma(\frac{1}{2})}{\Gamma(\frac{3}{4})\Gamma(1)} = \frac{1}{4\sqrt{\pi}} \left(\Gamma\left(\tfrac{1}{4}\right)\right)^2 \sin\frac{\pi}{4}$$
$$= \frac{1}{4\sqrt{2\pi}} \left(\Gamma\left(\tfrac{1}{4}\right)\right)^2,$$

for which

$$a_0 = 1, \quad a_{i+1} = \frac{\left(\frac{1}{2}+i\right)\left(\frac{1}{4}+i\right)}{(1+i)\left(\frac{5}{4}+i\right)} a_i = \left(\frac{1+i-\frac{1}{2}}{1+i}\right)\left(\frac{1+i-\frac{3}{4}}{1+i-\frac{3}{4}+1}\right) a_i$$

for $i = 0, 1, 2, \ldots$ with $z_0 = 1$, $m = \frac{1}{2}$, $n = \frac{3}{4}$. Stirling proceeds as in *Example 3*, the corresponding contribution from the new series being

$$\left(\tfrac{1}{2}\right)_9 \left(\tfrac{1}{4}\right)_9 \sum_{s=0}^{\infty} \frac{s!\,\left(\frac{1}{2}\right)_s \left(\left(\frac{3}{4}\right)_s\right)^2 \left((2s+\frac{3}{2})(2s+10) - (s+1)(s+\frac{3}{4})\right)}{\left(\frac{1}{2}\right)_{2s+2} (s+9)!\,\left(\frac{5}{4}\right)_{s+9}}$$

and the error

$$E = \sum_{i=9}^{\infty} a_i^{(10)} = \frac{9!\,\left(\frac{1}{2}\right)_9 \left(\left(\frac{3}{4}\right)_9\right)^2}{4\left(\frac{1}{2}\right)_{18}} \sum_{i=9}^{\infty} \frac{\left(\frac{1}{2}\right)_i}{(9+i)!\,(\frac{1}{4}+i)_{10}}.$$

Proceeding as before, we find that

$$\frac{4\left(\frac{1}{2}\right)_{18}\sqrt{\pi}}{9!\,\left(\frac{1}{2}\right)_9 \left(\frac{3}{4}\right)_9^2} E$$

lies between the values of the following two integrals:

$$\int_9^{\infty} \frac{dx}{\sqrt{x+.5}\,(x+1)(x+2)\ldots(x+9)(x+.25)(x+1.25)\ldots(x+9.25)};$$

$$\int_{8.5}^{\infty} \frac{dx}{\sqrt{x}\,(x+1)(x+2)\ldots(x+9)(x+.25)(x+1.25)\ldots(x+9.25)}.$$

From Maple we then obtain the error bounds

$$7.95 \times 10^{-18} < E < 1.7 \times 10^{-17}.$$

It is noted in [67] that Stirling's value is correct to 15DP. Quadruple precision arithmetic produces in the sixteenth to eighteenth places after the decimal point the digits 8, 9, 0; in the true value these are 9, 0, 5, so there is an error of about $1.5 \times 10^{-17}$. The value of $T_6$ in Stirling's tabulation should be $1,7922,56$.

*Example 3* and *Example 4* refer to Jakob Bernoulli's *elastic curve.*

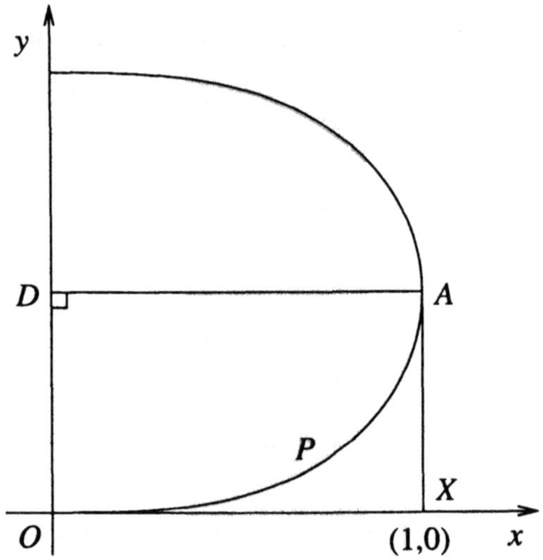


The 'Elastica'

If an elastic lamina is constrained so that its ends are perpendicular to the same straight line (the $y$-axis in the above diagram) and if coordinates are chosen so that $AD = 1$, then the lower portion of the curve, $OPA$, has equation

$$y = \int_0^x \frac{t^2}{\sqrt{1-t^4}}\,dt \qquad (0 \le x \le 1).$$

With $x = 1$ in this equation we have

$$AX = \int_0^1 t^2 \left(1 + \sum_{i=1}^{\infty} \frac{\left(-\frac{1}{2}\right)\left(-\frac{3}{2}\right)\ldots\left(-\frac{1}{2}-i+1\right)}{i!} \left(-t^4\right)^i\right) dt$$

$$= \frac{1}{3} + \sum_{i=1}^{\infty} \frac{\left(\frac{1}{2}\right)_i}{i!\,(4i+3)}$$

$$= \frac{1}{3}\left(1 + \sum_{i=1}^{\infty} \frac{\left(\frac{1}{2}\right)_i \left(\frac{3}{4}\right)_i}{i!\,\left(\frac{7}{4}\right)_i}\right) = \frac{1}{3} F\left(\frac{1}{2}, \frac{3}{4}; \frac{7}{4}; 1\right),$$

which is the series of *Example 3.* The length of the lower portion $OPA$ of the curve is

$$\int_0^1 \sqrt{1+\frac{t^4}{1-t^4}}\,dt = \int_0^1 \frac{dt}{\sqrt{1-t^4}}$$

$$= \int_0^1 1 + \sum_{i=1}^{\infty} \frac{\left(-\frac{1}{2}\right)\left(-\frac{3}{2}\right)\ldots\left(-\frac{1}{2}-i+1\right)}{i!}(-t^4)^i dt$$

$$= 1 + \sum_{i=1}^{\infty} \frac{\left(\frac{1}{2}\right)_i}{i!\,(4i+1)} = 1 + \sum_{i=1}^{\infty} \frac{\left(\frac{1}{2}\right)_i \left(\frac{1}{4}\right)_i}{i!\,\left(\frac{5}{4}\right)_i} = F\left(\tfrac{1}{2}, \tfrac{1}{4}; \tfrac{5}{4}; 1\right),$$

which is the series of *Example 4*.

These series representations were given by Bernoulli in 1694 [5]. In the same paper it is asserted that $AX$ lies strictly between 0.598 and 0.601, while the length is bounded strictly by 1.308 and 1.316 – these are not the exact bounds attributed by Stirling to Bernoulli. In a later dissertation[33] of 1704 [6] Bernoulli gave the limits (0.5983546, 0.6004034) and (1.3088173, 1.3152635).

Stirling notes that the sum of the quantites found in *Example 3* and *Example 4* is the semiperimeter of an ellipse with axes of length 1 and $\sqrt{2}$. The canonical equation of such an ellipse is $4x^2 + 2y^2 = 1$, so that the upper portion is given by $y = \sqrt{\frac{1}{2}(1-4x^2)}$ and has length

$$2\int_0^{1/2} \sqrt{1+\left(\frac{-4x}{\sqrt{2(1-4x^2)}}\right)^2}\,dx = 2\int_0^{1/2} \sqrt{\frac{1+4x^2}{1-4x^2}}\,dx = \int_0^1 \sqrt{\frac{1+t^2}{1-t^2}}\,dt$$

$$= \int_0^1 \frac{1+t^2}{\sqrt{1-t^4}}\,dt,$$

which is the sum of the two integrals considered above.

The two quantities are also known as the *lemniscate constants* (see [67]). Bernoulli was aware of quadrature relations between the elastic curve and his lemniscate; in particular, we have that the length of a quadrant of the lemniscate with polar equation $r^2 = \cos 2\theta$ is $2\int_0^1 (1-t^4)^{-1/2}dt$; moreover the product of the two quantities is $\pi/4$.

**Scholion to Proposition 11 (pp. 75–77).** Stirling begins by observing that Proposition 11 may be applied to evaluate $\int_0^x t^{\theta-1}\left(e+ft^{\eta}\right)^{\lambda-1} dt$ in the case where $e + fx^{\eta} = 0$. He has already noted that Propositions 7 and 8 may be applied for this purpose. The details are given in the note on Proposition 8, where it is shown that the integral may be expressed as a series whose terms satisfy the relation

[33] This dissertation was defended by Nikolaus Bernoulli, then a Masters candidate, before his uncle, Jakob Bernoulli, on 8 April 1704.

$$T' = \frac{(z-\lambda)\left(z-\left(1-\frac{\theta}{\eta}\right)\right)}{z\left(z-\left(1-\frac{\theta}{\eta}\right)+1\right)}T,$$

which is the relation of Proposition 11 with $m = \lambda$, $n = 1 - \frac{\theta}{\eta}$.

The transformation described in the scholion may be established using the method of Proposition 10, Case 1 (see the note on Proposition 10). We have

$$a_{i+1} = \frac{(z_0+i)^2+m}{(z_0+i)^2+n(z_0+i)+r}\,a_i \qquad (i=0,1,2,\ldots) \qquad (\gamma)$$

and we aim to determine with the same optimality criterion constants $p$, $q$, $a_i^{(2)}$ $(i = 0, 1, 2, \ldots)$ such that

$$\sum_{i=j}^{\infty} a_i = \frac{z_0+j+p}{q}\,a_j + \sum_{i=j}^{\infty} a_i^{(2)} \qquad (j=0,1,2,\ldots). \qquad (\delta)$$

The analogue of (6) in the note on Proposition 10 is

$$a_j^{(2)} = \left(1-\frac{z_0+j+p}{q}+\frac{z_0+j+1+p}{q}\cdot\frac{(z_0+j)^2+m}{(z_0+j)^2+n(z_0+j)+r}\right)a_j$$

$$= \frac{(1-n+q)(z_0+j)^2+(m-r-pn+qn)(z_0+j)+pm+m-pr+qr}{q\left((z_0+j)^2+n(z_0+j)+r\right)}\,a_j, \qquad (\epsilon)$$

so that we put $1 - n + q = 0$ and $m - r - pn + qn = 0$, which give

$$q = n-1 \quad\text{and}\quad p = \frac{m-r+qn}{n} = n-1-\frac{r-m}{n} = a-1, \qquad (\zeta)$$

where $a = n - \dfrac{r-m}{n}$. Then $(\delta)$ becomes

$$\sum_{i=j}^{\infty} a_i = \frac{z_0+j+n-\frac{r-m}{n}-1}{n-1}\,a_j + \sum_{i=j}^{\infty} a_i^{(2)}$$

$$= \frac{z_0+j+a-1}{n-1}\,a_j + \sum_{i=j}^{\infty} a_i^{(2)}, \qquad (\eta)$$

with, for $0 \le j \le i$,

$$a_i^{(2)} = \frac{m\left(n-\frac{r-m}{n}\right)+\frac{r-m}{n}r}{(n-1)\left((z_0+i)^2+n(z_0+i)+r\right)}\,a_i$$

$$= \frac{ma+(n-a)r}{(n-1)\left((z_0+i)^2+n(z_0+i)+r\right)}\,a_i; \qquad (\theta)$$

also from ($\theta$) and ($\gamma$)

$$\frac{a_{i+1}^{(2)}}{a_i^{(2)}} = \frac{(z_0+i)^2+n(z_0+i)+r}{(z_0+i+1)^2+n(z_0+i+1)+r}\cdot\frac{a_{i+1}}{a_i}$$

$$= \frac{(z_0+i)^2+m}{(z_0+i+1)^2+n(z_0+i+1)+r}$$

$$= \frac{(z_0+i)^2+m}{(z_0+i)^2+(n+2)(z_0+i)+r+n+1}\,. \qquad (\iota)$$

Now ($\iota$) is of the same form as ($\gamma$): in the denominator the coefficient of $z_0+i$ has been augmented by 2 and the constant term has had the original coefficient of $z+i$ and 1 added to it. We may now repeat the process with $\sum_{i=j}^{\infty} a_i^{(2)}$. The successive values of $a$ are Stirling's $a$, $b$, $c$, ... .[34]

In the *example* Stirling considers

$$\frac{1}{2}+\frac{1\times1}{2\times2\times4}+\frac{1\times1\times3\times3}{2\times2\times2\times4\times4\times6}+\frac{1\times1\times3\times3\times5\times5}{2\times2\times4\times4\times6\times6\times8}+\ldots$$

$$=\frac{1}{2}\left(1+\frac{\left(\frac{1}{2}\right)^2}{1\times2}+\frac{\left(\frac{1}{2}\cdot\frac{3}{2}\right)^2}{1\times2\times2\times3}+\frac{\left(\frac{1}{2}\cdot\frac{3}{2}\cdot\frac{5}{2}\right)^2}{1\times2\times3\times2\times3\times4}+\ldots\right)$$

$$=\tfrac{1}{2}F\left(\tfrac{1}{2},\tfrac{1}{2};2;1\right)=\frac{\Gamma(2)\,\Gamma(1)}{2\left(\Gamma(\frac{3}{2})\right)^2}=\frac{2}{\pi}\,.$$

Here we have

$$a_0=\tfrac{1}{2}\,,\quad a_{i+1}=\frac{\left(\frac{1}{2}+i\right)^2}{(i+1)(i+2)}\,a_i=\frac{\left(\frac{1}{2}+i\right)^2}{\left(\frac{1}{2}+i\right)^2+2\left(\frac{1}{2}+i\right)+\frac{3}{4}}\,a_i,$$

where $i=0,1,2,\ldots$ ; consequently $z_0=\frac{1}{2}$, $m=0$, $n=2$, $r=\frac{3}{4}$.

Stirling adds directly the first six terms of the series and then applies the above process to $\sum_{i=6}^{\infty} a_i$, so that $j=6$, and takes seven terms of the transformed series. The following tabulation shows the successive stages of the transformation; in order to illustrate the general series whose first four

[34] In his own copy of the *Methodus Differentialis* Stirling recorded the following alternative representations:

$$b=n-1-\frac{r-m-n}{n+2}\,;\qquad c=n+2-\frac{r-m-4}{n+4}\,;$$

$$d=n+3-\frac{r-m-9}{n+6}\,;\qquad e=n+4-\frac{r-m-16}{n+8}\,.$$

He also noted that $n$ should not be zero or a negative integer.

terms are stated by Stirling we keep $j$ arbitrary – recall that Stirling's $z$ is $\frac{1}{2}+j$ in our notation.

---

Stage I: $n = 2$, $r = \frac{3}{4}$;

$$a = 2 - \tfrac{3}{8} = \tfrac{13}{8};$$

$$\frac{z_0 + j + a - 1}{n-1} a_j = \frac{9 + 8j}{1 \times 8} a_j; \tag{I.1}$$

$$a_i^{(2)} = \frac{\frac{3}{8} \cdot \frac{3}{4}}{\left(\frac{1}{2}+i\right)^2 + 2\left(\frac{1}{2}+i\right) + \frac{3}{4}} a_i = \frac{1}{8} \cdot \frac{\left(\frac{3}{2}\right)^2}{(i+1)(i+2)} a_i. \tag{I.2}$$

---

Stage II: $n = 2 + 2 = 4$, $r = \frac{3}{4} + 2 + 1 = \frac{15}{4}$;

$$a = 4 - \tfrac{15}{16} = \tfrac{49}{16};$$

$$\frac{z_0 + j + a - 1}{n-1} a_j^{(2)} = \frac{41 + 16j}{3 \times 16} a_j^{(2)}; \tag{II.1}$$

$$a_i^{(3)} = \frac{\frac{15}{16} \cdot \frac{15}{4}}{3\left(\left(\frac{1}{2}+i\right)^2 + 4\left(\frac{1}{2}+i\right) + \frac{15}{4}\right)} a_i^{(2)} = \frac{3}{16} \cdot \frac{\left(\frac{5}{2}\right)^2}{(i+2)(i+3)} a_i^{(2)}. \tag{II.2}$$

---

Stage III: $n = 4 + 2 = 6$, $r = \frac{15}{4} + 4 + 1 = \frac{35}{4}$;

$$a = 6 - \tfrac{35}{24} = \tfrac{109}{24};$$

$$\frac{z_0 + j + a - 1}{n-1} a_j^{(3)} = \frac{97 + 24j}{5 \times 24} a_j^{(3)}; \tag{III.1}$$

$$a_i^{(4)} = \frac{\frac{35}{24} \cdot \frac{35}{4}}{5\left(\left(\frac{1}{2}+i\right)^2 + 6\left(\frac{1}{2}+i\right) + \frac{35}{4}\right)} a_i^{(3)} = \frac{5}{24} \cdot \frac{\left(\frac{7}{2}\right)^2}{(i+3)(i+4)} a_i^{(3)}. \tag{III.2}$$

---

---

Stage IV: $n = 6 + 2 = 8$, $r = \frac{35}{4} + 6 + 1 = \frac{63}{4}$;

$a = 8 - \frac{63}{32} = \frac{193}{32}$;

$$\frac{z_0 + j + a - 1}{n - 1} a_j^{(4)} = \frac{177 + 32j}{7 \times 32} a_j^{(4)}; \tag{IV.1}$$

$$a_i^{(5)} = \frac{\frac{63}{32} \cdot \frac{63}{4}}{7\left(\left(\frac{1}{2} + i\right)^2 + 8\left(\frac{1}{2} + i\right) + \frac{63}{4}\right)} a_i^{(4)} = \frac{7}{32} \cdot \frac{\left(\frac{9}{2}\right)^2}{(i+4)(i+5)} a_i^{(4)}. \tag{IV.2}$$

---

Stage V: $n = 8 + 2 = 10$, $r = \frac{63}{4} + 8 + 1 = \frac{99}{4}$;

$a = 10 - \frac{99}{40} = \frac{301}{40}$;

$$\frac{z_0 + j + a - 1}{n - 1} a_j^{(5)} = \frac{281 + 40j}{9 \times 40} a_j^{(5)}; \tag{V.1}$$

$$a_i^{(6)} = \frac{\frac{99}{40} \cdot \frac{99}{4}}{9\left(\left(\frac{1}{2} + i\right)^2 + 10\left(\frac{1}{2} + i\right) + \frac{99}{4}\right)} a_i^{(5)} = \frac{9}{40} \cdot \frac{\left(\frac{11}{2}\right)^2}{(i+5)(i+6)} a_i^{(5)}. \tag{V.2}$$

---

Stage VI: $n = 10 + 2 = 12$, $r = \frac{99}{4} + 10 + 1 = \frac{143}{4}$;

$a = 12 - \frac{143}{48} = \frac{433}{48}$;

$$\frac{z_0 + j + a - 1}{n - 1} a_j^{(6)} = \frac{409 + 48j}{11 \times 48} a_j^{(6)}; \tag{VI.1}$$

$$a_i^{(7)} = \frac{\frac{143}{48} \cdot \frac{143}{4}}{11\left(\left(\frac{1}{2} + i\right)^2 + 12\left(\frac{1}{2} + i\right) + \frac{143}{4}\right)} a_i^{(6)} = \frac{11}{48} \cdot \frac{\left(\frac{13}{2}\right)^2}{(i+6)(i+7)} a_i^{(6)}. \tag{VI.2}$$

---

Stage VII: $n = 12 + 2 = 14$, $r = \frac{143}{4} + 12 + 1 = \frac{195}{4}$;

$a = 14 - \frac{195}{56} = \frac{589}{56}$;

$$\frac{z_0 + j + a - 1}{n - 1} a_j^{(7)} = \frac{561 + 56j}{13 \times 56} a_j^{(7)}; \tag{VII.1}$$

$$a_i^{(8)} = \frac{\frac{195}{56} \cdot \frac{195}{4}}{13\left(\left(\frac{1}{2} + i\right)^2 + 14\left(\frac{1}{2} + i\right) + \frac{195}{4}\right)} a_i^{(7)} = \frac{13}{56} \cdot \frac{\left(\frac{15}{2}\right)^2}{(i+7)(i+8)} a_i^{(7)}. \tag{VII.2}$$

---

The first seven terms of the transformed series are (I.1)–(VII.1), which are evaluated with $j = 6$ ($z = 13/2$) and

$$a_6 = \frac{1}{2} \cdot \frac{\left(\left(\frac{1}{2}\right)_6\right)^2}{6!\,7!}$$

to complete the calculation. All Stirling's values for the terms are correctly rounded to the number of places shown, but his final value for the sum of the series is affected by the cumulative effects of rounding. Stirling's thirteenth and final digit is 3, but if we extend the calculation we find that the thirteenth and fourteenth digits are in fact 4 and 8, giving a value which is less than $2/\pi$ by about $1 \times 10^{-13}$. The actual error is $\sum_{i=6}^{\infty} a_i^{(8)}$ and the terms $a_i^{(8)}$ are determined by (VII.2)–(I.2) and the terms of the series. By an analysis similar to that used in the examples for the proposition we can show that

$$\begin{aligned} E &< \frac{(3 \times 5 \times \ldots \times 13)^3 \times 15^2}{7! \times 2^{36} \times \pi} \int_{5.5}^{\infty} \frac{dx}{x(x+1)^2(x+2)^2 \ldots (x+7)^2(x+8)} \\ &< 1.2 \times 10^{-13}. \end{aligned}$$

**Proposition 12 and Its Scholion (pp. 77–85).** Here Stirling applies Case 2 of Proposition 10 in conjunction with the procedures illustrated in the examples of Proposition 9 to develop a transformation process akin to that of Proposition 11. He does not present general formulae but demonstrates the procedure in three examples.

*Example 1.* The series[35] is

$$2\sqrt{3} \sum_{i=0}^{\infty} \frac{(-1)^i}{(2i+1)3^i} = 6 \sum_{i=0}^{\infty} \frac{(-1)^i}{(2i+1)(\sqrt{3})^{2i+1}} = 6 \tan^{-1} \frac{1}{\sqrt{3}} = \pi,$$

whose terms satisfy

$$a_0 = 2\sqrt{3}, \quad \frac{a_{i+1}}{a_i} = -\frac{1}{3}\left(\frac{2i+1}{2i+3}\right) = -\frac{1}{3}\left(\frac{\frac{1}{2}+i}{\frac{1}{2}+i+1}\right), \tag{1.1}$$

or equivalently,

$$a_0 = 2\sqrt{3}, \quad \left(\tfrac{1}{2}+i\right) a_i + 3\left(\tfrac{1}{2}+i+1\right) a_{i+1} = 0.$$

In order to apply identity (10) in the note on Proposition 10 we multiply[36] the last relation by $\frac{1}{2} + i$ to get

---

[35]Stirling ascribes this series to Halley who derived it and illustrated its use by calculating $\pi$ to 12DP in [25]. The calculation was extended by Abraham Sharp to 72DP [57].

[36]To fit in with the general relation (1.4) noted below it would have been more appropriate to multiply by $1 + i$ to get

$$\left(\tfrac{1}{2}+i\right)^2 a_i + 3\left(\left(\tfrac{1}{2}+i\right)^2 + \tfrac{1}{2} + i\right) a_{i+1} = 0.$$

Then from (10) (p. 205) we have

$$z_0 = \tfrac{1}{2}, \quad a = b = 0, \quad r = 3, \quad c = 1\,, \quad d = 0,$$

so that

$$p = \tfrac{3}{4}, \quad m = 1 - \tfrac{1}{4} = \tfrac{3}{4}, \quad n = \tfrac{3}{4} - \tfrac{1}{4} = \tfrac{1}{2},$$

and

$$\sum_{i=j}^{\infty} a_i = \frac{3}{4}\left(\frac{\frac{1}{2}+j+\frac{3}{4}}{\frac{1}{2}+j+\frac{1}{2}}\right) a_j + \sum_{i=j}^{\infty} a_i^{(2)}$$

$$= \frac{3}{4}\left(\frac{4j+5}{4j+4}\right) a_j + \sum_{i=j}^{\infty} a_i^{(2)} \qquad (j = 0, 1, 2, \ldots).$$

Now we have to determine the $a_i^{(2)}$. Replacing $j$ with $j+1$ we obtain

$$\sum_{i=j+1}^{\infty} a_i = \frac{3}{4}\left(\frac{4j+9}{4j+8}\right) a_{j+1} + \sum_{i=j+1}^{\infty} a_i^{(2)}$$

and subtraction of this from the previous identity produces

$$a_j = \frac{3}{4}\left(\frac{4j+5}{4j+4}\right) a_j - \frac{3}{4}\left(\frac{4j+9}{4j+8}\right) a_{j+1} + a_j^{(2)}.$$

Then by (1.1) we have

$$a_j^{(2)} = \left(1 - \frac{3}{4}\left(\frac{4j+5}{4j+4}\right) - \frac{1}{4}\left(\frac{4j+9}{4j+8}\right)\left(\frac{2j+1}{2j+3}\right)\right) a_j$$

$$= -\frac{12}{4(4j+4)(4j+8)(2j+3)} a_j$$

$$= -\frac{3!}{8(2j+2)(2j+3)(2j+4)} a_j, \tag{1.2}$$

and consequently,

$$a_{j+1}^{(2)} = -\frac{3!}{8(2j+4)(2j+5)(2j+6)} a_{j+1}.$$

Using (1.1) we now obtain

---

$$\left(\left(\tfrac{1}{2}+i\right)^2 + \tfrac{1}{2}\left(\tfrac{1}{2}+i\right)\right) a_i + 3\left(\left(\tfrac{1}{2}+i\right)^2 + \tfrac{3}{2}\left(\tfrac{1}{2}+i\right) + \tfrac{1}{2}\right) a_{i+1} = 0.$$

Then $z_0 = \frac{1}{2}$, $a = \frac{1}{2}$, $b = 0$, $r = 3$, $c = \frac{3}{2}$, $d = \frac{1}{2}$, giving $p = \frac{3}{4}$, $m = \frac{3}{2} - \frac{1}{4} - \frac{1}{2} = \frac{3}{4}$, $n = \frac{3}{4} - \frac{1}{4} = \frac{1}{2}$ as before.

$$\frac{a_{i+1}^{(2)}}{a_i^{(2)}} = \frac{(2i+2)(2i+3)}{(2i+5)(2i+6)}\left(-\frac{1}{3}\left(\frac{2i+1}{2i+3}\right)\right)$$

$$= -\frac{1}{3}\frac{(2i+1)(2i+2)}{(2i+5)(2i+6)},$$

so that

$$(2i+1)(2i+2)a_i^{(2)} + 3(2i+5)(2i+6)a_{i+1}^{(2)} = 0,$$

or, on dividing by 4 and expressing the result in terms of $\frac{1}{2}+i$,

$$\left(\left(\tfrac{1}{2}+i\right)^2 + \tfrac{1}{2}\left(\tfrac{1}{2}+i\right)\right)a_i^{(2)} + 3\left(\left(\tfrac{1}{2}+i\right)^2 + \tfrac{9}{2}\left(\tfrac{1}{2}+i\right)+5\right)a_{i+1}^{(2)} = 0. \quad (1.3)$$

Identities (1.2), (1.3) are equivalent to Stirling's equations relating $T_2$, $T$ and $T_2$, $T_2'$ (recall that $z = \frac{1}{2}+i$). We may now repeat the process with $\sum_{i=j}^{\infty} a_i^{(2)}$, for which

$$z_0 = \tfrac{1}{2}, \quad a = \tfrac{1}{2}, \quad b = 0, \quad r = 3, \quad c = \tfrac{9}{2}, \quad d = 5,$$

and consequently

$$p = \tfrac{3}{4}, \quad m = \tfrac{9}{2} - \tfrac{1}{4} - \tfrac{5}{4} = 3, \quad n = 3 - 1 = 2.$$

Thus

$$\sum_{i=j}^{\infty} a_i = \frac{3}{4}\left(\frac{4j+5}{4j+4}\right)a_j + \frac{3}{4}\left(\frac{2j+7}{2j+5}\right)a_j^{(2)} + \sum_{i=j}^{\infty} a_i^{(3)}.$$

In fact, we can show by induction that, for $s = 1, 2, \ldots$ with $a_i^{(1)} = a_i$,

$$\left(\left(\frac{1}{2}+i\right)^2 + \frac{1}{2}\left(\frac{1}{2}+i\right)\right)a_i^{(s)} + 3\left(\frac{1}{2}+i+\frac{3s-2}{2}\right)\left(\frac{1}{2}+i+\frac{3s-1}{2}\right)a_{i+1}^{(s)} = 0, \quad (1.4)$$

$$\sum_{i=j}^{\infty} a_i^{(s)} = \frac{3}{4}\left(\frac{4j+9s-4}{4j+6s-2}\right)a_j^{(s)} + \sum_{i=j}^{\infty} a_i^{(s+1)}, \quad (1.5)$$

$$a_i^{(s+1)} = -\frac{(3s-2)(3s-1)3s}{8(2i+3s-1)(2i+3s)(2i+3s+1)}a_i^{(s)}. \quad (1.6)$$

Stirling adds up the first ten terms of the series and then transforms $\sum_{i=10}^{\infty} a_i$, taking six terms of the transformed series, viz.

$$\sum_{s=1}^{6} \frac{3}{4}\left(\frac{36+9s}{38+6s}\right)a_{10}^{(s)}.$$

The error is

$$E = \sum_{i=10}^{\infty} a_i^{(7)}.$$

Now by (1.6)

$$a_i^{(7)} = \frac{18!}{8^6(2i+2)_{18}} a_i = \frac{18!\,6\,(-1)^i}{8^6(2i+1)_{19}\,3^{i+\frac{1}{2}}}$$

and because the terms alternate in sign and decrease in modulus to zero with $a_{10}^{(7)} > 0$,

$$a_{10}^{(7)} - a_{11}^{(7)} < E < a_{10}^{(7)}.$$

Thus the error is strictly less than

$$\frac{18!\,6}{8^6(21)_{19}\,3^{10.5}} < 1.71 \times 10^{-22}$$

and strictly greater than

$$\frac{18!\,6}{8^6(23)_{17}\,3^{10.5}} \left(\frac{1}{21 \times 22} - \frac{1}{3 \times 40 \times 41}\right) > 1.548 \times 10^{-22}.$$

Some of Stirling's terms are not correctly rounded, but repeating his calculation in quadruple precision arithmetic produces the same calculated value to the number of places given by Stirling. The actual error is about $1.6 \times 10^{-22}$.

Stirling's remark about six terms of the transformed series being as effective as thirty-two terms of the simple series refers to $\sum_{i=10}^{41} a_i$; the error involved in using this as an approximation to $\sum_{i=10}^{\infty} a_i$ is strictly less than

$$a_{42} = \frac{6}{85 \times 3^{42.5}} < 4 \times 10^{-22}$$

and strictly greater than

$$a_{42} - a_{43} = \frac{6}{3^{42.5}} \left(\frac{1}{85} - \frac{1}{3 \times 87}\right) > 2.5 \times 10^{-22}.$$

Thus we have the same order of accuracy in both cases.

*Example 2.* Here the series is

$$\sum_{i=0}^{\infty} \frac{(-1)^i}{(1+i)^2} = \sum_{k=1}^{\infty} \frac{1}{(2k-1)^2} - \sum_{k=1}^{\infty} \frac{1}{4k^2}.$$

Now

$$\sum_{k=1}^{\infty} \frac{1}{k^2} = \sum_{k=1}^{\infty} \frac{1}{(2k-1)^2} + \sum_{k=1}^{\infty} \frac{1}{4k^2}.$$

Hence

$$\sum_{k=1}^{\infty} \frac{1}{(2k-1)^2} = \frac{3}{4} \sum_{k=1}^{\infty} \frac{1}{k^2},$$

and so, as Stirling notes at the end of *Example 2*,[37]

[37] Concerning $\sum_{k=1}^{\infty} k^{-2}$ see Proposition 2, Example 6 and Proposition 11, Example 1.

$$\sum_{i=0}^{\infty} \frac{(-1)^i}{(1+i)^2} = \frac{1}{2}\sum_{k=1}^{\infty} \frac{1}{k^2} \quad \left(= \frac{\pi^2}{12}\right).$$

We have (cf. (2.1) below)

$$a_0 = 1, \quad (1+i)^2 a_i + (1+i+1)^2 a_{i+1} = 0,$$

from which

$$z_0 = 1, \quad r = 1, \quad a = b = 0, \quad c = 2, \quad d = 1.$$

Hence

$$p = \tfrac{1}{2}, \quad m = 2 - \tfrac{1}{2} - \tfrac{1}{2} = 1, \quad n = 1 - 1 = 0,$$

and

$$\sum_{i=j}^{\infty} a_i = \frac{1}{2}\left(\frac{j+2}{j+1}\right) a_j + \sum_{i=j}^{\infty} a_i^{(2)}.$$

Proceeding as above we deduce that

$$a_j^{(2)} = -\frac{2j+3}{2(j+1)(j+2)^3}\, a_j$$

and

$$(1+i)^3\left(1+i+\tfrac{3}{2}\right) a_i^{(2)} + \left(1+i+\tfrac{1}{2}\right)(1+i+2)^3 a_{i+1}^{(2)} = 0,$$

for which

$$r = 1, \quad a = \tfrac{3}{2}, \quad b = 0\,, \quad c = 6 + \tfrac{1}{2} = \tfrac{13}{2}, \quad d = 12 + 3 = 15.$$

We can show by induction that for $s = 1, 2, \ldots$

$$(1+i)^3\left(1+i+\frac{s+1}{2}\right) a_i^{(s)} + \left(1+i+\frac{s-1}{2}\right)(1+i+s)^3 a_{i+1}^{(s)} = 0, \tag{2.1}$$

$$\sum_{i=j}^{\infty} a_i^{(s)} = \frac{j+2s}{2j+s+1}\, a_j^{(s)} + \sum_{i=j}^{\infty} a_i^{(s+1)}, \tag{2.2}$$

$$a_i^{(s+1)} = -\left(\frac{2i+s+2}{2i+s+1}\right)\left(\frac{s}{i+s+1}\right)^3 a_i^{(s)}. \tag{2.3}$$

To facilitate the calculation of partial sums of the transformed series

$$\sum_{s} \frac{j+2s}{2j+s+1}\, a_j^{(s)},$$

Stirling introduces quantities $B$, $C$, $D$, ... whose definitions depend on the relationships

$$\frac{a_j^{(s+1)}}{2j+s+2} = \frac{1}{2j+s+2}\left(-\frac{2j+s+2}{2j+s+1}\right)\left(\frac{s}{j+s+1}\right)^3 a_j^{(s)}$$

$$= -\left(\frac{s}{j+s+1}\right)^3 \times \frac{a_j^{(s)}}{2j+s+1}.$$

Stirling finds $\sum_{i=0}^{9} a_i$ directly and then applies the transformation to $\sum_{i=10}^{\infty} a_i$, taking nine terms of the transformed series, viz. (see (2.2))

$$\sum_{s=1}^{9} \frac{10+2s}{21+s} a_{10}^{(s)}.$$

The error is (see (2.3))

$$\sum_{i=10}^{\infty} a_i^{(10)} = \frac{1}{2} \times 1^3 \times 2^3 \times \ldots \times 9^3 \sum_{i=10}^{\infty} \frac{(-1)^{i-1}(2i+11)}{(i+1)^3(i+2)^3\ldots(i+10)^3}.$$

This is negative and its modulus is strictly less than

$$\frac{1^3 \times 2^3 \times \ldots \times 9^3 \times 31}{2 \times 11^3 \times 12^3 \times \ldots \times 20^3} < 2.46 \times 10^{-18},$$

and strictly greater than

$$\frac{1^3 \times 2^3 \times \ldots \times 9^3}{2 \times 12^3 \times 13^3 \times \ldots \times 20^3}\left(\frac{31}{11^3} - \frac{33}{21^3}\right) > 2.08 \times 10^{-18}.$$

Stirling's calculated values of the multipliers $A - I$ are all correctly rounded to the number of places shown. However, his values for the terms contain some errors; correctly rounded to the number of places shown, they are

| | |
|---:|---:|
| .0495.8677.6859.5041.3 | .0000.3347.8726.6605.4 |
| 13.9322.3555.7 | 1542.2436.4 |
| 32.4950.1 | 1.0908.4 |
| 523.2 | 33.3 |
| 2.7 | |
| +.0495.8691.6214.4073.0 | −.0000.3348.0269.9983.5 |

The calculated value of the sum of the series should have 2 rather than 1 in the final place when rounded. The actual error is about $-2.1 \times 10^{-18}$.

*Example 3.* Stirling now applies his transformation to Leibniz's series

$$\sum_{i=0}^{\infty} \frac{(-1)^i}{2i+1} = \frac{\pi}{4},$$

whose terms satisfy the relation

$$a_0 = 1, \quad \left(\tfrac{1}{2} + i\right) a_i + \left(\tfrac{1}{2} + i + 1\right) a_{i+1} = 0,$$

or, equivalently, for the purposes of the proposition

$$\left(\tfrac{1}{2} + i\right)^2 a_i + \left(\left(\tfrac{1}{2} + i\right)^2 + \tfrac{1}{2} + i\right) a_{i+1} = 0.$$

With a view to a remark which Stirling makes in the *scholion* we will consider more generally

$$(z_0 + i)^2 a_i + \left((z_0 + i)^2 + z_0 + i\right) a_{i+1} = 0,$$

for which

$$r = 1, \quad a = b = 0, \quad c = 1, \quad d = 0.$$

Consequently

$$p = \tfrac{1}{2}, \quad m = 1 - \tfrac{1}{2} = \tfrac{1}{2}, \quad n = \tfrac{1}{2} - \tfrac{1}{2} = 0,$$

so that

$$\sum_{i=j}^{\infty} a_i = \frac{1}{2}\left(\frac{z_0 + j + \frac{1}{2}}{z_0 + j}\right) a_j + \sum_{i=j}^{\infty} a_i^{(2)} = \frac{2z_0 + 2j + 1}{4(z_0 + j)} a_j + \sum_{i=j}^{\infty} a_i^{(2)}.$$

Proceeding as before, we find that

$$a_j^{(2)} = -\frac{1}{4(z_0 + j)(z_0 + j + 1)^2} a_j$$

and

$$(z_0 + i)^2 a_i^{(2)} + (z_0 + i + 2)^2 a_{i+1}^{(2)} = 0,$$

for which

$$r = 1, \quad a = b = 0, \quad c = d = 4.$$

We can now show by induction that for $s = 1, 2, \ldots$

$$(z_0 + i)^2 \left(z_0 + i + \frac{1}{2}\right)\left(z_0 + i + \frac{s+1}{2}\right) a_i^{(s)}$$
$$+ \left(z_0 + i + s - \frac{1}{2}\right)\left(z_0 + i + \frac{s-1}{2}\right)\left(z_0 + i + s\right)^2 a_{i+1}^{(s)} = 0, \quad (3.1)$$

$$\sum_{i=j}^{\infty} a_i^{(s)} = \frac{2z_0 + 2j + 4s - 3}{2(2z_0 + 2j + s - 1)} a_j^{(s)} + \sum_{i=j}^{\infty} a_i^{(s+1)}, \quad (3.2)$$

$$a_i^{(s+1)} = -\frac{(2z_0 + 2i + s)s(2s - 1)^2}{2(2z_0 + 2i + s - 1)(2z_0 + 2i + 2s - 1)(z_0 + i + s)^2} a_i^{(s)}. \quad (3.3)$$

Note also that from (3.3)

$$\frac{a_j^{(s+1)}}{2z_0 + 2j + s} = -\frac{s(2s - 1)^2}{2(2z_0 + 2j + 2s - 1)(z_0 + j + s)^2} \times \frac{a_j^{(s)}}{2z_0 + 2j + s - 1};$$

Stirling's quantities $B$, $C$, $D$, ... are defined by these relationships (with $z_0 = \frac{1}{2}$) and, as in *Example 2*, they allow him to streamline the calculation of the partial sums of

$$\sum_s \frac{j+2s-1}{2j+s} a_j^{(s)}.$$

Stirling obtains $\sum_{i=0}^{11} a_i$ by direct addition and transforms $\sum_{i=12}^{\infty} a_i$, taking eight terms of the transformed series, viz. (from (3.2) with $z_0 = \frac{1}{2}$)

$$\sum_{s=1}^{8} \frac{11+2s}{24+s} a_{12}^{(s)}.$$

The error is (from (3.3) with $z_0 = \frac{1}{2}$)

$$\sum_{i=12}^{\infty} a_i^{(9)} = \frac{8! \times 1^2 \times 3^2 \times \ldots \times 15^2}{4^9} \sum_{i=12}^{\infty} \frac{(2i+9)(-1)^i}{(i+1)_8 \left((i+\frac{1}{2})_9\right)^2},$$

which is strictly less than

$$\frac{8! \times 1^2 \times 3^2 \times \ldots \times 15^2 \times 33}{4^9 (13)_8 \left((\frac{25}{2})_9\right)^2} < 6.3 \times 10^{-19}$$

and strictly greater than

$$\frac{8! \times 1^2 \times 3^2 \times \ldots \times 15^2}{4^9 (14)_7 \left((\frac{27}{2})_8\right)^2} \left( \frac{33}{13 \times \left(\frac{25}{2}\right)^2} - \frac{35}{21 \times \left(\frac{43}{2}\right)^2} \right) > 4.87 \times 10^{-19}.$$

Stirling's calculations of the multipliers $A$ – $H$ are all correctly rounded to the number of places shown, but there are some errors in the terms, which should be, correctly rounded to the number of places shown,

| | |
|---:|---:|
| .5200.0000.0000.0000.0 | .0000.6331.1174.4222.9 |
| 10.9694.9174.6 | 637.8782.6 |
| 7.9307.0 | 1689.2 |
| 54.1 | 2.4 |
| +.5200.0010.9702.8535.7 | −.0000.6331.1812.4697.1 |

The calculated value of the sum of the series should have 1 rather than 0 in the final place (when rounded). The actual error is about $5.1 \times 10^{-19}$. In fact, Stirling gives correctly the first seventeen digits of $\pi/4$. However, as the error bounds suggest, it is possible to extract the eighteenth digit from the calculation. Quadruple precision arithmetic produces 9 and 1 for the eighteenth and nineteenth digits of the calculated value, so the eighteenth digit of $\pi/4$ must be 9.

In the *scholion* Stirling notes that the procedure of *Example 3* also applies to the series

$$\sum_{i=0}^{\infty} \frac{(-1)^i}{m+in},$$

whose terms satisfy the relation

$$\left(\frac{m}{n}+i\right)a_i + \left(\frac{m}{n}+i+1\right)a_{i+1} = 0,$$

or equivalently

$$\left(\frac{m}{n}+i\right)^2 a_i + \left(\left(\frac{m}{n}+i\right)^2 + \frac{m}{n}+i\right)a_{i+1} = 0.$$

Thus we have $z_0 = \frac{m}{n}$ in the general case of *Example 3* described above. Unfortunately, both $m$ and $n$ are now playing two roles.

Stirling adds finally a few observations concerning power series with infinite radius of convergence, for which, he asserts, no transformations are required. The quoted series are the MacLaurin series for $e^x - 1$ and $\sin x$. As an example he notes in effect that

$$e^{12.3785} = e^{12}e^{.3785}$$

and that $e^{.3785}$ may be obtained from a few terms of the series. In a similar vein, since $\sin(\pi - \theta) = \sin\theta$, we may use $\pi - \theta$ rather than $\theta$ in the sine series if $\frac{\pi}{2} < \theta\,(< \pi)$, for the smaller value $\pi - \theta$ will produce a series which converges more rapidly.

**Proposition 13 (pp. 85–87).** Here we are concerned with a series $\sum a_k$ whose terms satisfy a linear recurrence relation of the form

$$\sum_{i=0}^{h} c_i a_{k+i} = 0 \qquad (k = k_0, k_0+1, \ldots). \tag{1}$$

If the series converges, we obtain for $m = k_0, k_0 + 1, \ldots$

$$0 = \sum_{k=m}^{\infty}\sum_{i=0}^{h} c_i a_{k+i} = \sum_{i=0}^{h} c_i \sum_{k=m}^{\infty} a_{k+i} = \sum_{i=0}^{h} c_i \sum_{j=m+i}^{\infty} a_j. \tag{2}$$

Thus, in Stirling's terminology, the successive sums of the series satisfy the same linear recurrence relation as the terms. Conversely, given (2), we can deduce (1) by taking the difference of the relations for $m$ and $m+1$.

As we see below, the MacLaurin series of a rational function which is analytic at the origin is of the above type for each $x$ in its domain of convergence. Such series are what Stirling means by "series which arise from division". We have an identity of the form

$$\frac{\alpha_0 + \alpha_1 x + \ldots + \alpha_l x^l}{\beta_0 + \beta_1 x + \ldots + \beta_h x^h} = \sum_{n=0}^{\infty} \gamma_n x^n,$$

which is valid for $|x|$ sufficiently small. Then if we consider the term of degree $k \geq k_0 = \max(h, l+1)$ on both sides of the identity

$$\alpha_0 + \alpha_1 x + \ldots + \alpha_l x^l = \left(\beta_0 + \beta_1 x + \ldots + \beta_h x^h\right) \sum_{n=0}^{\infty} \gamma_n x^n,$$

we obtain the relations

$$0 = \sum_{i=0}^{h} \left(\beta_{h-i} x^{h-i}\right) \left(\gamma_{k-h+i} x^{k-h+i}\right) \qquad (k = k_0, k_0+1, \ldots).$$

Stirling is particularly concerned with two- and three-term relations. From

$$r a_k + s a_{k+1} = 0 \quad (k = k_0, k_0+1, \ldots)$$

we obtain

$$r \sum_{k=m}^{\infty} a_k + s \sum_{k=m+1}^{\infty} a_k = 0;$$

thus

$$r \sum_{k=m}^{\infty} a_k + s \left( \sum_{k=m}^{\infty} a_k - a_m \right) = 0,$$

and so, if $r + s \neq 0$,

$$\sum_{k=m}^{\infty} a_k = \frac{s a_m}{r+s} \quad (m = k_0, k_0+1, \ldots).$$

The three-term relation

$$r a_k + s a_{k+1} + t a_{k+2} = 0 \quad (k = k_0, k_0+1, \ldots)$$

produces similarly

$$r \sum_{k=m}^{\infty} a_k + s \left( \sum_{k=m}^{\infty} a_k - a_m \right) + t \left( \sum_{k=m}^{\infty} a_k - a_m - a_{m+1} \right) = 0$$

if the series converges, in which case we have, provided[38] $r + s + t \neq 0$,

$$\sum_{k=m}^{\infty} a_k = \frac{(s+t) a_m + t a_{m+1}}{r+s+t} \qquad (m = k_0, k_0+1, \ldots). \tag{3}$$

[38]If $r + s + t = 0$ and $r$, $s$, $t$ are all nonzero, the recurrence relation $r a_k + s a_{k+1} + t a_{k+2} = 0$ has general solution

$$a_k = A + B \left(\frac{r}{t}\right)^k \quad (A, B \text{ arbitrary constants}).$$

Thus, to get a nontrivial convergent series, we require $A = 0$ and $\left|\frac{r}{t}\right| < 1$. In this case, the terms satisfy the two-term relation $r a_k - t a_{k+1} = 0$.

Note that (3) is not affected if we multiply the recurrence relation by a nonzero constant, since $r$, $s$, $t$ will be scaled by the same factor.

The series of *Example 1* is the geometric series $\sum\left(\frac{2}{x}\right)^k$, so that

$$\sum_{k=m}^{\infty}\left(\frac{2}{x}\right)^k=\left(\frac{2}{x}\right)^m\frac{1}{1-\frac{2}{x}}=\left(\frac{2}{x}\right)^m\frac{x}{x-2}\qquad(|x|>2,\ m=0,1,2,\ldots).$$

The terms of the series in *Example 2* satisfy the linear recurrence relation

$$x^2a_k-3xa_{k+1}+a_{k+2}=0,$$

which has general solution ($x>0$ or $x<0$)

$$a_k=A\left(\frac{(3+\sqrt{5})x}{2}\right)^k+B\left(\frac{(3-\sqrt{5})x}{2}\right)^k,$$

where $A$, $B$ are arbitrary constants. We also have $a_0=1$, $a_1=3x$, from which we obtain

$$A=\frac{3+\sqrt{5}}{2\sqrt{5}},\qquad B=-\frac{3-\sqrt{5}}{2\sqrt{5}}.$$

The series is therefore

$$\sum_{k\geq 0}\frac{1}{2^{k+1}\sqrt{5}}\left((3+\sqrt{5})^{k+1}-(3-\sqrt{5})^{k+1}\right)x^k.$$

Stirling introduces *Proposition 13* with a reference to the "principles of De Moivre", by which he means *recurrent series*, already mentioned in his Preface. De Moivre's main discussion of such series is in [43, pp. 26–42] (1730) but he had some of the ideas in the early 1710s and published relevant material in 1718 [41] and 1722 [42]. In fact recurrent series are just series whose terms satisfy a relation such as (1) above: "If some series is so formed that, if some number of terms in it are taken arbitrarily, each subsequent term always has given relations to the same number of preceding terms, I call such a series recurrent."[39] De Moivre's Theorem V [43, p. 35] is particularly relevant to Stirling's *Proposition 13*.

**Scholion to Proposition 13 (pp. 87–88).** In the *scholion* Stirling introduces an important concept which he uses to great effect in Proposition 14. He is concerned with a series $\sum a_k$ whose terms satisfy a recurrence relation of the form

$$\sum_{i=0}^{h}c_ip_i(k)a_{k+i}=0\qquad(k=k_0,k_0+1,\ldots),\tag{4}$$

[39] Si Series aliqua ita sit constituta, ut assumptis in ea ad libitum terminis quotlibet, terminus quisque subsequens ad eundem semper antecedentium numerum habeat rationes datas, talem seriem voco recurrentem; ... [43, p. 27].

where the $p_i(k)$ are polynomials of the same degree $m$ with 1 as the coefficient of $k^m$. We have equivalently

$$\sum_{i=0}^{h} c_i k^{-m} p_i(k) a_{k+i} = 0 \qquad (k = k_0, k_0+1, \ldots);$$

also $k^{-m}p_i(k) \to 1$ as $k \to \infty$. Thus for large $k$ we expect the terms to be given approximately by

$$\sum_{i=0}^{h} c_i a_{k+i} = 0.$$

This is what Stirling means by the ultimate relation of the terms (*ultima relatio terminorum*).

In the case $h = 3$ we can write (4) in the form

$$rp_0(k)a_k + sp_1(k)a_{k+1} + tp_2(k)a_{k+2} = 0 \qquad (k = k_0, k_0+1, \ldots), \tag{5}$$

for which the ultimate relation is

$$ra_k + sa_{k+1} + ta_{k+2} = 0 \qquad (k = k_0, k_0+1, \ldots).$$

Now, assuming that this ultimate relation defines a convergent series, we can obtain the successive sums of this series by means of (3) in the note on Proposition 13 provided $r + s + t \neq 0$ and we may expect that for large $m$ these sums

$$\frac{(s+t)a_m + ta_{m+1}}{r+s+t}$$

will approximate closely to the corresponding sums for the series defined by (5). In Proposition 14 Stirling uses this approximation iteratively to transform and sum suitable series.

**Proposition 14 (pp. 88–91).** This proposition appears to have caught the interest of Stirling's contemporaries. In his review [1] of the *Methodus Differentialis* Castel describes Proposition 14 as the most powerful and complicated result.[40] And Euler in his letter to Stirling of 8 June 1736 ([74, p. 179; 70, p. 141; 33, p. 125]) includes the following in some remarks about the *Methodus Differentialis*:

> But especially pleasing to me was Prop. 14 of Part I, in which you present a method for summing so easily series whose law of progression is not even established, using only the relation of the last terms; certainly this method applies very widely and has the greatest use. But the demonstration of this proposition, which you seem to have concealed from study, caused me immense difficulty, until at last with

[40] "La quatorziéme proposition est tout ce qu'il y a de plus fort et de plus compliqué."

the greatest pleasure I obtained it from the things which had gone before ....

Indeed, Stirling provided no proof apart from a few remarks relating to the case where the series can be summed as a rational function. Perhaps Stirling and Euler proceeded along the following lines.

Let $S = \sum_{k=1}^{\infty} a_k$, where in the notation of the preceding note the terms $a_k$ satisfy the recurrence relation

$$rp_1(k)a_k + sp_2(k)a_{k+1} + tp_3(k)a_{k+2} = 0,$$

for which the ultimate relation of the terms is

$$ra_k + sa_{k+1} + ta_{k+2} = 0$$

and $r + s + t \neq 0$. Define sequences $(s_k)$, $(d_k^{(1)})$ as follows:

$$\begin{aligned} s_1 &= \frac{(s+t)a_1 + ta_2}{r+s+t}, \\ s_k &= a_1 + \ldots + a_k + \frac{(s+t)a_{k+1} + ta_{k+2}}{r+s+t} \qquad (k = 1, 2, \ldots), \\ d_k^{(1)} &= s_{k+1} - s_k = \frac{ra_k + sa_{k+1} + ta_{k+2}}{r+s+t} \qquad (k = 1, 2, \ldots). \end{aligned}$$

Note that in defining $(s_k)$ we have used the procedure which Stirling gives in the preceding scholion and which I have described in more detail in the previous note for approximating to the sums $a_{k+1} + a_{k+2} + \ldots$ and that the $d_k^{(1)}$ are Stirling's $A_2/n$, $B_2/n$, $C_2/n$, ....

Clearly $s_k \to S$ as $k \to \infty$ and

$$\begin{aligned} S &= s_1 + (s_2 - s_1) + (s_3 - s_2) + \ldots \\ &= \frac{(s+t)a_1}{r+s+t} + \frac{ta_2}{r+s+t} + \sum_{k=1}^{\infty} d_k^{(1)}. \end{aligned}$$

Thus we have the first terms for the two parts of the transformed series: $(s+t)A/n$ and $tB/n$.

Now it can be shown that the terms of $\sum_{k=1}^{\infty} d_k^{(1)}$ satisfy the recurrence relation[41]

$$\begin{aligned} &rp_1(k)\{rt(p_3(k+2) - p_1(k+2))(p_3(k+1) - p_1(k+1)) \\ &\qquad - s^2(p_3(k+2) - p_2(k+2))(p_2(k+1) - p_1(k+1))\}\, d_k^{(1)} \\ &+ s\,\{rt\,(p_3(k+1)(p_3(k+2) - p_1(k+2))(p_2(k) - p_1(k)) \\ &\qquad + p_1(k+1)(p_3(k) - p_1(k))(p_3(k+2) - p_2(k+2))) \end{aligned}$$

[41] I found this relation by lengthy but elementary manipulations, which do not seem to justify reproducing here.

$$\begin{aligned} &\quad - s^2(p_3(k+2) - p_2(k+2))p_2(k+1)(p_2(k) - p_1(k))\big\}\, d_{k+1}^{(1)} \\ &+ tp_3(k+2)\left\{rt(p_3(k) - p_1(k))(p_3(k+1) - p_1(k+1))\right. \\ &\quad \left. - s^2(p_3(k+1) - p_2(k+1))(p_2(k) - p_1(k))\right\} d_{k+2}^{(1)} = 0. \qquad (*) \end{aligned}$$

It follows easily from this that if

$$p_1(k) = k^m + ak^{m-1} + \dots, \quad p_2(k) = k^m + bk^{m-1} + \dots,$$
$$p_3(k) = k^m + ck^{m-1} + \dots,$$

then the ultimate relation of the terms is

$$\left(rt(c-a)^2 - s^2(c-b)(b-a)\right)\left(rd_k^{(1)} + sd_{k+1}^{(1)} + td_{k+2}^{(1)}\right) = 0,$$

that is, provided $rt(c-a)^2 - s^2(c-b)(b-a) \neq 0$,

$$rd_k^{(1)} + sd_{k+1}^{(1)} + td_{k+2}^{(1)} = 0,$$

which is the same as that for the $a_k$. We may therefore treat $\sum d_k^{(1)}$ in the same way to get

$$\sum_{k=1}^{\infty} d_k^{(1)} = \frac{(s+t)d_1^{(1)}}{r+s+t} + \frac{td_2^{(1)}}{r+s+t} + \sum_{k=1}^{\infty} d_k^{(2)},$$

where

$$d_k^{(2)} = \frac{rd_k^{(1)} + sd_{k+1}^{(1)} + td_{k+2}^{(1)}}{r+s+t} \qquad (k = 1, 2, \dots);$$

the $d_k^{(2)}$ are in fact Stirling's $A_3/n^2$, $B_3/n^2$, $C_3/n^2$, .... Moreover,

$$\frac{(s+t)d_1^{(1)}}{r+s+t} + \frac{td_2^{(1)}}{r+s+t} = \frac{(s+t)(ra_1 + sa_2 + ta_3)}{(r+s+t)^2} + \frac{t(ra_2 + sa_3 + ta_4)}{(r+s+t)^2},$$

which provides the second terms in the two parts of the transformed series: $(s+t)A_2/n^2$ and $tB_2/n^2$.

Likewise, except in degenerate cases, the terms of $\sum d_k^{(2)}$ will have the same ultimate relation of terms as before and the same procedure applied to this series will produce the next pair of terms for the transformed series. The transformation process continues in this way.

Stirling presents three examples to illustrate Proposition 14. In *Example 1* and *Example 2* only two terms are involved in the recurrence relation ($t = 0$)[42]

[42]Stirling attributes this case to De Montmort, but I have not been able to find an exact equivalent in his work. De Montmort does discuss certain series of the form

$$\frac{M}{h} + \frac{N}{h^2} + \frac{O}{h^3} + \frac{P}{h^4} + \dots$$

in [45, pp. 669–675], where $h$ is of the form $1+q$. In the two-term case we have $rT + sT' = 0$, or $T + \frac{s}{r}T' = 0$; for the latter version the $n$ in Stirling's transformation, which corresponds to De Montmort's $h$, becomes $1 + \frac{s}{r}$.

but in *Example 3* all three coefficients $r$, $s$ and $t$ are nonzero. Let us examine the last example in more detail. The recurrence relation for the terms is[43]

$$x^2 a_k(k+5) - 2x a_{k+1}(k+3) - a_{k+2}(k+1) = 0 \qquad (k = 1, 2, \ldots)$$

with ultimate relation

$$x^2 a_k - 2x a_{k+1} - a_{k+2} = 0.$$

From (*) we find that the terms of $\sum d_k^{(1)}$ satisfy

$$x^2 d_k^{(1)}(k+5) - 2x d_{k+1}^{(1)}(k+4) - d_{k+2}^{(1)}(k+3) = 0$$

and those of $\sum d_k^{(2)}$ satisfy

$$x^2 d_k^{(2)} - 2x d_{k+1}^{(2)} - d_{k+2}^{(2)} = 0,$$

which is the same as the ultimate relation. Consequently the process terminates at this stage since the $d_k^{(3)}$ will all be zero – Stirling finds $A_4 = 0$ and $B_4 = 0$ by direct calculation. The series may now be summed in closed form as Stirling shows.

In all three examples the process terminates. Stirling asserts, however, that where this does not happen a rapidly converging series will be produced. In fact this is not true in general, as the following simple example shows. Consider the series

$$\sum_{k=1}^{\infty} \frac{x^k}{k} = -\ln(1-x) \qquad (-1 \le x < 1).$$

The terms $a_k = x^k/k$ satisfy the recurrence relation

$$xk a_k - (k+1)a_{k+1} = 0,$$

for which $r = x$, $s = -1$, $t = 0$, $p_1(k) = k$, $p_2(k) = k+1$, $p_3(k) = k+\alpha$, where $\alpha$ is an arbitrary constant, and the ultimate relation of the terms is $xa_k - a_{k+1} = 0$. Since $t = 0$ the second part of the transformed series drops out. The first term in the first part is $-x/(x-1)$ and

$$d_k^{(1)} = \left(x \cdot \frac{x^k}{k} - \frac{x^{k+1}}{k+1}\right)\frac{1}{x-1} = \frac{x^{k+1}}{k(k+1)(x-1)} \qquad (k = 1, 2, \ldots).$$

Consequently the second term in the first part of the transformed series is

$$-\frac{x^2}{2(x-1)} \cdot \frac{1}{x-1} = -\frac{x^2}{2(x-1)^2}$$

[43] Stirling's own version $x^2T(z+4) - 2xT'(z+2) - T''z = 0$ requires $z = 2, 3, \ldots$, contrary to his usual $z = 1, 2, \ldots$.

and

$$d_k^{(2)} = \left(x \cdot \frac{x^{k+1}}{k(k+1)(x-1)} - \frac{x^{k+2}}{(k+1)(k+2)(x-1)}\right) \cdot \frac{1}{x-1}$$
$$= \frac{2x^{k+2}}{k(k+1)(k+2)(x-1)^2} \quad (k = 1, 2, \ldots).$$

The third term of the first part of the transformed series is

$$-\frac{2x^3}{2.3(x-1)^3} = -\frac{x^3}{3(x-1)^3}.$$

We can show by induction that this pattern continues, so that the transformed series is

$$\sum_{k=1}^{\infty} -\frac{x^k}{k(x-1)^k} = \ln\left(1 - \frac{x}{x-1}\right)$$
$$= \ln\left(\frac{1}{1-x}\right) = -\ln(1-x)$$

provided $-1 \leq \dfrac{x}{x-1} < 1$, that is $x \leq \frac{1}{2}$. Thus the region of convergence of the transformed series neither includes nor is included in that of the original series.

The series

$$\sum_{k=1}^{\infty} (-1)^k x^k \left\{\frac{1}{2^k} + \frac{1}{2^{2k}}\right\} \qquad (|x| < 2)$$

is given in [33, p. 153] as a counterexample to the general validity of the *proposition*. I believe, however, that this example would not be admissible in Stirling's intended scheme because its terms do not satisfy a recurrence relation of the type considered above.

**Proposition 15 (pp. 92–95).** This consists of five examples in which Stirling demonstrates techniques for finding equations satisfied by given series. In *Example 1* he simply sums the series, but in each of *Examples 2–5* he obtains a differential equation which has the function defined by the given power series as a solution.

The series of *Example 2* is

$$\sum_{n=0}^{\infty} \frac{(-1)^n n!\, x^{2n}}{(\frac{3}{2})_n} = F(1, 1\,; \tfrac{3}{2}\,; -x^2),$$

for which Stirling finds the differential equation

$$x(1+x^2)\frac{dy}{dx} + (1+2x^2)y = 1$$

or in linear form

$$\frac{dy}{dx} + \frac{1+2x^2}{x(1+x^2)}\,y = \frac{1}{x(1+x^2)}\,.$$

The integrating factor for this equation is $x\sqrt{1+x^2}$, which leads to the general solution

$$yx\sqrt{1+x^2} = \ln\left(x + \sqrt{1+x^2}\right) + c.$$

The particular solution given by the series is

$$y = \frac{\ln\left(x+\sqrt{1+x^2}\right)}{x\sqrt{1+x^2}}\,,$$

since we require $c = 0$ to produce a finite limit for $y$ as $x \to 0_+$.

In *Example 3* the series is $F(-\frac{1}{2}, \frac{1}{2}\,; 1\,; x^2)$ with differential equation

$$x(1-x^2)\frac{d^2y}{dx^2} + (1-x^2)\frac{dy}{dx} + xy = 0$$

and in *Example 4* we have $F(\frac{1}{2}, \frac{1}{2}\,; 1\,; x^2)$ for which Stirling obtains the differential equation

$$x(1-x^2)\frac{d^2y}{dx^2} + (1-3x^2)\frac{dy}{dx} - xy = 0.$$

For both these equations 0 is a regular singular point and if we attempt to solve by the method of Frobenius we find that the indicial equation has repeated root $\sigma = 0$, so that just one linearly independent series solution, an ordinary power series, emerges in both cases.

The initial series of *Example 5* is

$$\sum_{n=1}^{\infty} \frac{x^n}{n^2} = -\int_0^x \frac{\ln(1-t)}{t}\,dt \qquad (|x| \le 1).$$

The associated differential equation is

$$x\frac{d^2y}{dx^2} + \frac{dy}{dx} = \frac{1}{1-x}\,,$$

whose general solution for $0 < x < 1$ is given by

$$y = -\int_0^x \frac{\ln(1-t)}{t}\,dt + c_1 \ln x + c_2.$$

**Scholion to Proposition 15 (pp. 95–100).** Following some general observations concerning manipulations with series, Stirling moves on to some specific remarks about fluxional or differential equations.

First he notes that for given $a$, $b$, $c$, the algebraic equation

$$y^2 = a^2 + bx + \frac{x^4}{c^2} \tag{$*$}$$

provides solutions for each of the following differential equations:

(i) $2y\dfrac{dy}{dx} = b + \dfrac{4x^3}{c^2}$;

(ii) $2xy\dfrac{dy}{dx} - y^2 = -a^2 + \dfrac{3x^4}{c^2}$;

(iii) $2xy\dfrac{dy}{dx} - 4y^2 = -4a^2 - 3bx$;

(iv) $\left(2a^2 + 2bx + \dfrac{2x^4}{c^2}\right)\dfrac{dy}{dx} = \left(b + \dfrac{4x^3}{c^2}\right)y.$

The first of these has general solution

$$y^2 = bx + \frac{x^4}{c^2} + k_1;$$

multiplying the second by $1/x^2$ we obtain

$$\frac{d}{dx}\left(\frac{y^2}{x}\right) = -\frac{a^2}{x^2} + \frac{3x^2}{c^2},$$

so that

$$y^2 = a^2 + \frac{x^4}{c^2} + k_2x;$$

multiplication of the third by $1/x^5$ leads to

$$\frac{d}{dx}\left(\frac{y^2}{x^4}\right) = -\frac{4a^2}{x^5} - \frac{3b}{x^4}$$

and thence

$$y^2 = a^2 + bx + k_3x^4;$$

the fourth equation can be expressed as

$$\frac{1}{y}\frac{dy}{dx} = \frac{b + \frac{4x^3}{c^2}}{2a^2 + 2bx + \frac{2x^4}{c^2}},$$

from which it follows that for $y > 0$ and $a^2 + bx + \frac{x^4}{c^2} > 0$

$$\ln y = \frac{1}{2}\ln\left(a^2 + bx + \frac{x^4}{c^2}\right) + c$$

and so

$$y = k_4\left(a^2 + bx + \frac{x^4}{c^2}\right)^{1/2}.$$

The quantities $k_1$, $k_2$, $k_3$ are arbitrary constants and $k_4$ is an arbitrary positive constant. Thus, as Stirling points out, in cases (i)–(iii) exactly one of the coefficients in the original form $(*)$ is not uniquely determined by the differential equation, although of course $(*)$ does result from the solution if we give the arbitrary constant an appropriate value; moreover, while $a$, $b$, $c$ all appear explicitly in equation (iv), the form $(*)$ still does not come out uniquely upon solution.

Stirling's remark about certain coefficients in the series not being determined by the equation may be explained as follows: consider, for example, case (iii); from the general solution we can develop the non-negative square root for $|x|$ sufficiently small by means of the binomial series as

$$(a^2 + bx + k_3x^4)^{1/2} = |a|\left(1 + \frac{1}{a^2}(bx + k_3x^4)\right)^{1/2}$$

$$= |a|\left(1 + \frac{1}{2a^2}(bx + k_3x^4) - \frac{1}{8a^4}(bx + k_3x^4)^2 + \frac{1}{16a^3}(bx + k_3x^4)^3 + \dots\right);$$

on expanding the powers and combining terms of the same degree we see that the coefficients of $x^0$, $x^1$, $x^2$, $x^3$ are uniquely determined in terms of $a$ and $b$, but the coefficient of $x^4$ (as also the coefficient of each higher power) depends on the arbitrary constant $k_3$. Stirling also points out that the solution of (i) (with $k_1 = 0$) does not take the form of an ordinary power series: in fact

$$\left(bx + \frac{x^4}{c^2}\right)^{1/2} = (bx)^{1/2}\left(1 + \frac{x^3}{bc^2}\right)^{1/2}$$

$$= b^{1/2}x^{1/2}\left(1 + \frac{x^3}{2bc^2} + \dots\right)$$

(assuming $b > 0$ with $x > 0$ and sufficiently small).

Stirling now considers the differential equation

$$r^2\left(\frac{dy}{dx}\right)^2 = r^2 - y^2,$$

for which he obtains two power series solutions which he identifies as a sine and a cosine; in modern notation they are

$$r\sin\frac{x}{r} \quad \text{and} \quad r\cos\frac{x}{r}.$$

Now the equation is equivalent to

$$\frac{1}{\sqrt{r^2 - y^2}} \frac{dy}{dx} = \pm\frac{1}{r},$$

from which it follows that

$$\sin^{-1}\frac{y}{r} = \pm\frac{x}{r} + c \quad \text{and hence that} \quad y = r\sin\left(\pm\frac{x}{r} + c\right).$$

Thus we have the general solution[44]

$$y = r\sin\left(\frac{x}{r} + c\right) = r\sin\frac{x}{r}\cos c + r\cos\frac{x}{r}\sin c, \qquad (**)$$

where $c$ is an arbitrary constant. Stirling's two solutions are of course contained in this form ($c = 0$, $c = \pi/2$ respectively); however, Stirling appears to go further and assert that there are no solutions other than the two he has found. In a letter dated 1 April 1733 (see [74, pp. 141–150]) Nikolaus Bernoulli told Stirling quite curtly that he had not examined the matter with sufficient accuracy and informed him of the more general solution

$$y = A + Bx + Cx^2 + Dx^3 + Ex^4 + \dots,$$

where $B^2 = \dfrac{r^2 - A^2}{r^2}$ and

$$C = -\frac{A}{1 \times 2r^2}, \quad D = -\frac{B}{2 \times 3r^2}, \quad E = -\frac{C}{3 \times 4r^2}, \quad F = -\frac{D}{4 \times 5r^2}, \quad \dots;$$

this corresponds to the power series for (**).

Finally, Stirling deals with the equation

$$(1 - x^2)\frac{d^2y}{dx^2} - x\frac{dy}{dx} + a^2y = 0,$$

which is now known as Chebyshev's equation. This has solutions of the form $\sum_{n=0}^{\infty} a_n x^n$ which are determined by the recurrence relation

$$a_{n+2}(n + 2)(n + 1) = (n^2 - a^2)a_n.$$

Setting $a_0 = A$, $a_1 = 0$, we obtain Stirling's second solution, while $a_0 = 0$, $a_1 = A$ produce his first solution. When $a$ is a positive integer $n$, one of these solutions will reduce to a polynomial, namely a multiple of the Chebyshev polynomial $T_n(x)$. Stirling remarks that the first two series relate to the multiplication or division of an arc – in this connection we have, in particular, the trigonometric identities[45]

[44]Note that the same set of solutions will be generated by the form $r\sin(-\frac{x}{r} + c)$.

[45]See [70, Chapter 3, Section 1], where some other writings of Stirling on this topic are discussed.

$$\cos 2n\theta = (-1)^n T_{2n}(\sin\theta)$$
$$= 1 - \frac{4n^2 \sin^2\theta}{2!} + \frac{4n^2(4n^2-4)\sin^4\theta}{4!} - \frac{4n^2(4n^2-4)(4n^2-16)\sin^6\theta}{6!} + \dots$$

and for $n$ odd

$$\sin n\theta = (-1)^{(n-1)/2} T_n(\sin\theta)$$
$$= n\left(\sin\theta - \frac{(n^2-1)\sin^3\theta}{3!} + \frac{(n^2-1)(n^2-9)\sin^5\theta}{5!} - \dots\right).$$

His other two solutions correspond to solutions near infinity. If we put $z = x^{-1}$, the equation transforms to

$$z^2(z^2-1)\frac{d^2y}{dz^2} + z(2z^2-1)\frac{dy}{dz} + a^2 y = 0,$$

for which the point $z = 0$ is a regular singular point. Solutions of the form $\sum_{n=0}^{\infty} a_n z^{n+\sigma}$ are determined by

$$\sigma = \pm a, \quad a_{2n+1} = 0 \quad (n = 0, 1, \dots),$$
$$a_{2n} = \frac{(2n+\sigma-2)(2n+\sigma-1)}{(2n+\sigma+a)(2n+\sigma-a)} a_{2n-2} \qquad (n = 1, 2, \dots).$$

The case $\sigma = a$ produces Stirling's fourth solution, while $\sigma = -a$ gives the third. The third solution is in fact

$$Ax^a F\left(-\frac{a}{2}, -\frac{a}{2} + \frac{1}{2}; -a+1; \frac{1}{x^2}\right)$$
$$= A\left(\frac{x}{2}\left(1 + \sqrt{1 - \frac{1}{x^2}}\right)\right)^a = 2^{-a} A\left(x + \sqrt{x^2-1}\right)^a \quad (x > 1),$$

while the fourth is

$$Ax^{-a} F\left(\frac{a}{2}, \frac{a}{2} + \frac{1}{2}; a+1; \frac{1}{x^2}\right)$$
$$= A\left(\frac{x}{2}\left(1 + \sqrt{1 - \frac{1}{x^2}}\right)\right)^{-a} = 2^a A\left(x + \sqrt{x^2-1}\right)^{-a} \quad (x > 1).$$

The connection with the area of a hyperbola may be seen as follows: for positive $\alpha$, $\beta$ the equations

$$x = \left(u + \sqrt{u^2-1}\right)^{1/\alpha}, \quad y = \left(u - \sqrt{u^2-1}\right)^{1/\beta} \quad (u \geq 1)$$

parameterize the general hyperbola $x^\alpha y^\beta = 1$ for $x \geq 1$, $0 < y \leq 1$ and areas will be determined from

$$\int y(u)\dot{x}(u)\,du = \frac{1}{\alpha}\int \left(u - \sqrt{u^2-1}\right)^{1/\beta} \frac{\left(u+\sqrt{u^2-1}\right)^{1/\alpha}}{\sqrt{u^2-1}}\,du$$

$$= \frac{1}{\alpha}\int \frac{\left(u+\sqrt{u^2-1}\right)^{\frac{1}{\alpha}-\frac{1}{\beta}}}{\sqrt{u^2-1}}\,du$$

$$= \frac{\beta}{\beta-\alpha}\left(u+\sqrt{u^2-1}\right)^{\frac{1}{\alpha}-\frac{1}{\beta}} + c.$$

Again Bernoulli had a few critical observations to make:

(i) he pointed out in effect that any linear combination of Stirling's first two or last two solutions is also a solution;

(ii) Stirling had erroneously put minus signs between the terms in the fourth solution (corrected here);

(iii) Stirling has the letter $A$ playing two distinct roles, which is significant except in the first solution – it is the *coefficient* in the first term, but in the second term it stands for the *whole* of the first term.

**Stirling's Introductory Remarks in Part II (p. 101).** Generally, Stirling deals with the situation where he has a sequence of function values $f(k)$ with $k = 0, 1, 2, \ldots$ or $k = 0, \pm 1, \pm 2, \ldots$ together with a rule by means of which $f(k+1)$ is determined from $f(k)$. The aim is to determine $f(z)$ for $z \in [0, \infty)$, or $\mathbb{R}$, or some appropriate subset of $\mathbb{R}$ which contains the initial set of integral $z$, under the assumption that the rule extends to the determination of $f(z+1)$ from $f(z)$. He also indicates that the initial values might be given more generally at a sequence of points $x_0 + kh$ ($h$ constant).

**Proposition 16 and Its Scholion (pp. 102–103).** The idea here is that, if we have a sequence $\{a_k\}$ defined by a relation[46]

$$a_{k+1} = F(a_k, k),$$

then an interpolating function $f(z)$ should satisfy

$$f(z+1) = F(f(z), z).$$

Consequently, if we know $f(z_0)$ for some $z_0$ (which is not one of the suffices in the sequence), then we can deduce $f(z_0+1)$, $f(z_0+2)$, ....

[46]In a handwritten note Stirling has added to the statement of the proposition: "provided the relation is only between two terms."

In *Example 1* we have $a_k = x^k$ $(k = 0, 1, \ldots)$, so that $a_{k+1} = xa_k$. We therefore want $f(z+1) = xf(z)$. The natural interpolating function is of course $f(z) = x^z$, which clearly satisfies this relation.

In *Example 2* the sequence is $a_k = k! = \Gamma(k+1)$ $(k = 0, 1, \ldots)$. Since $a_{k+1} = (k+1)a_k$ we require $f(z+1) = (z+1)f(z)$, which is certainly satisfied by $f(z) = \Gamma(z+1)$. In particular, if $a = f(\frac{1}{2})$ the relation $f(z+1) = (z+1)f(z)$ produces the sequence

$$a, \quad f(\tfrac{3}{2}) = \tfrac{3}{2}\,a, \quad f(\tfrac{5}{2}) = \tfrac{5}{2} \times \tfrac{3}{2}\,a, \quad \ldots,$$

while if $a = f(\frac{1}{3})$ the sequence is

$$a, \quad f(\tfrac{4}{3}) = \tfrac{4}{3}\,a, \quad f(\tfrac{7}{3}) = \tfrac{7}{3} \times \tfrac{4}{3}\,a, \quad \ldots .$$

In Example 2 of Proposition 21 Stirling calculates $f(\frac{1}{2}) = \Gamma(\frac{3}{2}) = \sqrt{\pi}/2$ by means of an interpolating series.

The sequence of *Example 3* is defined by

$$a_0 = 1, \quad a_{k+1} = \frac{2k+1}{2(k+1)}\,a_k \quad (k = 0, 1, \ldots).$$

Thus

$$f(z+1) = \frac{2z+1}{2(z+1)}\,f(z) = \frac{z+\frac{1}{2}}{z+1}\,f(z)$$

and from $a = f(\frac{1}{2})$ we therefore obtain the sequence

$$a, \quad f(\tfrac{3}{2}) = \tfrac{2}{3}\,a, \quad f(\tfrac{5}{2}) = \tfrac{4}{5} \times \tfrac{2}{3}\,a, \quad \ldots .$$

Series expressions for this $f(\frac{1}{2})$ are developed in Example 1 of Proposition 21. The sequence $\{a_k\}$ is given explicitly by

$$a_k = \frac{1 \times 3 \times \ldots \times (2k-1)}{2 \times 4 \times \ldots \times 2k} \qquad (k = 1, 2, \ldots)$$

$$= \frac{(2k)!}{2^{2k}(k!)^2} = \frac{\Gamma(2k+1)}{2^{2k}(\Gamma(k+1))^2} \qquad (k = 0, 1, \ldots).$$

Its natural interpolating function is therefore

$$f(z) = \frac{\Gamma(2z+1)}{2^{2z}(\Gamma(z+1))^2},$$

for which

$$f(\tfrac{1}{2}) = \frac{1}{2(\Gamma(\frac{3}{2}))^2} = \frac{2}{(\Gamma(\frac{1}{2}))^2} = \frac{2}{\pi}.$$

In the *scholion* Stirling refers to Newton's Proposition 7 of *De Quadratura Curvarum* in [47]. In it Newton compares the areas of curves whose ordinates are of the form

$$z^{\theta+\eta\sigma}\left(e + fz^{\eta} + gz^{2\eta} + \ldots\right)^{\lambda+\tau},$$

where $\theta$, $\eta$, $\lambda$, $e$, $f$, $g$, ... are constants and $\sigma$ and $\tau$ are integers; he points out in particular that all the areas will be determined if a certain number of them, depending on the number of nonzero coefficients $e$, $f$, $g$, ..., are known. This is the connection with Stirling's observation about the number of terms which have to be given to determine the sequence (series).

**Proposition 17 and Its Scholion (pp. 103–106).** Stirling states some more or less self-evident principles which he applies later to great effect.

(i) (*Examples 1 & 2.*) Suppose that

$$a_k = \frac{\prod_{i=1}^{m} n_k^{(i)}}{\prod_{j=1}^{n} d_k^{(j)}} \qquad (m, n \text{ fixed}).$$

If $\{n_k^{(i)}\}$ is interpolated by $f_i(z)$ $(i = 1, \ldots, m)$ and $\{d_k^{(j)}\}$ is interpolated by $g_j(z)$ $(j = 1, \ldots, n)$, then $\{a_k\}$ is interpolated by

$$f(z) = \frac{\prod_{i=1}^{m} f_i(z)}{\prod_{j=1}^{n} g_j(z)}.$$

In particular, we may interpolate separately the sequence of numerators and the sequence of denominators. (See Proposition 21, Example 3 and Proposition 22, Example 1.)

(ii) (*Scholion*, first part.) Given $a_k$ $(k = 0, \pm 1, \pm 2, \ldots)$ form the sequence $a_k a_{-k+1}$ $(k = 0, \pm 1, \pm 2, \ldots)$. Then the term midway between $a_0 a_1$ and $a_1 a_0$ in the new sequence $(k = 0, 1)$ will be the square of the term midway between $a_0$ and $a_1$ in the original sequence. (See Proposition 22, Example 1.)

(iii) (*Scholion*, last part.) Interpolate the logarithms of the terms of a given sequence and then deduce the required interpolated value of the given sequence. Stirling subsequently employs logarithms to base 10 (tabular logarithms) for this purpose and advocates their use when the sequence increases very rapidly. (See Proposition 21, Example 2.)

**Proposition 18 (pp. 106–110).** The two sequences are

$$r_k = \frac{\Gamma(r+k)\,\Gamma(p)}{\Gamma(p+k)\,\Gamma(r)}\,a, \qquad s_k = \frac{\Gamma(r+k)\,\Gamma(q)}{\Gamma(q+k)\,\Gamma(r)}\,a,$$

which are interpolated respectively by

$$f(z) = \frac{\Gamma(r+z)\,\Gamma(p)}{\Gamma(p+z)\,\Gamma(r)}\,a, \qquad g(z) = \frac{\Gamma(r+z)\,\Gamma(q)}{\Gamma(q+z)\,\Gamma(r)}\,a.$$

Now, assuming that $p$, $q$, $r$ are such that all the required quantities are defined, we obtain

$$f(q-r) = \frac{\Gamma(q)\,\Gamma(p)}{\Gamma(p+q-r)\,\Gamma(r)}\,a = g(p-r).$$

This is the content of *Proposition 18.* Stirling's proof of his assertion consists of verification in a few cases where $p-r$ and $q-r$ are positive integers. However, he does apply the result in situations where this is not the case.

In *Example 2* Stirling introduces an important sequence which he discusses in detail in Example 1 of Proposition 22 and Proposition 23, namely the sequence $(r_k)$ defined by

$$r_0 = 1, \quad r_{k+1} = \frac{2(k+1)}{2k+1} r_k = \frac{k+1}{k+\frac{1}{2}} r_k \qquad (k = 0, 1, 2, \ldots).$$

In fact,

$$r_k = \frac{k!}{\frac{1}{2} \times \frac{3}{2} \times \ldots \frac{2k-1}{2}} = \frac{\Gamma(k+1)\,\Gamma(\frac{1}{2})}{\Gamma(k+\frac{1}{2})} = \frac{\Gamma(k+1)\sqrt{\pi}}{\Gamma(k+\frac{1}{2})}$$

or, without introducing the Gamma function,

$$r_k = 2^{2k} / \binom{2k}{k}.$$

The proposition is applied for given $m$ to $(r_k)$ and the sequence $(s_k)$ defined by

$$s_0 = 1, \quad s_{k+1} = \frac{k+1}{k+m+1} s_k \qquad (k = 0, 1, 2, \ldots).$$

The "difference of the factors" in $(r_k)$ is $(k+\frac{1}{2}) - (k+1) = -\frac{1}{2}$ and the corresponding quantity for $(s_k)$ is $m$. Thus the term determined by $(r_k)$ which comes $m$ units after the initial term 1 is equal to the term determined by $(s_k)$ which is half a unit before its initial term 1. This observation is applied in Example 1 of Proposition 22. Stirling goes on to consider the sequence of reciprocals in the same way and again his conclusions are relevant to this example. *Example 1* and *Example 3* provide similar illustrations of the proposition.

**On the Differences of Quantities (pp. 110–111).** The material here is quite standard. Stirling points out that the $n$-th-order differences of a polynomial of degree $n$ are constant; he also warns that we cannot expect the sequence of differences to converge in general.

**On the Description of Curves Through Given Points (pp. 111–112).** Here we are concerned with interpolating polynomials or series. Newton had presented his interpolating formulae in Lemma V of Book III of the *Principia* [48] and in his *Methodus Differentialis* which appears on pp. 93–101 of [47]. As its title suggests, a large part of Stirling's 1719 paper [61] is devoted to discussing and illustrating Newton's results, which are given there with some minor variations.[47] Most of the relevant parts of the paper are reproduced in Propositions 19, 20, 31, 32 and 33, again with some variations.

Stirling begins with a quotation from Newton's *epistola posterior* to Oldenburg dated 24 October 1676 at Cambridge.[48] He then refers to Proposition 60 in Newton's *Arithmetica Universalis* [46]; in fact the requirement of this proposition is that the conic section should pass through four given points *and* be tangential to a given line through one of them. Concerning the reference to curves of the third order[49] we should note that Newton had described these curves in [47] (see [77, Vol. VII, pp. 609–610] about missing types) and, as with the *Methodus Differentialis*, Stirling wrote a commentary on the work [60]. The organic description of curves, which Stirling tells us is not required here, was another development of Newton's subsequently taken up by others, notably Colin MacLaurin [38].

**Proposition 19 and Its Scholion (pp. 112–119).** Here Stirling presents Newton's forward difference formula [27, p. 95]

$$f_s = f_0 + s\nabla f_0 + \frac{s(s-1)}{2!}\nabla^2 f_0 + \frac{s(s-1)(s-2)}{3!}\nabla^3 f_0 + \dots .$$

He gives two proofs, both of which involve identifying the values of the coefficients in the formal expression

$$A + Bz + C\frac{z(z-1)}{2!} + D\frac{z(z-1)(z-2)}{3!} + \dots$$

which will generate the required ordinates when $z = 0, 1, 2, \dots$. Stirling had already given this formula as the first case of the proposition in his 1719 paper [61]. Newton's version was given in the first case of Lemma V in Book III of the *Principia* (see [48]).

In the *scholion* Stirling indicates that Taylor's Theorem may be obtained as a limiting case of the interpolation formula. For this he obtains approximations to the ordinates $A_1$, $A_2$, $A_3$, ... by means of linear approximations and their derivatives:

$$\begin{aligned} A_1 &\approx A + \dot{A}n \\ A_2 &\approx A_1 + \dot{A}_1 n \approx A + \dot{A}n + (\dot{A} + \ddot{A}n)n = A + 2\dot{A}n + \ddot{A}n^2 \\ A_3 &\approx A_2 + \dot{A}_2 n \approx A + 2\dot{A}n + \ddot{A}n^2 + (\dot{A} + 2\ddot{A}n + \dddot{A}n^2)n \\ &= A + 3\dot{A}n + 3\ddot{A}n^2 + \dddot{A}n^3, \end{aligned}$$

[47] The other part of Stirling's paper has been discussed in detail in [71].

[48] This is Letter 188 in Volume II of [68] (pp. 110–161). The original manuscript refers to a curve of three dimensions through *eight* points rather than *seven* as quoted by Stirling. This and other textual variations are discussed in [68]. Stirling's source for the quotation was pp. 75–76 of the first edition or pp. 156–157 of the second edition of [12]. (See the note on Stirling's Preface.)

[49] This is a curve defined by a third-degree equation in two variables.

and so on.

The formula of Johann Bernoulli to which Stirling refers in connection with his expression for the area *BEDA* may be obtained from Taylor's Theorem as follows:

$$\begin{aligned}\int_a^b f(x)\,dx &= \int_a^b f(b) + \frac{(x-b)}{1!}f'(b) + \frac{(x-b)^2}{2!}f''(b) + \dots\,dx\\ &= (b-a)f(b) - \frac{(a-b)^2}{2!}f'(b) - \frac{(a-b)^3}{3!}f''(b) - \dots\\ &= (b-a)f(b) - \frac{(b-a)^2}{2!}f'(b) + \frac{(b-a)^3}{3!}f''(b) - \dots.\end{aligned}$$

It has been argued on the basis of this formula, which Bernoulli gave in [7], that he is the true discoverer of Taylor's Theorem, a view which is disputed by Gibson in [17]. Gibson also queries Stirling's reference in the *scholion* to Hermann as having given Taylor's Theorem in the Appendix to [26] – Gibson's opinion is that Hermann only produced Bernoulli's formula.

**Proposition 20 and Its Scholion (pp. 119–123).** The first case is what has become known as "Stirling's interpolation formula" and is usually expressed as [27, p. 100]

$$f_s = f_0 + s\mu\delta f_0 + \frac{s^2}{2!}\delta^2 f_0 + \frac{s(s^2-1^2)}{3!}\mu\delta^3 f_0 + \frac{s^2(s^2-1^2)}{4!}\delta^4 f_0 + \dots.$$

This is essentially how Newton gives it in Case I of Proposition III in his *Methodus Differentialis* [47]. In Stirling's version the terms involving $\mu\delta^{2n-1}f_0$ and $\delta^{2n}f_0$ are combined. In the second case we have "Bessel's interpolation formula" [27, p. 101]:

$$f_s = \mu f_{\frac{1}{2}} + (s-\tfrac{1}{2})\delta f_{\frac{1}{2}} + \frac{s(s-1)}{2!}\mu\delta^2 f_{\frac{1}{2}} + \frac{s(s-1)(s-\frac{1}{2})}{3!}\delta^3 f_{\frac{1}{2}} + \dots,$$

which is equivalent to the formula of Case II of Proposition III in Newton's *Methodus Differentialis*. Again, pairs of terms are combined in Stirling's version; we also have to note that $z = 2s - 1$. Thus

$$\mu f_{\frac{1}{2}} + (s-\tfrac{1}{2})\delta f_{\frac{1}{2}} = \frac{1}{2}({}_1A + A_1) + \frac{z}{2}({}_1A - A_1) = \frac{A+az}{2},$$

and

$$\begin{aligned}&\frac{s(s-1)}{2!}\mu\delta^2 f_{\frac{1}{2}} + \frac{s(s-1)(s-\frac{1}{2})}{3!}\delta^3 f_{\frac{1}{2}}\\ &= \frac{1}{2!}\left(\frac{z+1}{2}\right)\left(\frac{z-1}{2}\right)\frac{1}{2}({}_1B + B_1) + \frac{1}{3!}\left(\frac{z+1}{2}\right)\left(\frac{z-1}{2}\right)\frac{z}{2}b\\ &= \frac{1}{2!\,2^3}(z^2-1)B + \frac{1}{3!\,2^3}(z^2-1)zb = \frac{3B+bz}{2}\times\frac{z^2-1}{4\times 6},\ \text{etc.}\end{aligned}$$

Stirling had already given these formulae in the second and third cases of the proposition in his 1719 paper [61] but with certain notational differences; in particular, in the case of Bessel's formula the sums of differences are also divided by 2 and the variable $z$ is half that used in the second case of Proposition 20. Braunmühl [10] interprets Stirling's remarks in the *scholion* as an acknowledgement that he was only giving variants of Newton's results. The formulae were also investigated by Cotes (see *De Methodo Differentiali Newtoniana* in [13] and [21, Chapter 6]).

In the first formula the given ordinates are generated by putting $z = 0, \pm 1, \pm 2, \ldots$, while in the second they correspond to $z = \pm 1, \pm 2, \ldots$. Stirling indicates that the formulae are to be established by the method used in Proposition 19, by which he means presumably that one assumes a series of the required form and then identifies the coefficients by giving $z$ the appropriate values required to generate the given ordinates. However, as noted by Braunmühl, this is not altogether trivial.

**Proposition 21 (pp. 123–130).** This proposition consists of four examples to illustrate the application of Propositions 19 and 20 to interpolation problems.

*Example 1.* Here the given sequence is

$$t_n = \frac{(2n)!}{2^{2n}(n!)^2} = \frac{\Gamma(2n+1)}{2^{2n}(\Gamma(n+1))^2} \qquad (n = 0, 1, 2, \ldots),$$

for which the differences may be shown to be

$$\Delta_n^k = \frac{(-1)^k (2k)!(2n)!}{2^{2(k+n)} k!\, n!(k+n)!}\,.$$

The series which emerges from Proposition 19 on taking for $A$ the first term 1 ($n = 0$) may be expressed as

$$1 + \sum_{k=1}^{\infty} (-1)^k \frac{(2k)!}{2^{2k}(k!)^3} z(z-1)\ldots(z-k+1).$$

Stirling requires the term midway between the first two terms 1 and 1/2, for which $z = 1/2$. The series then becomes

$$1 - \sum_{k=1}^{\infty} \frac{k(2k)!(2k-2)!}{2^{4k-1}(k!)^4}$$

and its value should be

$$\frac{\Gamma(2)}{2(\Gamma(\frac{3}{2}))^2} = \frac{2}{\pi}\,.$$

Now for this series the ratio of the $(k+1)$-th term to the $k$-th term is[50]

$$\frac{4k^2-1}{4(k+1)^2}=\frac{k^2-\frac{1}{4}}{k^2+2k+1},$$

which fits the general relation required in the scholion to Proposition 11, viz.

$$T'=\frac{z^2+m}{z^2+nz+r}T$$

(note that $z$ is playing a different role here and corresponds to our $k$). Thus, as Stirling asserts, the procedure described in the scholion may be applied to sum the series. He provides no details but let us illustrate the procedure by using two terms of the transformed series from $k=3\,(=z)$: we have for the scholion $m=-\frac{1}{4}$, $n=2$, $r=1$ and so

$$\begin{aligned} a=2-\frac{1+\frac{1}{4}}{2}&=\frac{11}{8},\\ b=2+2-\frac{1+\frac{1}{4}+2+1}{2+2}&=\frac{47}{16},\\ \frac{z+a-1}{n-1}=\frac{3+\frac{11}{8}-1}{1}&=\frac{27}{8},\\ \frac{z+b-1}{n+1}=\frac{3+\frac{47}{16}-1}{3}&=\frac{79}{48},\\ \frac{ma+(n-a)r}{z^2+nz+r}=\frac{-\frac{11}{32}+\frac{5}{8}}{9+6+1}&=\frac{9}{512}; \end{aligned}$$

hence

$$1-\frac{1}{4}-\frac{3}{64}-\frac{5}{256}-\ldots\approx 1-\frac{1}{4}-\frac{3}{64}-\frac{5}{256}\left(\frac{27}{8}+\frac{79}{48}\times\frac{9}{512}\right)=.63664\ldots,$$

whereas

$$1-\frac{1}{4}-\frac{3}{64}-\frac{5}{256}=.684 \text{ (to 3DP)}$$

In fact $2/\pi=.636619772\ldots.$

The second series, where $\frac{1}{2}$ is used as the initial ordinate, may be dealt with similarly.

---

[50]This ratio can be expressed as

$$1-\frac{2}{k}+O\left(\frac{1}{k^2}\right),$$

from which we see by Gauss's test that the series converges.

*Example 2.* This calculation has attained a certain celebrity – in modern terminology Stirling indentifies $\Gamma(\frac{1}{2})$ as $\sqrt{\pi}$ on the basis of numerical evidence. He first applies Bessel's interpolation formula (second case of Proposition 20) to the twelve quantities

$$\log_{10}(5!),\ \log_{10}(6!),\ \ldots\ ,\ \log_{10}(16!),$$

that is,

$$\log_{10}\{(\tfrac{1}{2}(z+21))!\} = \log_{10}\Gamma(\tfrac{1}{2}(z+23)) \quad (z = \pm1, \pm3, \pm5, \pm7, \pm9, \pm11).$$

With $z = 0$ he obtains an approximation to the term midway between $\log_{10}(10!)$ and $\log_{10}(11!)$;[51] thus in modern terminology he has an approximation to $\log_{10}\Gamma(11.5)$ and then from tables an approximation to $\Gamma(11.5)$. Next Stirling invokes Proposition 16 to observe that, since the factorials are built up according to the rule $n \times (n-1)! = n!$, the same rule must carry over to the function produced by interpolating the factorials – this is equivalent to $z\Gamma(z) = \Gamma(z+1)$, which is applied in reverse to deduce values for $\Gamma(10.5)$, $\Gamma(9.5)$, ... , $\Gamma(1.5)$. Having determined $\Gamma(1.5)$ numerically, Stirling asserts that its square is the area of a circle whose diameter is 1; in other words, he claims that $(\Gamma(1.5))^2 = \pi/4$. He also appends to his list of values the "term which stands before the first principal term by half the common interval". This is $\Gamma(\frac{1}{2})$, which should come out directly by the assumed rule:

$$\Gamma(\tfrac{1}{2}) = \frac{\Gamma(1.5)}{.5} = \sqrt{\pi}\,.$$

However, it seems that he wishes to justify this separately by considering a corresponding interpolation of the squares of the factorials. Presumably $\sqrt{\pi}$ would have been a familiar number to Stirling and he would simply have recognized its occurrence in his list of values – he certainly offers no proof here for the introduction of $\pi$.

Stirling's stated aim in this example is to find the term which is midway between the first two terms $1 = 0! = \Gamma(1)$ and $1 = 1! = \Gamma(2)$; this he has found to be .8862269251, which agrees with $\sqrt{\pi}/2$ except in the last place, which should be 5 (correctly rounded). The accuracy of the final result is remarkable, but it appears to be difficult to justify this by elementary error analysis. In the notation of [27, p. 102], Stirling's initial interpolation to find $\log_{10}\Gamma(11.5)$ is equivalent to finding $f_{\frac{1}{2}}$ for $f(t) = \dfrac{\ln\Gamma(t+11)}{\ln 10}$ using Bessel's interpolation formula with differences up to order 10 (see also [55, p. 32]). In this case the error can be represented in the form[52]

$$\frac{(1 \times 3 \times \ldots \times 11)^2}{2^{12} \times 12!} f^{12}(\xi)$$

[51] Note that for this purpose Stirling could have used equivalently the formula which he gives in Proposition 33.

[52] The sign is wrong in the version in [27]: $(-1)^m$ should be $(-1)^{m+1}$.

for some $\xi$ between $-5$ and 6. Now, in terms of the psi (digamma) function,

$$\frac{d^{12}}{dt^{12}}(\ln\Gamma(t+11)) = \psi^{(11)}(t+11) = \int_0^\infty \frac{u^{11}e^{-(t+11)u}}{1-e^{-u}}\,du$$

$$= \sum_{n=0}^{\infty}\int_0^\infty u^{11}e^{-(n+11)u}e^{-tu}\,du = 11!\sum_{n=0}^{\infty}\frac{1}{(t+n+11)^{12}},$$

so that the error is

$$\frac{(1\times 3\times\ldots\times 11)^2}{2^{12}\times 12\ln 10}\sum_{n=0}^{\infty}\frac{1}{(\xi+n+11)^{12}}$$

for some $\xi$ between $-5$ and 6. However, this sum takes on a wide range of values over the interval $[-5,6]$: the maximum value is attained at $\xi=-5$, where we have

$$2.5\times 10^{-10} < \int_6^\infty \frac{1}{x^{12}}\,dx < \sum_{n=0}^{\infty}\frac{1}{(n+6)^{12}} < \int_{5.5}^\infty \frac{1}{x^{12}}\,dx < 6.53\times 10^{-10};$$

for the minimum $\xi = 6$ and

$$2.65\times 10^{-15} < \int_{17}^\infty \frac{1}{x^{12}}\,dx < \sum_{n=0}^{\infty}\frac{1}{(n+17)^{12}} < \int_{16.5}^\infty \frac{1}{x^{12}}\,dx < 3.684\times 10^{-15}.$$

Thus all we can deduce from the error term is that the calculated value of $\log_{10}\Gamma(11.5)$ is an underestimate by an amount which does not exceed

$$\frac{(1\times 3\times\ldots\times 11)^2}{2^{12}\times 12\ln 10}\times 6.53\times 10^{-10} < 6.235\times 10^{-7}$$

and is at least

$$\frac{(1\times 3\times\ldots\times 11)^2}{2^{12}\times 12\ln 10}\times 2.65\times 10^{-15} > 2.53\times 10^{-12}.$$

Stirling's ordinates are all correctly rounded to 10 decimal places. Repeating his calculation in double precision arithmetic, I find agreement except in the last decimal place, where Stirling's 9 should be 8. The value of $\log_{10}\Gamma(11.5)$ rounded to 11 decimal places is 7.07552590614, which means that Stirling has an underestimate by about $4.5\times 10^{-10}$.

*Example 3.* Stirling interpolates the sequence

$$A_n = (e+fx^\eta)^{-n}\int_0^x t^{\theta-1}(e+ft^\eta)^n\,dt \qquad (n=0,1,2,\ldots)$$

with a view to obtaining a series for $\int_0^x t^{\theta-1}(e+ft^\eta)^\lambda\,dt$ ($\lambda$ now replaces $z$). Although he does not say so explicitly, he is presumably using Proposition 17 to deduce the desired series by multiplying by $(e+fx^\eta)^\lambda$ the series obtained by applying Proposition 19 to the $A_n$. We can show inductively that the successive differences for the $A_n$ are

$$\Delta_n^k = \frac{f^k}{(e+fx^\eta)^{n+k}} \int_0^x t^{\theta-1}(t^\eta - x^\eta)^k (e+ft^\eta)^n \, dt.$$

For Proposition 19 we require

$$\Delta_0^k = \frac{f^k}{(e+fx^\eta)^k} \int_0^x t^{\theta-1}(t^\eta - x^\eta)^k \, dt;$$

on putting $t = xu^{1/\eta}$ this becomes

$$\begin{aligned}\Delta_0^k &= \frac{(-1)^k f^k x^{k\eta+\theta}}{\eta(e+fx^\eta)^k} \int_0^1 u^{\frac{\theta}{\eta}-1}(1-u)^k \, du = \frac{(-1)^k f^k x^{k\eta+\theta}}{\eta(e+fx^\eta)^k} B\Big(\frac{\theta}{\eta}, k+1\Big) \\ &= \frac{(-1)^k f^k x^{k\eta+\theta} k!}{\eta(e+fx^\eta)^k(\frac{\theta}{\eta}+k)(\frac{\theta}{\eta}+k-1)\dots\frac{\theta}{\eta}} \\ &= \frac{(-1)^k k!\, \eta^k f^k x^{k\eta+\theta}}{\theta(\theta+\eta)(\theta+2\eta)\dots(\theta+k\eta)(e+fx^\eta)^k}.\end{aligned}$$

Consequently the interpolating series is

$$\frac{x^\theta}{\theta} + \sum_{k=1}^{\infty} \frac{(-1)^k \eta^k f^k x^{k\eta+\theta}}{\theta(\theta+\eta)(\theta+2\eta)\dots(\theta+k\eta)(e+fx^\eta)^k} \lambda(\lambda-1)\dots(\lambda-k+1),$$

or equivalently,

$$\frac{x^\theta}{\theta}\left(1 + \sum_{k=1}^{\infty} \frac{(-\lambda)(-\lambda+1)\dots(-\lambda+k-1)}{(\theta+\eta)(\theta+2\eta)\dots(\theta+k\eta)} \left(\frac{\eta f x^\eta}{e+fx^\eta}\right)^k\right).$$

For this series the ratio of the $(k+1)$-th term to the $k$-th term is

$$\begin{aligned}\frac{k-\lambda}{\theta+(k+1)\eta} \times \frac{\eta f x^\eta}{e+fx^\eta} &= \frac{k-\lambda}{k+\frac{\theta+\eta}{\eta}} \times \frac{fx^\eta}{e+fx^\eta} \\ &= \frac{z-\left(\frac{\theta+\eta}{\eta}+\lambda\right)}{z} \times \frac{fx^\eta}{e+fx^\eta},\end{aligned}$$

where $z = k + \frac{\theta+\eta}{\eta}$. Comparing this with the relation required in Proposition 7, viz.

$$\frac{T'}{T} = \frac{z-n}{(1-m)z},$$

we see that Proposition 7 may be applied with

$$n = \frac{\theta+\eta}{\eta} + \lambda \quad \text{and} \quad m = 1 - \frac{e+fx^\eta}{fx^\eta} = -\frac{e}{fx^\eta}.$$

Transforming the series from its first term $x^\theta/\theta$ ($k=0$, $z=(\theta+\eta)/\eta$) we then obtain

$$\left(\frac{e+fx^\eta}{e}\right)\frac{x^\theta}{\theta} - \frac{(\theta+(\lambda+1)\eta)(e+fx^\eta)fx^{\eta+\theta}}{(\theta+\eta)e^2\theta} + \frac{(\theta+(\lambda+2)\eta)(\theta+(\lambda+1)\eta)(e+fx^\eta)f^2x^{2\eta+\theta}}{(\theta+2\eta)(\theta+\eta)e^3\theta} - \cdots.$$

When this is multiplied by $(e+fx^\eta)^\lambda$ we obtain, as Stirling notes, a series of Newton for $\int_0^x t^{\theta-1}(e+ft^\eta)^\lambda\,dt$ (cf. p. 33 where Stirling quotes the series but with $\lambda$ replaced by $\lambda-1$).

*Example 4.* Stirling applies the interpolation formulae of Proposition 20 to the binomial coefficients $\binom{n}{k}$ ($k=0,1,\ldots,n$). Where $n$ is an even positive integer he asserts that from the first case ("Stirling's interpolation formula") with middle term $\binom{n}{n/2}$ we obtain

$$\binom{n}{\frac{n}{2}\pm k} = \binom{n}{\frac{n}{2}} \times \left(1 - \frac{r^2}{2(n+2)} + \frac{r^2(r^2-4)}{2\times 4(n+2)(n+4)} - \frac{r^2(r^2-4)(r^2-16)}{2\times 4\times 6(n+2)(n+4)(n+6)} + \cdots\right),$$

where $r=2k$ ($k=0,1,\ldots,n/2$). The series will of course terminate. Calculation of the table of differences is obviously tedious, but we can easily verify the result by means of Gauss's formula for $F(a,b;c;1)$. For *any* $r$

$$1 - \frac{r^2}{2(n+2)} + \frac{r^2(r^2-4)}{2\times 4(n+2)(n+4)} - \frac{r^2(r^2-4)(r^2-16)}{2\times 4\times 6(n+2)(n+4)(n+6)} + \cdots$$

$$= 1 + \frac{\frac{r}{2}\left(-\frac{r}{2}\right)}{1!\left(\frac{n}{2}+1\right)} + \frac{\frac{r}{2}\left(\frac{r}{2}+1\right)\left(-\frac{r}{2}\right)\left(-\frac{r}{2}+1\right)}{2!\left(\frac{n}{2}+1\right)\left(\frac{n}{2}+2\right)} + \frac{\frac{r}{2}\left(\frac{r}{2}+1\right)\left(\frac{r}{2}+2\right)\left(-\frac{r}{2}\right)\left(-\frac{r}{2}+1\right)\left(-\frac{r}{2}+2\right)}{3!\left(\frac{n}{2}+1\right)\left(\frac{n}{2}+2\right)\left(\frac{n}{2}+3\right)} + \cdots$$

$$= F\left(\tfrac{r}{2}, -\tfrac{r}{2}; \tfrac{n}{2}+1; 1\right)$$

$$= \frac{\Gamma(\frac{n}{2}+1)\,\Gamma(\frac{n}{2}+1)}{\Gamma(\frac{n}{2}-\frac{r}{2}+1)\,\Gamma(\frac{n}{2}+\frac{r}{2}+1)}.$$

When $r$ is an even integer such that $-n \le r \le n$ this is

$$\frac{(\frac{n}{2})!\,(\frac{n}{2})!}{(\frac{n}{2}-\frac{r}{2})!\,(\frac{n}{2}+\frac{r}{2})!},$$

so that multiplication by $\binom{n}{n/2}$ produces

$$\frac{n!}{(\frac{n}{2}-\frac{r}{2})!\,(\frac{n}{2}+\frac{r}{2})!}=\binom{n}{\frac{n}{2}\pm\frac{r}{2}}.$$

When $n$ is odd we can apply Bessel's interpolation formula with "middle terms" $\binom{n}{\frac{n}{2}\pm\frac{1}{2}}$ to obtain Stirling's stated expression

$$\binom{n}{\frac{n}{2}\pm(k+\frac{1}{2})}=\binom{n}{\frac{n}{2}\pm\frac{1}{2}}\left(1-\frac{r^2-1}{2(n+3)}+\frac{(r^2-1)(r^2-9)}{2\times 4(n+3)(n+5)}\right.$$
$$\left.-\frac{(r^2-1)(r^2-9)(r^2-25)}{2\times 4\times 6(n+3)(n+5)(n+7)}+\ldots\right),$$

where $r=2k+1$ $(k=0,1,\ldots,\frac{n}{2}-\frac{1}{2})$. We see as above that for *any* $r$

$$1-\frac{r^2-1}{2(n+3)}+\frac{(r^2-1)(r^2-9)}{2\times 4(n+3)(n+5)}-\frac{(r^2-1)(r^2-9)(r^2-25)}{2\times 4\times 6(n+3)(n+5)(n+7)}+\ldots$$
$$=F\left(\tfrac{r}{2}+\tfrac{1}{2},-\tfrac{r}{2}+\tfrac{1}{2};\tfrac{n}{2}+\tfrac{3}{2};1\right)$$
$$=\frac{\Gamma(\frac{n}{2}+\frac{3}{2})\,\Gamma(\frac{n}{2}+\frac{1}{2})}{\Gamma(\frac{n}{2}-\frac{r}{2}+1)\,\Gamma(\frac{n}{2}+\frac{r}{2}+1)}.$$

Then if $r$ is an odd integer in the range $-n\le r\le n$ this is

$$\frac{(\frac{n}{2}+\frac{1}{2})!\,(\frac{n}{2}-\frac{1}{2})!}{(\frac{n}{2}-\frac{r}{2})!\,(\frac{n}{2}+\frac{r}{2})!},$$

which on multiplication by $\binom{n}{\frac{n}{2}\pm\frac{1}{2}}$ produces the quantity

$$\frac{n!}{(\frac{n}{2}-\frac{r}{2})!\,(\frac{n}{2}+\frac{r}{2})!}=\binom{n}{\frac{n}{2}\pm\frac{r}{2}}.$$

**Proposition 22 (pp. 131–133).** Proposition 22 consists of two examples whose purpose is to illustrate a particular interpolation technique. In *Example 1* Stirling is concerned with the sequence $(r_k)$ defined by

$$r_0=1,\quad r_{k+1}=\frac{2(k+1)}{2k+1}r_k=\frac{k+1}{k+\frac{1}{2}}r_k\qquad(k=0,1,2,\ldots).$$

We saw in the note on Proposition 18 that

$$r_k = \frac{\Gamma(k+1)\sqrt{\pi}}{\Gamma(k+\frac{1}{2})} = 2^{2k}/\binom{2k}{k}.$$

Stirling shows here by interpolation methods that

$$\begin{aligned}
&\frac{\Gamma(m+1)\sqrt{\pi}}{\Gamma(m+\frac{1}{2})} \\
&= \sqrt{\left\{\pi m\left(1+\frac{\left(\frac{1}{2}\right)^2}{m+1}+\frac{\left(\frac{1}{2}\cdot\frac{3}{2}\right)^2}{2!(m+1)(m+2)}\right.\right.} \\
&\qquad\qquad \left.\left. + \frac{\left(\frac{1}{2}\cdot\frac{3}{2}\cdot\frac{5}{2}\right)^2}{3!(m+1)(m+2)(m+3)}+\dots\right)\right\} \\
&= \sqrt{\left\{\pi m F\left(\tfrac{1}{2},\tfrac{1}{2};m+1;1\right)\right\}}.
\end{aligned}$$

We can see that this is correct by using Gauss's formula,

$$F(a,b;c;1) = \frac{\Gamma(c)\Gamma(c-a-b)}{\Gamma(c-a)\Gamma(c-b)} \qquad (\mathrm{Re}(c-a-b)>0).$$

In the case in which $m$ is a positive integer, the sum is $2^{2m}/\binom{2m}{m}$ and we have one of Stirling's solutions to De Moivre's middle-ratio problem,[53] which is discussed in greater detail in Proposition 23. It is difficult to justify the arguments presented by Stirling in this example, but fortunately the proofs given in Proposition 23 are much more satisfactory and can easily be made precise. I will therefore simply describe what Stirling is doing in this example without attempting to justify the steps.

Stirling applies a result from Example 2 of Proposition 18 (see the note on Proposition 18): if the sequence $(s_k)$ is defined for given $m$ by

$$s_0 = 1, \quad s_{k+1} = \frac{k+1}{k+m+1}s_k \qquad (k=0,1,2,\dots),$$

then the term determined by $(r_k)$ which comes $m$ units after the initial term 1 is equal to the term determined by $(s_k)$ which is half a unit before its initial term 1. Now according to Proposition 17 we may interpolate the numerators and denominators separately in $(s_k)$. But Stirling has already found in Example 2 of Proposition 21 that the term which is half a unit before the initial term 1 in the sequence of numerators $1, 1, 1\times 2, 1\times 2\times 3, \dots$ is $\sqrt{\pi}$; it therefore remains to find the corresponding term for the sequence $(d_k)$ defined by

$$d_0 = 1, \quad d_{k+1} = \frac{1}{k+m+1}d_k \qquad (k=0,1,2\dots).$$

[53]See the Preface and its note and the Appendix.

Putting $k = -1, -2, -3, \ldots$ in the recurrence relation for $(d_k)$ yields

$$d_{-1} = m, \quad d_{-2} = m(m-1), \quad d_{-3} = m(m-1)(m-2), \ \ldots .$$

By this means Stirling produces a sequence which goes off to infinity on both sides – clearly the extended terms will eventually be zero if $m$ is a positive integer. He then forms the sequence $(d_k d_{-k-1})_{-\infty}^{\infty}$, some of whose terms are given in the following table:

| $k$ | $-3$ | $-2$ | $-1$ | $0$ | $1$ | $2$ |
|---|---|---|---|---|---|---|
| $d_k d_{-k-1}$ | $\frac{m(m-1)(m-2)}{(m+1)(m+2)}$ | $\frac{m(m-1)}{m+1}$ | $m$ | $m$ | $\frac{m(m-1)}{m+1}$ | $\frac{m(m-1)(m-2)}{(m+1)(m+2)}$ |

The term determined by this sequence for $k = -\frac{1}{2}$ should be equal to the square of the required quantity and that term is now obtained by means of the second case of Proposition 20, for which the two terms $m$ $(k = -1, 0)$ are the two middle ordinates and $z = 0$. To illustrate the calculation consider the table of differences below, in which the differences up to order 4 are given for the six terms listed above.

| | | | | |
|---|---|---|---|---|
| $\frac{m(m-1)(m-2)}{(m+1)(m+2)}$ | | | | |
| | $\frac{-4m(m-1)}{(m+1)(m+2)}$ | | | |
| $\frac{m(m-1)}{m+1}$ | | $\frac{-2m(m-4)}{(m+1)(m+2)}$ | | |
| | $\frac{-2m}{m+1}$ | | $\frac{12m}{(m+1)(m+2)}$ | |
| $m$ | | $\frac{-2m}{m+1}$ | | $\frac{12m}{(m+1)(m+2)}$ |
| | $0$ | | $0$ | |
| $m$ | | $\frac{-2m}{m+1}$ | | $\frac{12m}{(m+1)(m+2)}$ |
| | $\frac{2m}{m+1}$ | | $\frac{-12m}{(m+1)(m+2)}$ | |
| $\frac{m(m-1)}{m+1}$ | | $\frac{-2m(m-4)}{(m+1)(m+2)}$ | | |
| | $\frac{4m(m-1)}{(m+1)(m+2)}$ | | | |
| $\frac{m(m-1)(m-2)}{(m+1)(m+2)}$ | | | | |

From this table we see that for the second case of Proposition 20 we have

$$A = 2m, \quad B = \frac{-4m}{m+1}, \quad C = \frac{24m}{(m+1)(m+2)}$$

and the series is

$$\frac{2m}{2} + \frac{3}{2}\left(\frac{-4m}{m+1}\right)\left(\frac{-1}{4\times 6}\right)$$

$$+\frac{5}{2}\left(\frac{24m}{(m+1)(m+2)}\right)\left(\frac{-1}{4\times 6}\right)\left(\frac{-9}{8\times 10}\right)+\dots$$

$$= m + \frac{m}{4(m+1)} + \frac{9m}{32(m+1)(m+2)} + \dots,$$

as Stirling gives it.

In his numerical illustration Stirling finds $2^{200}/\binom{200}{100}$. The case $m = 99.5$ which he mentions would produce

$$\frac{\Gamma(101.5)\sqrt{\pi}}{\Gamma(100)} = \frac{\Gamma(101.5)\sqrt{\pi}}{99!}.$$

Finally in *Example 1* Stirling indicates that the term determined by the sequence of reciprocals at distance $m/2$ from the initial term 1 is given by

$$\sqrt{\left\{\frac{2}{\pi(m+1)}\left(1+\frac{1}{2(m+3)}+\frac{9}{2\times 4(m+3)(m+5)}+\dots\right)\right\}},$$

which is

$$\sqrt{\left\{\frac{2}{\pi(m+1)}F\left(\tfrac{1}{2},\tfrac{1}{2};\tfrac{m}{2}+\tfrac{3}{2};1\right)\right\}}.$$

Again we can verify the correctness of this assertion by using Gauss's formula quoted above. Moreover, when $m$ is an even positive integer the value is $\binom{m}{m/2}/2^m$; this gives another of Stirling's solutions to the middle-ratio problem which is also discussed in greater detail in Proposition 23.

In *Example 2* Stirling only indicates how to proceed. He gives an alternative treatment for both the first series of *Example 1* and the series of *Example 2* in Examples 1 and 2 of Proposition 26, where the solutions obtained are in fact asymptotic series.

**Proposition 23 (pp. 134–139).** In Proposition 23 Stirling gives four series from which he asserts the middle ratio may be determined; he also gives two expressions for approximating this quantity. The first series of his *First Solution* and the first series of his *Second Solution* are in fact the series given in Example 1 of Proposition 22, viz.

$$S_1 = nF\left(\tfrac{1}{2},\tfrac{1}{2};\tfrac{n}{2}+1;1\right) \quad\text{and}\quad S_3 = \frac{1}{n+1}F\left(\tfrac{1}{2},\tfrac{1}{2};\tfrac{n}{2}+\tfrac{3}{2};1\right).$$

When $n$ is an even positive integer their sums, according to Gauss's formula, are respectively $\frac{2}{\pi}\left(2^n/\binom{n}{n/2}\right)^2$ and its reciprocal. As Stirling points out, in an odd power of the binomial there are two middle coefficients, $\binom{n}{(n-1)/2}$ and $\binom{n}{(n+1)/2}$, and

$$\frac{2^n}{\binom{n}{(n-1)/2}} = \frac{2^{n+1}}{\binom{n+1}{(n+1)/2}};$$

consequently we have to replace $n$ by $n+1$ in $S_1$ and $S_3$ to get the required quantity when $n$ is an *odd* positive integer.

In the *Analysis of the First Solution* Stirling is concerned with $S_1$. For

$$y_n = \left(2^n / \binom{n}{n/2}\right)^2 \quad (n = 0, 2, 4, \ldots)$$

he obtains the recurrence relation

$$2y_{n+2} + (n+2)(y_n - y_{n+2}) - \frac{y_{n+2}}{n+2} = 0,$$

which he solves by assuming that $y_n$ can be expressed in the form

$$y_n = \alpha_0 n + \frac{\alpha_1 n}{n+2} + \frac{\alpha_2 n}{(n+2)(n+4)} + \frac{\alpha_3 n}{(n+2)(n+4)(n+6)} + \cdots.$$

Substitution of this expression into the recurrence relation leads to the following equations for the coefficients:

$$\alpha_k = \frac{(2k-1)^2}{2k}\alpha_{k-1} \quad (k = 1, 2, 3, \ldots).$$

These determine $y_n$ in terms of $\alpha_0$ as

$$y_n = \alpha_0 \left(n + \frac{n}{2(n+2)} + \frac{9n}{2 \times 4(n+2)(n+4)} + \frac{9 \times 25n}{2 \times 4 \times 6(n+2)(n+4)(n+6)} + \cdots\right). \quad (*)$$

The determination of $\alpha_0$ is achieved by an application of Wallis's product[54] which Stirling indicates after his *Analysis of the Second Solution.* Assuming the validity of $(*)$, we can see that $y_n/n \to \alpha_0$ as $n$ goes off to infinity through even values. But

$$\frac{y_n}{n} = \frac{2^{2n}}{n\binom{n}{n/2}^2} = \frac{2^2 \times 4^2 \times \ldots \times n^2}{3^2 \times 5^2 \times \ldots \times (n-1)^2 \times (n+1)} \times \frac{n+1}{n} \to \frac{\pi}{2}$$

[54]See the note on Proposition 27.

as $n \to \infty$; thus $\alpha_0 = \pi/2$ as required. The various steps in this development of $S_1$ can be justified by using properties of hypergeometric series. Stirling deals similarly with $S_3$ in his *Analysis of the Second Solution.*

Before turning to the other two series and the approximations, let me comment on Stirling's letter of 19 June 1729 to De Moivre, in which he communicated the two series just discussed as well as the approximations but gave no proofs (see Appendix). De Moivre was quite astonished by Stirling's solutions to the middle-ratio problem and in particular by the occurrence of $\pi$ in these. On pages 172–173 of [43] he writes:

> Indeed there is no one who having seen this solution of the above problem may refuse to acknowledge that it is marvellous in every aspect; but perhaps nothing in it will be seen as more extraordinary than by what means the quadrature of the circle could have been brought in to it; ...

He goes on to tell us how he puzzled over the occurrence of $\pi$ in Stirling's solutions and eventually, after coming upon a reference to Wallis's product, discovered for himself the relevance of this result for the middle-ratio problem. It is asserted in [55, p. 273] that Stirling in fact learned of this application from De Moivre. I am not aware of any historical record to substantiate this assertion, although it is quite conceivable that, when Stirling wrote to De Moivre, he had only found his solutions by the interpolation methods discussed in Example 1 of Proposition 22 and that the more satisfactory proof given in Proposition 23 was indeed inspired by De Moivre. Be that as it may, the results communicated by Stirling enabled De Moivre to improve and extend his own work on this and related topics (see the note on Proposition 28).

The other two series of Proposition 23 are stated without proof, although Stirling indicates that the same method of proof applies. The second series of the *First Solution*, whose sum is claimed to be $\frac{2}{\pi}\left(2^n/\binom{n}{n/2}\right)^2$ when $n$ is an even positive integer, is

$$\begin{aligned}
S_2 &= (n+1)\left(1 - \frac{1}{2(n-1)} + \frac{9}{2\times 4(n-1)(n-3)}\right.\\
&\qquad\left. - \frac{9\times 25}{2\times 4\times 6(n-1)(n-3)(n-5)} + \ldots\right)\\
&= (n+1)\left(1 + \frac{(\frac{1}{2})^2}{1!(\frac{1}{2}-\frac{n}{2})} + \frac{(\frac{1}{2})^2(\frac{3}{2})^2}{2!(\frac{1}{2}-\frac{n}{2})(\frac{1}{2}-\frac{n}{2}+1)}\right.\\
&\qquad\left. + \frac{(\frac{1}{2})^2(\frac{3}{2})^2(\frac{5}{2})^2}{3!(\frac{1}{2}-\frac{n}{2})(\frac{1}{2}-\frac{n}{2}+1)(\frac{1}{2}-\frac{n}{2}+2)} + \ldots\right)\\
&= (n+1)F\left(\tfrac{1}{2}, \tfrac{1}{2}; \tfrac{1}{2}-\tfrac{n}{2}; 1\right)
\end{aligned}$$

and the second series of the *Second Solution* for $\frac{\pi}{2}\left(\binom{n}{n/2}/2^n\right)^2$ is

$$S_4 = \frac{1}{n}\left(1 - \frac{1}{2(n-2)} + \frac{9}{2\times 4(n-2)(n-4)} - \frac{9\times 25}{2\times 4\times 6(n-2)(n-4)(n-6)} + \ldots\right)$$

$$= \frac{1}{n}\left(1 + \frac{(\frac{1}{2})^2}{1!(1-\frac{n}{2})} + \frac{(\frac{1}{2})^2(\frac{3}{2})^2}{2!(1-\frac{n}{2})(1-\frac{n}{2}+1)} + \frac{(\frac{1}{2})^2(\frac{3}{2})^2(\frac{5}{2})^2}{3!(1-\frac{n}{2})(1-\frac{n}{2}+1)(1-\frac{n}{2}+2)} + \ldots\right)$$

$$= \frac{1}{n}F\left(\tfrac{1}{2},\tfrac{1}{2};1-\tfrac{n}{2};1\right).$$

At first sight $S_3$ and $S_4$ appear to be quite wrong – it is easy to see that $S_3$ diverges, while only the first $n/2$ terms of $S_4$ are defined (there is a zero factor in the denominator of each subsequent term). Although Stirling did not include $S_2$ or $S_4$ in his letter to De Moivre, he did remark, "There are also other series for the solution of this problem which are just as simple as those presented so far, but a little less convergent when the index of the binomial is a small number." It seems reasonable to infer that Stirling is referring here to $S_2$ and $S_4$ and that he had carried out some calculations with them. In fact $S_2$ and $S_4$ have certain asymptotic properties which are illustrated in the following two tables:[55]

**Partial sums of $S_1$ and $S_2$ with $n = 100$.**

| # | $S_1$ | $S_2$ | $S_1 - S_2$ |
|---|---|---|---|
| 1 | 100 | 101 | $-1$ |
| 2 | 100.490196 | 100.489899 | $3\times 10^{-4}$ |
| 3 | 100.500801 | 100.501731 | $-9\times 10^{-4}$ |
| 4 | 100.501218 | 100.501212 | $6\times 10^{-6}$ |
| 25 | 100.501243711840822 | 100.501243711840829 | $-7.2\times 10^{-15}$ |
| 26 | 100.501243711840822 | 100.501243711840815 | $7\times 10^{-15}$ |
| 27 | 100.501243711840822 | 100.501243711840830 | $-7.5\times 10^{-15}$ |
| 50 | 100.501243711840822 | 100.42 | $8\times 10^{-2}$ |
| 51 | 100.501243711840822 | 108.46 | $-8$ |
| 52 | 100.501243711840822 | 912.39 | $-812$ |

[55] These tables have appeared in [70]. Except for the first partial sums all entries are rounded to the number of places shown.

**Partial sums of $S_3$ and $S_4$ with $n = 100$.**

| # | $S_3$ | $S_4$ | $S_3 - S_4$ |
|---|---|---|---|
| 1 | $0.\dot{0}09\dot{9}$ | 0.01 | $-9.9 \times 10^{-5}$ |
| 2 | 0.0099490532 | 0.0099489796 | $7 \times 10^{-8}$ |
| 3 | 0.0099500831 | 0.0099501754 | $-9 \times 10^{-8}$ |
| 4 | 0.0099501232 | 0.0099501224 | $8 \times 10^{-10}$ |
| 25 | 0.009950125621004453 | 0.009950125621004454 | $-9.97 \times 10^{-19}$ |
| 26 | 0.009950125621004453 | 0.009950125621004452 | $9.97 \times 10^{-19}$ |
| 27 | 0.009950125621004453 | 0.009950125621004454 | $-1.08 \times 10^{-18}$ |
| 48 | 0.009950125621004453 | 0.00995007 | $5 \times 10^{-8}$ |
| 49 | 0.009950125621004453 | 0.009951 | $-1 \times 10^{-6}$ |
| 50 | 0.009950125621004453 | 0.009887 | $6 \times 10^{-5}$ |

I have discussed the relationships between $S_3$ and $S_4$ in detail in [70, pp. 44–49] and a corresponding discussion for $S_1$ and $S_2$ may be based on the same techniques. Here I will summarise the discussion for $S_1$ and $S_2$ and state the corresponding results for $S_3$ and $S_4$.

Let $n$ be an even positive integer and let the terms of the corresponding series $S_2$ be $v_0, v_1, v_2, \ldots$. Then for $k = 0, 1, 2, \ldots$

$$v_k = \frac{(n+1)\Gamma\left(\frac{1}{2} - \frac{n}{2}\right)\left(\Gamma\left(k + \frac{1}{2}\right)\right)^2}{\pi\Gamma\left(k - \frac{n}{2} + \frac{1}{2}\right)k!}$$

and

$$\frac{v_{k+1}}{v_k} = -\frac{(2k+1)^2}{2(k+1)(n-2k-1)}. \tag{**}$$

The ratio on the right-hand side of (**) is negative as long as $n - 2k - 1 > 0$, i.e., $k = 0, 1, \ldots, \frac{n}{2} - 1$, so that $v_0, v_1, \ldots, v_{n/2}$ are alternately positive and negative, while the sign of $v_k$ is fixed for $k \geq n/2$ (positive if $4|n$, negative otherwise). We can also deduce from (**) that

$$\left|\frac{v_{k+1}}{v_k}\right| < 1 \quad \text{if and only if} \quad k < \left[\frac{n+2}{4}\right],$$

from which it follows that $v_{[(n+2)/4]}$ has the least modulus of all the $v_k$.

Now let $\tau_m$ denotes the sum of the first $m$ terms of $S_2$. If we apply Whipple's result (see pp. 5–6) that the sum of the first $m$ terms of $F(a, b; c; 1)$ is

$$\frac{\Gamma(1+a-c)\Gamma(1+b-c)}{\Gamma(1-c)\Gamma(1-c+a+b)}\left\{1 - \frac{(a)_m(b)_m}{m!(c-1)_m}\,{}_3F_2\begin{bmatrix}1-a,\ 1-b,\ m\\ 2-c,\ m+1\end{bmatrix}\right\}$$

provided the expressions are all defined and $a + b - c > -1$, we see that

$$\tau_m = \frac{(n+1)\left(\Gamma\left(\frac{n}{2}+1\right)\right)^2}{\Gamma\left(\frac{n}{2}+\frac{1}{2}\right)\Gamma\left(\frac{n}{2}+\frac{3}{2}\right)}\left\{1-\frac{\left(\left(\frac{1}{2}\right)_m\right)^2}{m!\left(-\frac{1}{2}-\frac{n}{2}\right)_m}\,{}_3F_2\left[\begin{matrix}\frac{1}{2},\,\frac{1}{2},\,m\\ \frac{n}{2}+\frac{3}{2},\,m+1\end{matrix}\right]\right\}$$

$$= \frac{2}{\pi}\left(\frac{2^n}{\binom{n}{n/2}}\right)^2\left\{1-\frac{\left(\left(\frac{1}{2}\right)_m\right)^2}{m!\left(-\frac{1}{2}-\frac{n}{2}\right)_m}\,{}_3F_2\left[\begin{matrix}\frac{1}{2},\,\frac{1}{2},\,m\\ \frac{n}{2}+\frac{3}{2},\,m+1\end{matrix}\right]\right\}.$$

Now for $m \le \dfrac{n}{2}+1$,

$${}_3F_2\left[\begin{matrix}\frac{1}{2},\,\frac{1}{2},\,m\\ \frac{n}{2}+\frac{3}{2},\,m+1\end{matrix}\right] > 0 \quad \text{while} \quad \left(-\tfrac{1}{2}-\tfrac{n}{2}\right)_m \begin{cases} <0 & \text{if } m \text{ is odd,}\\ >0 & \text{if } m \text{ is even.}\end{cases}$$

Thus the $\tau_m$ oscillate about $\dfrac{2}{\pi}\left(2^n/\binom{n}{n/2}\right)^2$ for $m = 1, 2, \ldots, \frac{n}{2}+1$. It follows in particular that for $m = 1, 2, \ldots, [(n+2)/4]$ the $\tau_m$ oscillate about this quantity with *decreasing* deviation, so that the best approximation to it that we can obtain from $S_2$ is that it lies between $\tau_m$ and $\tau_{m+1}$ for $m = [(n+2)/4]$; this provides an interval of length[56]

$$\left|v_{[(n+2)/4]}\right| < \frac{n+1}{2^{n/2-3/2}\sqrt{\pi n}}.$$

For example, in the case $n = 100$ the best bounds from the corresponding $S_2$ are given by $\tau_{25}$ and $\tau_{26}$, which provide an interval of length less than $1.432 \times 10^{-14}$ (cf. first table). We can also show, however, that the partial sums of $S_1$ always give better approximations than the corresponding partial

[56]If $4|n$ we have by direct calculation

$$v_{[(n+2)/4]} = v_{n/4} = \frac{(-1)^{n/4}(n+1)\binom{n/2}{n/4}^2}{2^{n/2}\binom{n}{n/2}}.$$

The required inequality follows on applying the inequalities (see [65, p. 256, Qu. 13])

$$\binom{2h}{h} < \frac{2^{2h}}{\sqrt{\pi h}}\exp\left(-\frac{1}{12h+1}\right)$$

with $h = n/4$ in the numerator and

$$\frac{2^{2h}}{\sqrt{\pi h}}\exp\left(-\frac{1}{6h}\right) < \binom{2h}{h}$$

with $h = n/2$ in the denominator. If $n$ has remainder 2 on division by 4, then $v_{[(n+2)/4]} = v_{(n+2)/4}$; a similar calculation leads to the required inequality in this case.

sums of $S_2$ except where only the first terms are taken. It is convenient to note at this point the following simple inequalities which were used in error analysis in the note on Proposition 11:

$$\sqrt{\frac{2}{\pi(n+1)}} < 2^{-n}\binom{n}{n/2} < \sqrt{\frac{2}{\pi n}}.$$

The upper bound comes immediately from the first partial sum of $S_1$, while the lower bound comes from the first partial sum of $S_2$ on noting the oscillatory nature of the $\tau_m$.

Now let the first $n/2$ terms of $S_4$ (i.e., its finite terms) be $u_0, u_1, u_2, \ldots, u_{\frac{n}{2}-1}$ and let $t_m$ be the sum of the first $m$ of these. Again the terms alternate in sign and the $t_m$ oscillate about $\frac{\pi}{2}\left(\binom{n}{n/2}/2^n\right)^2$; also $\left|\frac{u_{k+1}}{u_k}\right| < 1$ if and only if $k < [n/4]$. Thus the best bounds we can obtain for the desired quantity from $S_4$ are $t_m$ and $t_{m+1}$ where $m = [n/4]$; this provides an interval of length $\left|u_{[n/4]}\right|$, which again tends to zero rapidly as the order $n$ of the binomial coefficient tends to infinity. However, each $t_m$ provides a poorer approximation than the corresponding partial sum of $S_3$.

Finally we consider the approximations stated by Stirling. The first of these is that

$$2^n/\binom{n}{n/2} \approx \sqrt{\frac{\pi}{2}\left(n+\frac{1}{2}\right)}.$$

We can easily see this from $S_1$: for large $n$

$$\begin{aligned}
2^n/\binom{n}{n/2} &= \sqrt{\frac{\pi}{2}\left(n+\frac{n}{2(n+2)}+\frac{9n}{8(n+2)(n+4)}+\ldots\right)} \\
&> \sqrt{\frac{\pi}{2}\left(n+\frac{n}{2(n+2)}+\frac{9n}{8(n+2)(n+4)}\right)} \\
&= \sqrt{\frac{\pi}{2}\left(n+\frac{1}{2}-\frac{1}{n+2}+\frac{9n}{8(n+2)(n+4)}\right)} \\
&= \sqrt{\frac{\pi}{2}\left(n+\frac{1}{2}+\frac{n-32}{8(n+2)(n+4)}\right)} \qquad (***) \\
&> \sqrt{\frac{\pi}{2}\left(n+\frac{1}{2}\right)}.
\end{aligned}$$

The error term given by Stirling relates to the *reciprocal*, as is clear from the application which he makes of it;[57] from $(***)$ for large $n$

[57]Note that in Stirling's discussion $c = 4/\pi$.

$$\binom{n}{n/2}/2^n < \left\{\frac{\pi}{2}\left(n+\frac{1}{2}+\frac{n-32}{8(n+2)(n+4)}\right)\right\}^{-1/2}$$

$$= \left(\frac{2}{\pi(n+\frac{1}{2})}\right)^{\frac{1}{2}}\left(1+\frac{n-32}{8(n+2)(n+4)(n+\frac{1}{2})}\right)^{-1/2}$$

$$\approx \sqrt{\frac{4}{\pi(2n+1)}}\left(1+\frac{1}{8n^2}\right)^{-1/2}$$

$$\approx \sqrt{\frac{4}{\pi(2n+1)}}\left(1-\frac{1}{16n^2}\right).$$

Thus $\sqrt{\dfrac{4}{\pi(2n+1)}}$ is the approximation to $\binom{n}{n/2}/2^n$ and this *exceeds* the required quantity by about $\dfrac{1}{16n^2}\sqrt{\dfrac{4}{\pi(2n+1)}}$ for large $n$. For further discussion of this approximation see [70, pp. 50–52].

Stirling's second approximation asserts that[58]

$$\ln\left(2^n/\binom{n}{n/2}\right) \approx \frac{1}{16}\ln\frac{n+2}{n-2}+\frac{1}{2}\ln n+\frac{1}{2}\ln\frac{\pi}{2}.$$

From the first approximation we obtain

$$\ln\left(2^n/\binom{n}{n/2}\right) \approx \ln\sqrt{\frac{\pi}{2}(n+\frac{1}{2})}$$

$$= \frac{1}{2}\ln\frac{\pi}{2}+\frac{1}{2}\ln n+\frac{1}{2}\ln\left(1+\frac{1}{2n}\right)$$

$$= \frac{1}{2}\ln\frac{\pi}{2}+\frac{1}{2}\ln n+\frac{1}{2}\left(\frac{1}{2n}-\frac{1}{8n^2}+\ldots\right)$$

$$\approx \frac{1}{2}\ln\frac{\pi}{2}+\frac{1}{2}\ln n+\frac{1}{4n};$$

but

$$\frac{1}{16}\ln\frac{n+2}{n-2} = \frac{1}{16}\ln\frac{1+\frac{2}{n}}{1-\frac{2}{n}}$$

$$= \frac{1}{8}\left(\frac{2}{n}+\frac{8}{3n^3}+\ldots\right) \approx \frac{1}{4n}.$$

[58] As usual, Stirling works with common rather than natural logarithms.

**Proposition 24 (pp. 139–142).** Here we are concerned with the sequence

$$A_n = \int_0^1 x^{r+n-1}(1-x)^{p-r-1}dx = B(r+n, p-r) \quad (n = 0, 1, 2, \ldots),$$

which is defined for $p > r > 0$. By integrating by parts we can easily obtain the relations which Stirling obtains from Newton's Proposition 7 of *De Quadratura Curvarum* in [47] (see the note on Proposition 16 and its scholion):

$$\begin{aligned} A_n &= \int_0^1 x^{r+n-1}(1-x)^{p-r-1}dx \\ &= \left[-\frac{x^{r+n-1}(1-x)^{p-r}}{p-r}\right]_0^1 + \frac{r+n-1}{p-r}\int_0^1 x^{r+n-2}(1-x)^{p-r}dx \\ &= \frac{r+n-1}{p-r}\int_0^1 x^{r+n-2}(1-x)(1-x)^{p-r-1}dx \\ &= \frac{r+n-1}{p-r}\left\{\int_0^1 x^{r+n-2}(1-x)^{p-r-1}dx - \int_0^1 x^{r+n-1}(1-x)^{p-r-1}dx\right\} \\ &= \frac{r+n-1}{p-r}(A_{n-1} - A_n), \end{aligned}$$

from which it follows that

$$A_n = \frac{r+n-1}{p+n-1}A_{n-1} \qquad (n = 1, 2, \ldots);$$

thus we have the sequence

$$a, \quad \frac{r}{p}a, \quad \frac{r+1}{p+1}b, \quad \ldots \quad \text{with} \quad a = A_0.$$

The *corollary* asserts that the function which interpolates the integrals is a constant multiple of the function which interpolates the sequence

$$a, \quad \frac{r}{p}a, \quad \frac{r+1}{p+1}b, \quad \ldots.$$

In fact the latter sequence is interpolated by

$$\frac{\Gamma(r+z)\,\Gamma(p)}{\Gamma(r)\,\Gamma(p+z)}a,$$

while for the integrals we have

$$B(r+z, p-r) = \frac{\Gamma(r+z)\,\Gamma(p-r)}{\Gamma(p+z)}.$$

In *Example 1* Stirling considers the sequence

$$\binom{2n}{n}/2^{2n} = \frac{\Gamma(2n+1)}{(\Gamma(n+1))^2 2^{2n}} \qquad (n = 0, 1, 2, \ldots),$$

which can be put in the required form by setting $a = 1$, $r = \frac{1}{2}$, $p = 1$ (see Proposition 22, Example 1 (sequence of reciprocals)). Then according to the proposition,

$$\frac{2^{2z}(\Gamma(z+1))^2}{\Gamma(2z+1)} = \frac{\int_0^1 x^{-1/2}(1-x)^{-1/2}dx}{\int_0^1 x^{z-1/2}(1-x)^{-1/2}dx}.$$

In particular, with $z = \frac{1}{2}$ we have

$$2(\Gamma\left(\tfrac{3}{2}\right))^2 = \frac{\int_0^1 x^{-1/2}(1-x)^{-1/2}dx}{\int_0^1 (1-x)^{-1/2}dx}.$$

Apparently, Stirling evaluates these integrals directly:

$$\int_0^1 x^{-1/2}(1-x)^{-1/2}dx = \int_0^1 \frac{dx}{\sqrt{x-x^2}} = \left[\sin^{-1}(2x-1)\right]_0^1 = \pi$$

and

$$\int_0^1 (1-x)^{-1/2}dx = \left[-2(1-x)^{1/2}\right]_0^1 = 2.$$

Thus we have

$$2(\Gamma\left(\tfrac{3}{2}\right))^2 = \frac{\pi}{2}$$

(cf. Proposition 21, Example 2).

Stirling's remarks relating to the interpolation of the reciprocals appear to be invalid as the integrals involved do not converge – in particular,

$$\int_0^x (1-t)^{-3/2}dt = 2(1-(1-x)^{-1/2}) \not\to 2 \quad \text{as} \quad x \to 1_-,$$

contrary to what he seems to be asserting.

Note that in *Example 2*

$$\int_0^1 x^{-2/3}(1-x)^{-2/3}dx = B\left(\tfrac{1}{3}, \tfrac{1}{3}\right) = \frac{\left(\Gamma\left(\frac{1}{3}\right)\right)^2}{\Gamma\left(\frac{2}{3}\right)} = \frac{\sqrt{3}}{2\pi}\left(\Gamma\left(\tfrac{1}{3}\right)\right)^3.$$

**Proposition 25 (pp. 142–143).** Now we have the sequence

$$B_m = \int_0^1 x^{p-m}(1-x)^{r+m-1}dx = B(p-m+1, r+m),$$

for which we require $p - m + 1 > 0$ and $r + m > 0$; consequently $B_m$ is defined (by the integral) for only finitely many integers $m$. Where meaningful, integration by parts yields

$$B_m = \left[\frac{x^{p-m+1}(1-x)^{r+m-1}}{p-m+1}\right]_0^1$$
$$+ \frac{r+m-1}{p-m+1}\int_0^1 x^{p-(m-1)}(1-x)^{r+(m-1)-1}dx$$
$$= \frac{r+m-1}{p-m+1}B_{m-1}.$$

Further,

$$B(p-m+1, r+m) = \frac{\Gamma(p-m+1)\,\Gamma(r+m)}{\Gamma(p+r+1)},$$

so that the finite sequence $B_m$ is interpolated by

$$B(p-z+1, r+z) = \frac{\Gamma(p-z+1)\,\Gamma(r+z)}{\Gamma(p+r+1)},$$

which is defined except where $p-z+1$ or $r+z$ is a non-positive integer.

In *Example 1* Stirling considers the finite sequence of reciprocals of binomial coefficients for given order $n$, for which $a=1$, $r=1$ and $p=n$; these are interpolated by

$$\frac{1}{n!}\Gamma(n-z+1)\Gamma(1+z) = (n+1)B(n-z+1, 1+z) \qquad (*)$$

or

$$(n+1)\int_0^1 x^{n-z}(1-x)^z dz \quad \text{if} \quad n-z+1>0,\ z+1>0.$$

Now for $r=1$ and $p=n$ we have

$$B_0 = \int_0^1 x^n dx = \frac{1}{n+1},$$

so that, according to the proposition, for $m = 0, 1, \ldots, n$,

$$\binom{n}{m} = \frac{1}{1/\binom{n}{m}} = \frac{B_0}{B_m} = \frac{1}{(n+1)\int_0^1 x^{n-m}(1-x)^m dx}.$$

Stirling illustrates this formula by using it to calculate $\binom{9}{4}$.

*Example 2* deals with the special case of *Example 1* in which $n=1$. There are just two coefficients, $\binom{1}{0}=1$ and $\binom{1}{1}=1$. The function $(*)$ becomes $\Gamma(2-z)\Gamma(1+z)$, whose value at $z=\frac{1}{2}$ is $(\Gamma(\frac{3}{2}))^2 = \pi/4$; consequently the term $t_{1/2}$ midway between $\binom{1}{0}$ and $\binom{1}{1}$ is $4/\pi$. Alternatively, in Stirling's version this comes from

$$\frac{1}{1/t_{1/2}} = \frac{B_0}{\int_0^1 x^{1/2}(1-x)^{1/2}dx},$$

so that

$$t_{1/2} = \frac{1}{2}\left(\int_0^1 x^{1/2}(1-x)^{1/2}dx\right)^{-1} = \frac{1}{2}\left(\int_{-1/2}^{1/2}\sqrt{\frac{1}{4}-u^2}\,du\right)^{-1} = \frac{4}{\pi},$$

where we have substituted $x = u + \frac{1}{2}$.

**Proposition 26 and Its Scholion (pp. 143–147).** Here Stirling is concerned with the sequence

$$a_k = \frac{\Gamma(r+k)\,\Gamma(p)}{\Gamma(p+k)\,\Gamma(r)} \qquad (k = 0,1,2,\ldots),$$

which he claims may be interpolated by a series of the form $z^n \sum_{m=0}^{\infty} \alpha_m z^{-m}$, where $n = r - p$ and $z - p$ is the distance of the term from the beginning; in other words, he asserts that

$$\frac{\Gamma(z+r-p)\,\Gamma(p)}{\Gamma(z)\,\Gamma(r)} = z^n \sum_{m=0}^{\infty} \frac{\alpha_m}{z^m}$$

for suitable coefficients $\alpha_m$. The coefficients are determined by substituting the desired form in the recurrence relation for the terms $za_{k+1} - (z+n)a_k = 0$ $(z = p+k)$, expanding all factors of the form $(z+1)^a = z^a\left(1+z^{-1}\right)^a$ by means of the binomial theorem, combining all terms of the same degree in $z$ and finally setting their coefficients equal to zero. The resulting series is generally asymptotic. Using the general version of "Stirling's formula",

$$\Gamma(az+b) \sim \sqrt{2\pi}e^{-az}(az)^{az+b-1/2},$$

we see that

$$a_k \sim \frac{\Gamma(p)}{\Gamma(r)}z^{r-p} = \frac{\Gamma(p)}{\Gamma(r)}z^n \qquad (z = p+k),$$

from which it follows that $\alpha_0$ (Stirling's initial coefficient $A$) is in fact $\Gamma(p)/\Gamma(r)$.

In *Example 1* Stirling returns to the middle-ratio sequence $\binom{2k}{k}/2^{2k}$ (see the note on Proposition 22). Here $r = \frac{1}{2}$ and $p = 1$; consequently

$$n = -\tfrac{1}{2} \quad \text{and} \quad A = \frac{\Gamma(1)}{\Gamma(\frac{1}{2})} = \frac{1}{\sqrt{\pi}}.$$

The function being interpolated is $\dfrac{\Gamma(z-\frac{1}{2})}{\Gamma(z)\sqrt{\pi}}$.

It is curious that, although Stirling calculates in this example a value for $A$ which coincides with $1/\sqrt{\pi}$ rounded to 12DP, he does not actually identify $A$ as $1/\sqrt{\pi}$, which he could easily have done by comparison with the series of Proposition 23 (or Example 1 of Proposition 22). However, he did make this identification later in a handwritten addendum and also in one of his notebooks.[59]

*Example 2* deals with the sequence

$$\frac{\Gamma(\frac{2}{3}+k)\,\Gamma(\frac{1}{3})}{\Gamma(\frac{1}{3}+k)\,\Gamma(\frac{2}{3})}, \quad \text{for which} \quad A = \frac{\Gamma(\frac{1}{3})}{\Gamma(\frac{2}{3})} = \frac{\Gamma(\frac{1}{3})^2\sqrt{3}}{2\pi}.$$

The values given in the *Methodus Differentialis* for the coefficients of $z^{-3}$, $z^{-4}$ and $z^{-5}$ in the series are wrong and these errors affect the subsequent calculations. While he was working on his translation of the *Methodus Differentialis* in 1749, Holliday queried the figures given in *Example 2* with Stirling, who provided him with corrections. The corrected figures appear in Holliday's translation and also replace the originals here. Stirling's revised calculated value for $A$ is correct to the number of decimal places given.

In the *scholion* Stirling indicates how the proposition might be extended to the situation discussed in Proposition 6. Note that the $S$, $S'$ of Proposition 6 correspond to the $T$, $T'$ of the *scholion* since it is intended to interpolate the sequence of $T$ values by means of a series and that the $m$ of Proposition 6 is now 1. In his rather imprecise proof of Proposition 6 Stirling assumes that the sums can be expressed in the form

$$\frac{z^n}{p^z} \times \left(A + \frac{B}{z} + \frac{C}{z^2} + \frac{D}{z^3} + \dots\right)$$

and deduces that $p = m$ ($= 1$ in the present case) and $n = a - c$. According to this the appropriate form to take for the terms of the *scholion* is, as Stirling has it,

$$z^{a-c} \times \left(A + \frac{B}{z} + \frac{C}{z^2} + \frac{D}{z^3} + \dots\right).$$

As Stirling notes in his proof, the terms of the sequence considered in Proposition 26 satisfy the relation

$$T' = \frac{z+n}{z}T \qquad (n = r - p).$$

In a handwritten addendum on three pages at the back of his copy of the *Methodus Differentialis*, Stirling discusses interpolation of the related sequence defined by

[59] I have reproduced the latter and given a more precise derivation of the asymptotic series of *Example 1* in [70, pp. 53–56]. The tables on pp. 57–58 of [70] contain some comparative middle-ratio calculations using various series, in particular the present one and those of Proposition 23.

$$T' = \frac{z+n}{z-n}T,$$

where $n$ is a constant and successive values of $z$ differ by 2. Here we have

$$f(z) = \alpha\,\frac{\Gamma\left(\frac{1}{2}(z+n)\right)}{\Gamma\left(\frac{1}{2}(z-n)\right)},$$

where $\alpha$ is a constant. I have discussed this material in [70, pp. 95–98].

**Proposition 27 (pp. 147–149).** In this proposition Stirling finds an inverse factorial series to represent and interpolate the terms of the sequence defined by

$$a_1 = 1, \quad a_{k+1} = \frac{k^2}{k^2+r}\,a_k \qquad (k = 1, 2, \ldots).$$

The series is

$$A\left(1 + \sum_{k=1}^{\infty} \frac{r(r+1^2)(r+2^2)\ldots(r+(k-1)^2)}{k!\,z(z+1)\ldots(z+k-1)}\right).$$

Now if $r < 0$, so that $r = -s$ with $s > 0$, we may write the series as

$$A\left(1 + \sum_{k=1}^{\infty} \frac{(-\sqrt{s})\sqrt{s}(1-\sqrt{s})(1+\sqrt{s})\ldots(k-1-\sqrt{s})(k-1+\sqrt{s})}{k!\,z(z+1)\ldots(z+k-1)}\right)$$

$$= AF(-\sqrt{s}, \sqrt{s}\,; z\,; 1),$$

and by Gauss's formula this is

$$\frac{A(\Gamma(z))^2}{\Gamma(z+\sqrt{s})\,\Gamma(z-\sqrt{s})}$$

provided $\operatorname{Re} z > 0$. If $r > 0$ we obtain similarly

$$AF(-i\sqrt{r}, i\sqrt{r}\,; z\,; 1) = \frac{A(\Gamma(z))^2}{\Gamma(z+i\sqrt{r})\,\Gamma(z-i\sqrt{r})}$$

for the value of the series, provided $\operatorname{Re} z > 0$. We see easily from these expressions that Stirling's solution does in fact generate the required sequence. Moreover, it follows from the asymptotic properties of the Gamma function (generalized Stirling's formula) or properties of hypergeometric series that the value of the series tends to $A$ as $z \to \infty$.

In the *example* Stirling refers to Wallis's product in the form[60]

[60]See *Arithmetica Infinitorum*, Proposition 191, in particular p. 469 in Vol. II of [75]. Stirling has already applied this result in Proposition 23.

$$\lim_{m\to\infty}\prod_{k=2}^{m}\frac{(2k-1)^2-1}{(2k-1)^2}=\frac{\pi}{4}$$

and proposes to find the value of

$$\lim_{m\to\infty}\prod_{k=1}^{m}\frac{(2k)^2}{(2k)^2-1}$$

by means of the proposition. In fact he could have deduced this directly from Wallis's result since

$$\prod_{k=1}^{m}\frac{(2k)^2}{(2k)^2-1}=\prod_{k=1}^{m}\frac{(2k)^2}{(2k-1)(2k+1)}$$

$$=\frac{2^2\times 4^2\times\ldots(2m)^2}{(1\times 3)(3\times 5)(5\times 7)\ldots((2m-3)(2m-1))((2m-1)(2m+1))}$$

$$=2\left(\frac{2m}{2m+1}\right)\frac{(2\times 4)(4\times 6)\ldots((2m-2)\times 2m)}{(3\times 3)(5\times 5)\ldots(2m-1)^2}$$

$$=2\left(\frac{2m}{2m+1}\right)\prod_{k=2}^{m}\frac{(2k-2)2k}{(2k-1)^2}$$

$$=2\left(\frac{2m}{2m+1}\right)\prod_{k=2}^{m}\frac{(2k-1)^2-1}{(2k-1)^2}\to 2\times\frac{\pi}{4}=\frac{\pi}{2}\quad\text{as}\quad m\to\infty.$$

For the application of the proposition

$$a_1=1,\quad a_{k+1}=\frac{(2k)^2}{(2k)^2-1}a_k=\frac{k^2}{k^2-\frac{1}{4}}a_k\qquad(k=1,2,\ldots),$$

so that $r=-\frac{1}{4}$ and the resulting series is

$$\frac{\pi}{2}F\left(-\tfrac{1}{2},\tfrac{1}{2};z;1\right)=\frac{\pi(\Gamma(z))^2}{2\,\Gamma(z+\frac{1}{2})\,\Gamma(z-\frac{1}{2})},$$

which holds for $\operatorname{Re} z>0$. Stirling determines the initial constant $A=\frac{\pi}{2}$ numerically from the tenth term of the sequence, viz.

$$\frac{2^2\times 4^2\times 6^2\times\ldots\times 18^2}{(1\times 3)(3\times 5)\ldots(17\times 19)}.$$

He quotes .9740 3924 54 for the sum of the corresponding series

$$F\left(-\tfrac{1}{2},\tfrac{1}{2};10;1\right)=1-\frac{1}{4.10}-\frac{1.3}{4.8.10.11}-\frac{1.3.15}{4.8.12.10.11.12}-\ldots,$$

but does not tell us how he obtained it. In fact, this sum is .9740 3924 84 to 10 decimal places, which suggests that Stirling's penultimate figure 5 is a misprint for 8. It would require direct addition of at least 23 terms to obtain this value.

**Proposition 28 (pp. 149–152).** Here we have the original versions of "Stirling's formula". The proposition deals with the sum $\sum_{r=1}^{h}\log_{10}(x+(2r-1)n)$, where $x+(2h-1)n = z-n$, so that $h=(z-x)/(2n)$. This is

$$\log_{10}\prod_{r=1}^{k}(x+(2r-1)n) = \frac{\ln\prod_{r=1}^{k}(x+(2r-1)n)}{\ln 10} = a\ln\prod_{r=1}^{k}(x+(2r-1)n).$$

Once more Stirling offers little more than a hint of his proof, but we can see formally what he is doing from the following. Put

$$S(z) = \frac{z\log_{10}z}{2n} - \frac{az}{2n} + a\sum_{r=1}^{\infty}a_r\left(\frac{n}{z}\right)^{2r-1}$$

and consider

$$S(z)-S(z-2n) = \frac{z\log_{10}z}{2n} - \frac{(z-2n)\log_{10}(z-2n)}{2n} - a + a\sum_{r=1}^{\infty}a_r\left\{\left(\frac{n}{z}\right)^{2r-1} - \left(\frac{n}{z-2n}\right)^{2r-1}\right\}.$$

The appropriate procedure now is to express everything in terms of $z-n$. We have

$$\begin{aligned}
&\frac{z\log_{10}z}{2n} - \frac{(z-2n)\log_{10}(z-2n)}{2n} - a\\
&= \frac{z}{2n}\log_{10}\left\{(z-n)\left(1+\frac{n}{z-n}\right)\right\}\\
&\qquad - \frac{(z-2n)}{2n}\log_{10}\left\{(z-n)\left(1-\frac{n}{z-n}\right)\right\} - a\\
&= \log_{10}(z-n) + \frac{a(z-n)}{2n}\left\{\ln\left(1+\frac{n}{z-n}\right) - \ln\left(1-\frac{n}{z-n}\right)\right\}\\
&\qquad + \frac{a}{2}\left\{\ln\left(1+\frac{n}{z-n}\right) + \ln\left(1-\frac{n}{z-n}\right)\right\} - a\\
&= \log_{10}(z-n) + a\sum_{r=1}^{\infty}\frac{1}{2r-1}\left(\frac{n}{z-n}\right)^{2r-2} - a\sum_{r=1}^{\infty}\frac{1}{2r}\left(\frac{n}{z-n}\right)^{2r} - a\\
&= \log_{10}(z-n) - a\sum_{r=1}^{\infty}\left(\frac{n}{z-n}\right)^{2r}\left(\frac{1}{2r}-\frac{1}{2r+1}\right)
\end{aligned}$$

and

$$\left(\frac{n}{z}\right)^{2r-1} - \left(\frac{n}{z-2n}\right)^{2r-1}$$
$$= \left(\frac{n}{z-n}\right)^{2r-1} \left\{ \left(1+\frac{n}{z-n}\right)^{-(2r-1)} - \left(1-\frac{n}{z-n}\right)^{-(2r-1)} \right\}$$
$$= -2\left(\frac{n}{z-n}\right)^{2r-1} \sum_{s=1}^{\infty} \binom{2r+2s-3}{2r-2} \left(\frac{n}{z-n}\right)^{2s-1}.$$

Consequently

$$S(z) - S(z-2n)$$
$$= \log_{10}(z-n) - a\sum_{r=1}^{\infty} \left(\frac{n}{z-n}\right)^{2r} \frac{1}{2r(2r+1)}$$
$$- 2a\sum_{t=1}^{\infty} a_t \left(\frac{n}{z-n}\right)^{2t-1} \sum_{s=1}^{\infty} \binom{2t+2s-3}{2t-2} \left(\frac{n}{z-n}\right)^{2s-1}.$$

Stirling wants $S(z) - S(z-2n) = \log_{10}(z-n)$, so we equate the coefficient of each power of $z-n$ on the right-hand side to zero, that is we put

$$-\frac{1}{2r(2r+1)} - 2\sum a_t \binom{2t+2s-3}{2t-2} = 0,$$

where the sum is over all positive integers $s$, $t$ such that $(2t-1)+(2s-1) = 2r$. Thus we have for $r = 1, 2, \ldots$

$$-\frac{1}{4r(2r+1)} = \sum_{t=1}^{r} a_t \binom{2r-1}{2t-2}. \qquad (1)$$

These are the equations that Stirling gives for his coefficients $A$, $B$, $C$, .... The above analysis does not depend on the value of $z$, so we may replace $z$ by $z-2n$, $z-4n$, .... Stirling's proof is completed by noting that, if the terms are taken in reverse order, then the required sum is

$$(S(z) - S(z-2n)) + (S(z-2n) - S(z-4n)) + \ldots + (S(x+2n) - S(x))$$
$$= S(z) - S(x)$$

and $S(z)$, $S(x)$ are the two series in the proposition.

Using properties of the Bernoulli numbers we can show that the equations (1) give

$$a_r = -\frac{(2^{2r-1}-1)B_{2r}}{2r(2r-1)} \qquad (r = 1, 2, \ldots).$$

There is no evidence of Stirling's being aware when he wrote the *Methodus Differentialis* that his coefficients could be expressed in terms of the Bernoulli numbers, although he may have discovered this connection at a later date (see below).

In *Example 2* Stirling takes up the problem of determining $\log_{10} m!$. For this we want $2n = 1$, $z - n = m$ and $x + n = 1$ (or 2 since $\log_{10} 1 = 0$), so that $z = m + \frac{1}{2}$ and $x = \frac{1}{2}$. According to Stirling the $x$ series becomes $-\frac{1}{2}\log_{10}(2\pi)$, although he does not explain why; thus he has

$$\log_{10} m! = (m + \tfrac{1}{2})\log_{10}(m + \tfrac{1}{2}) + \tfrac{1}{2}\log_{10}(2\pi) - a(m + \tfrac{1}{2}) - \frac{a}{24(m + \frac{1}{2})} + \frac{7a}{2880(m + \frac{1}{2})^3}\cdots,$$

or equivalently

$$\ln m! = (m + \tfrac{1}{2})\ln(m + \tfrac{1}{2}) + \tfrac{1}{2}\ln(2\pi) - (m + \tfrac{1}{2}) - \frac{1}{24(m + \frac{1}{2})} + \frac{7}{2880(m + \frac{1}{2})^3}\cdots. \tag{2}$$

If we neglect all the terms involving reciprocal powers of $m + \frac{1}{2}$ and exponentiate, we obtain formally

$$m! \approx \sqrt{2\pi}\left(\frac{m + \frac{1}{2}}{e}\right)^{m+1/2} = s_m. \tag{3}$$

Formulae (2) and (3) are therefore Stirling's own versions of the expressions which are generally labelled "Stirling's formula" nowadays.

How did Stirling come up with the $\frac{1}{2}\log_{10}(2\pi)$ term? One possibility is direct calculation. Evaluation of

$$\ln m! - \left\{(m + \tfrac{1}{2})\ln(m + \tfrac{1}{2}) - (m + \tfrac{1}{2}) - \frac{1}{24(m + \frac{1}{2})} + \frac{7}{2880(m + \frac{1}{2})^3}\cdots\right\}$$

using just a few terms of the series in conjunction with a sufficiently large value of $m$ would produce a good approximation to $\frac{1}{2}\log_{10}(2\pi)$, which would probably have been a familiar number to Stirling – we have seen in Example 2 of Proposition 21 how Stirling identified a certain quantity as $\sqrt{\pi}/2$ apparently on the basis of a calculated value. However, we have also seen in his proof of Proposition 23 that Stirling could make effective use of Wallis's product, so that he may well have applied this just as in modern texts to obtain the $\sqrt{2\pi}$ in (2).

De Moivre tells us in [44, p. 1] how a few days after the publication of [43] Stirling had written to him to advise him of some errors in a table of sums of logarithms in this work and at the same time communicated his series[61]

$$\log_{10} m! = (m + \tfrac{1}{2})\log_{10}(m + \tfrac{1}{2}) + \tfrac{1}{2}\log_{10}(2\pi) - a(m + \tfrac{1}{2}) - \frac{a}{2 \times 12(m + \frac{1}{2})} + \frac{7a}{8 \times 360(m + \frac{1}{2})^3} - \frac{31a}{32 \times 1260(m + \frac{1}{2})^5} + \frac{127a}{128 \times 1680(m + \frac{1}{2})^7}\cdots.$$

[61]In place of 127 in the numerator of the last term quoted De Moivre has in error 63.

Struck by the appearance in the denominators of the numbers 12, 360, 1260, 1680, which had occurred in his own work on the middle ratio (see [43, p. 127]), De Moivre set about applying his own methods to determine the logarithm of a factorial and obtained the expression

$$\ln(m-1)! = (m-\tfrac{1}{2})\ln m + \tfrac{1}{2}\ln(2\pi) - m \\ + \frac{1}{12m} - \frac{1}{360m^3} + \frac{1}{1260m^5} - \frac{1}{1680m^7} + \cdots. \tag{4}$$

Here the general term is

$$\frac{B_{2r}}{2r(2r-1)m^{2r-1}} \qquad (r = 1, 2, \ldots).$$

Note that De Moivre deals with natural logarithms (hence the absence of Stirling's quantity $a$); moreover De Moivre was aware that his coefficients involved the Bernoulli numbers [44, pp. 18–21]. If we neglect the reciprocal powers of $m$ and exponentiate we obtain formally

$$(m-1)! \approx \sqrt{\frac{2\pi}{m}}\left(\frac{m}{e}\right)^m$$

or, on multiplying by $m$,

$$m! \approx \sqrt{2\pi m}\left(\frac{m}{e}\right)^m = t_m. \tag{5}$$

Unfortunately (4) and (5) are the expressions which posterity has adopted and labelled as "Stirling's formula", probably because they are a little simpler than Stirling's own versions.

De Moivre also found a simpler analogue of Stirling's Proposition 28 to deal with $\ln\{(m-d)(m-2d)\ldots(m-ld)\}$ (see [44, pp. 11–15]) and Stirling in turn included his own version of De Moivre's result along with two systems of equations for the coefficients in a handwritten addendum in his own copy of the *Methodus Differentialis* (see [70, pp. 12–14]).

The common link in these various series is the general Euler–MacLaurin summation formula. Putting $f(s) = \ln(x + 2sn)$ and $h = (z-x)/(2n)$ in the second (midpoint) form of the Euler–MacLaurin summation formula

$$\sum_{s=0}^{h-1} f(s+\tfrac{1}{2}) = \int_0^h f(t)\,dt - \sum_{r=1}^{k-1} \frac{(2^{2r-1}-1)B_{2r}}{2^{2r-1}(2r)!}\left(f^{(2r-1)}(h) - f^{(2r-1)}(0)\right) \\ - \frac{h(2^{2k-1}-1)B_{2k}}{2^{2k-1}(2k)!} f^{(2k)}(\eta) \quad \text{for some } \eta \in [0, h],$$

we obtain

$$\ln(x+n) + \ln(x+3n) + \ldots + \ln(z-n)$$
$$= \frac{z\ln z - x\ln x}{2n} - \frac{z-x}{2n} + \sum_{r=1}^{k-1} \frac{(2^{2r-1}-1)B_{2r}n^{2r-1}}{2r(2r-1)} \left\{ \frac{1}{z^{2r-1}} - \frac{1}{x^{2r-1}} \right\}$$
$$+ \frac{h(2^{2k-1}-1)B_{2k}n^{2k}}{k(x+2\eta n)^{2k}} \quad \text{for some } \eta \in [0,h],$$

which is the formula of Proposition 28 expressed in terms of natural logarithms and with remainder term. If we use the more familiar first (endpoint) form of the formula

$$\sum_{s=0}^{h-1} f(s) = \int_0^h f(t)\,dt - \frac{1}{2}(f(h)-f(0)) + \sum_{r=1}^{k-1} \frac{B_{2r}}{(2r)!}(f^{(2r-1)}(h) - f^{(2r-1)}(0))$$
$$+ \frac{hB_{2k}}{(2k)!} f^{(2k)}(\eta) \quad \text{for some } \eta \in [0,h],$$

with $f(s) = \ln(x+sn)$ and $h = (z-x)/n$, we obtain

$$\ln x + \ln(x+n) + \ln(x+2n) + \ldots + \ln(z-n)$$
$$= \frac{(z-\frac{n}{2})\ln z - (x-\frac{n}{2})\ln x}{n} - \frac{z-x}{n} + \sum_{r=1}^{k-1} \frac{B_{2r}n^{2r-1}}{2r(2r-1)} \left\{ \frac{1}{z^{2r-1}} - \frac{1}{x^{2r-1}} \right\}$$
$$- \frac{hB_{2k}n^{2k}}{2k(x+\eta n)^{2k}} \quad \text{for some } \eta \in [0,h];$$

this corresponds to Stirling's version of De Moivre's result mentioned above. The asymptotic series of Stirling (2) and De Moivre (4) for the logarithm of a factorial can be developed from these applications of the Euler–MacLaurin summation formula (see [70, pp. 16–19]). Stirling does not tell us how many terms he used in his three *examples*, so I have not included any error analysis for these calculations. All his answers are correct to the number of places given except that in *Example 2* the value of $\log_{10}(11 \times 12 \times 13 \times \ldots \times 1000)$ should have 3 and not 2 as the final digit when correctly rounded.

In his first letter[62] to Stirling dated 8 June 1736, Euler included a statement of his version of the summation formula which he had found about 1732 [15]; this is a variant of the first form in which all the terms involving values at zero are combined into a single "Euler–MacLaurin constant". In 1737 MacLaurin sent Stirling some printed pages of his proposed *Treatise of Fluxions* (published 1742) [39] in which Maclaurin gave both the first and the second forms of the summation formula.[63] When Stirling replied to Euler on

[62]For the Stirling–Euler correspondence see [74, 70, 33].

[63]Of course neither Euler nor Stirling gave an error term and apparently neither was aware at this time that the coefficients could be expressed in terms of Bernoulli numbers.

16 April 1738 he pointed out MacLaurin's independent discovery of the summation formulae and remarked that his theorem for summing logarithms is a special case of Euler's summation formula.[64] In fact it is De Moivre's version which comes readily from the formula communicated by Euler. MacLaurin derived the results of Stirling and De Moivre from his versions of the summation formula in Articles 838, 839 and 840 of [39].

Stirling's versions are generally more accurate than the corresponding formulae of De Moivre (see [70, pp. 18–19]). In particular, Stirling's $s_m$ (3), which exceeds $m!$, and De Moivre's $t_m$ (5), which is less than $m!$, have the property that

$$\frac{s_m - m!}{m! - t_m} \to \frac{1}{2} \quad \text{as } m \to \infty;$$

in this sense $s_m$ is twice as good as $t_m$ (see [69]).

**Note on the Scholion to Proposition 28 (pp. 152–153).** The remarks in the *scholion* do not relate specifically to Proposition 28 but are directed rather to the technique of using suitable series for purposes of interpolation or solving difference equations.

Stirling notes in passing that the sequence

$$u_n = \int_0^1 (1-x)^n dx = \frac{1}{n+1} \qquad (n = 0, 1, 2, \ldots)$$

is of the type considered in Proposition 26:

$$\frac{u_{n+1}}{u_n} = \frac{n+1}{n+2} \quad \text{so that} \quad r = 1, p = 2.$$

He then considers the sequence defined by

$$u_0 = a\,, \quad u_{k+1} = \frac{r+n+k}{r+k} u_k \qquad (k = 0, 1, 2, \ldots),$$

where $r$, $n$ are constants. For this he has the difference equation

$$(k+r)(u_{k+1} - u_k) - n u_k = 0,$$

which he proceeds to solve by fitting a Newton series[65]

$$\alpha_0 + \sum_{m=1}^{\infty} \alpha_m z(z-1)\ldots(z-m+1);$$

---

[64] "Perhaps you have not noticed that my theorem for summing logarithms is nothing more than a special case of your general theorem."

[65] For properties of such series see [40, Chapter X]. Note in particular that either a Newton series converges only at the non-negative integers or it converges in a half-plane of the form $\{z \in \mathbb{C} : \mathrm{Re}\, z > \lambda\}$ $(\lambda \geq -\infty)$.

setting $z$ equal to $k$ and $k+1$ to give $u_k$ and $u_{k+1}$ determines the coefficients and then the series provides the intermediate values of the function, in this case

$$\frac{\Gamma(z+n+r)\,\Gamma(r)}{\Gamma(z+r)\,\Gamma(n+r)}\,a. \tag{$*$}$$

Stirling notes that the same series could have been obtained from Proposition 19 (Newton's forward difference formula). To obtain in this way the five terms given by Stirling we need to consider the first five terms of the sequence,

$$a,\ \frac{(r+n)a}{r},\ \frac{(r+n+1)(r+n)a}{(r+1)r},\ \frac{(r+n+2)(r+n+1)(r+n)a}{(r+2)(r+1)r},$$

$$\frac{(r+n+3)(r+n+2)(r+n+1)(r+n)a}{(r+3)(r+2)(r+1)r},$$

and form their differences. The appropriate differences $A, B, C, D, E$ required in Proposition 19 turn out to be respectively

$$a\,,\ \frac{na}{r}\,,\ \frac{n(n-1)a}{(r+1)r}\,,\ \frac{n(n-1)(n-2)a}{(r+2)(r+1)r}\,,\ \frac{n(n-1)(n-2)(n-3)a}{(r+3)(r+2)(r+1)r}\,,$$

and substitution of these values in the formula of Proposition 19 produces the required series:

$$a\biggl(1+\frac{nz}{r\,1!}+\frac{n(n-1)z(z-1)}{(r+1)r\,2!}+\frac{n(n-1)(n-2)z(z-1)(z-2)}{(r+2)(r+1)r\,3!}$$

$$+\frac{n(n-1)(n-2)(n-3)z(z-1)(z-2)(z-3)}{(r+3)(r+2)(r+1)r\,4!}+\dots\biggr).$$

In fact the series is $a\,F(-n,-z\,;r\,;1)$, which by Gauss's formula converges to the correct value $(*)$ provided $r$ is not a negative integer or zero and $\mathrm{Re}(z+n+r)>0$.

**Proposition 29 (pp. 153–156).** This is the general divided difference interpolation formula which is the second case of Lemma V in Book III of Newton's *Principia* [48]. Once again we are left to verify that giving $z$ appropriate values, in this case $a, b, c, \dots$, produces the given ordinates in turn. It is curious that Stirling did not discuss this formula in his 1719 paper, although Cotes dealt with it in his *De Methodo Differentiali Newtoniana* (see [13] and [21, Chapter 6]).

In *Example 2* four zenith-distance measurements of the sun's position at noon, given as chords corresponding to radius $10^5$ and taken at Nürnberg on 8, 9, 12, 16 June 1500, are used to calculate the time of the summer solstice in that year. The measurements were made by Bernard Walther (1430–1504), patron, student and collaborator of Regiomontanus.[66] Stirling finds the interpolating polynomial for these chords as ordinates with respective $z$-values determined from the dates as $0 = 8 - 8$, $1 = 9 - 8$, $4 = 12 - 8$, $8 = 16 - 8$. At the summer solstice the zenith-distance is a minimum, so we would expect that a reasonable approximation to the corresponding $z$-value would be obtained by setting the derivative of the polynomial equal to zero and extracting the appropriate root. This produces $z = 3.889355$ with $44882.92395$ as the corresponding value of the polynomial. Referring to Walther's data Tycho Brahe [8, Vol. II p. 38] found 44884 for this quantity (before correction for parallax).

In the notation of the diagram the chord of angle $\alpha$ is $PQ = 2R\sin(\alpha/2)$, so that $\alpha = 2\sin^{-1}(\frac{1}{2}PQ/R)$. For $PQ = 44884$ and $R = 10^5$ we obtain $\alpha = 25° 56' 15''$, which is the observed angle from the vertical to the sun. Walther gave the latitude of Nürnberg as $49° 24'$ and the maximum declination of the sun as $23° 28'$, whose difference is approximately this $\alpha$.

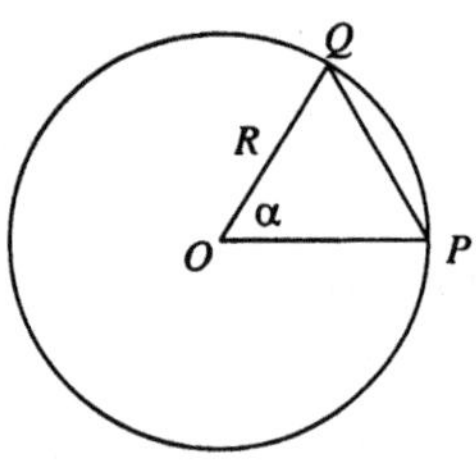

Chord of angle

**Scholion to Proposition 29 (pp. 156–157).** Newton described the application of the general divided-difference interpolation formula to the determination of the orbit of a comet in Lemma VI of Book III of the *Principia* [48].

Halley's paper [24], which is reproduced in the 1707 edition of Newton's *Arithmetica Universalis* [46], deals with approximations to the roots of polynomial equations. Other important contributions to the solution of such equations are contained in his four lectures delivered in October and November 1704. These are included as an appendix in [32]. In particular, the third lecture deals with the application of the conical parabola to solve equations of degrees three and four. These items are probably what Stirling intends by his reference to Halley's solutions of equations both here and at the end of Example 2 of the proposition.

**Proposition 30 (pp. 157–160).** Here Stirling assumes a function of the form

$$f(z) = a - br^z - cr^{2z} - dr^{3z} - \dots,$$

where $r$ is a positive constant with $r > 1$, and he wishes to determine

$$PQ = a = \lim_{z \to -\infty} f(z).$$

[66]For an interesting account of Walther's astronomical work see [4]. Stirling may have obtained the data from [9], whose pages 42–67 contain the observations of Regiomontanus and Walther.

This he proposes to do by solving for $a$ the infinite system of equations

$$\begin{aligned} a - b - c - d - e - \ldots &= f(0)\ (= A) \\ a - br - cr^2 - dr^3 - er^4 - \ldots &= f(1)\ (= B) \\ a - br^2 - cr^4 - dr^6 - er^8 - \ldots &= f(2)\ (= C) \\ a - br^3 - cr^6 - dr^9 - er^{12} - \ldots &= f(3)\ (= D) \\ a - br^4 - cr^8 - dr^{12} - er^{16} - \ldots &= f(4)\ (= E) \quad \text{(and so on)} \end{aligned}$$

which can be expressed in augmented matrix form as

$$\left.\begin{array}{cccccc|c} 1 & -1 & -1 & -1 & -1 & \ldots & g(1) \\ 1 & -r & -r^2 & -r^3 & -r^4 & \ldots & g(r) \\ 1 & -r^2 & -r^4 & -r^6 & -r^8 & \ldots & g(r^2) \\ 1 & -r^3 & -r^6 & -r^9 & -r^{12} & \ldots & g(r^3) \\ 1 & -r^4 & -r^8 & -r^{12} & -r^{16} & \ldots & g(r^4) \\ & & & \ldots & & & \end{array}\right.$$

where $g(x) = f(\log_r x)$. The columns are just the values at $x = 1, r, r^2, r^3, \ldots$ of the constant function 1, the functions $-x^{k-1}$ $(k = 2, 3, \ldots)$ and the function $g$. I have shown in [71, pp. 115–117] how this system may be solved formally to produce Stirling's expression for $PQ$ by formation of the divided differences of these functions at the points in question.

Stirling's formulation and explanation are perhaps not well-suited to the kind of application he intends. Apparently he has a sequence of ordinates $f(k)$ $(k = 0, 1, 2, \ldots)$, yet he wants to determine $\lim_{z \to -\infty} f(z)$. In fact, he has an arbitrarily large finite number of ordinates going off to the left from some initial ordinate, but he chooses his origin at the point corresponding to the left-most ordinate. Thus if he introduces further ordinates to the left, as would happen if the process were applied iteratively, the origin and therefore also the series for the function would change.[67] It therefore seems preferable to fix the representation of the function

$$f(z) = a - br^z - cr^{2z} - dr^{3z} - \ldots$$

so that the ordinates are $f(-l)$, $f(-l-1)$, ..., $f(-h+1)$, $f(-h)$, where $l$ is some initial non-negative integer and $h$ is an arbitrarily large positive integer greater than $l$. Putting $g(x) = f(\log_r x)$ as before, we have

(i) $g(1/r^m) = f(-m) \qquad (m = l, l+1, \ldots, h)$,

[67] Note, however, that changing the origin alters only the coefficients $b, c, d, \ldots$ : if $\zeta = z - z_0$ then

$$a - br^z - cr^{2z} - dr^{3z} - \ldots = a - (br^{z_0})r^\zeta - (cr^{2z_0})r^{2\zeta} - (dr^{3z_0})r^{3\zeta} - \ldots.$$

(ii) $g(x) = a - bx - cx^2 - dx^3 - \ldots$, so that $g$ is analytic on some domain containing the origin,

(iii) $a = \lim_{z \to -\infty} f(z) = \lim_{x \to 0} g(x)$.

I.J. Schoenberg [56] has put Stirling's process on a firm base, both theoretical and practical, by showing that the formula of Proposition 30 is obtained if we find the interpolating polynomial for $g$ at the points $r^{-m}$ ($m = l, l+1, \ldots, h$), which are in geometric progression, and use the value at zero of this polynomial to estimate the value of $a$. For example, with three points

$$g(1/r^l) = C, \quad g(1/r^{l+1}) = B, \quad g(1/r^{l+2}) = A,$$

and the table of divided differences is

$$A$$

$$\frac{A-B}{1/r^{l+2} - 1/r^{l+1}} = \frac{r^{l+2}(A-B)}{1-r}$$

$$\frac{r^{l+1}(r(A-B) - (B-C))}{(1-r)(1/r^{l+2} - 1/r^l)}$$

$$B$$

$$= \frac{r^{2l+3}(rA - (r+1)B + C)}{(1-r)(1-r^2)}$$

$$\frac{B-C}{1/r^{l+1} - 1/r^l} = \frac{r^{l+1}(B-C)}{1-r}$$

$$C$$

Thus the required interpolating polynomial is

$$p(x) = A + (x - 1/r^{l+2})\frac{r^{l+2}(A-B)}{1-r} + (x - 1/r^{l+2})(x - 1/r^{l+1})\frac{r^{2l+3}(rA - (r+1)B + C)}{(1-r)(1-r^2)},$$

so that

$$p(0) = A + \frac{A-B}{r-1} + \frac{rA - (r+1)B + C}{(r-1)(r^2-1)};$$

this coincides with the sum of the first three terms in Stirling's expression.

Now let us consider the *example* which Stirling presents with very little explanation as an illustration of *Proposition 30*. The area of a regular polygon of $2^{n+1}$ sides inscribed in a circle of radius 1 is $A_n = 2^n \sin(\pi/2^n)$, which tends to $\pi$ as $n \to \infty$. These values are generated by putting $z = 0, -1, -2, \ldots$ in

$$f(z) = \frac{\sin(\pi 2^{z-1})}{2^{z-1}} = \frac{2\sin(\frac{\pi}{2}.4^{z/2})}{4^{z/2}} = \sum_{k=0}^{\infty} \frac{(-1)^k \pi^{2k+1} 4^{kz}}{(2k+1)!\, 2^{2k}},$$

for which $r = 4$ and

$$g(x) = \frac{2\sin(\frac{\pi}{2}\sqrt{x})}{\sqrt{x}} = \sum_{k=0}^{\infty} \frac{(-1)^k \pi^{2k+1} x^k}{(2k+1)!\, 2^{2k}}.$$

Stirling tabulates the areas $A_n$ for $n = 1, 2, \ldots, 6$ and uses the five terms given explicitly in Proposition 30 to calculate $\pi$. Here

$$A = A_6 = g(4^{-5}), \quad B = A_5 = g(4^{-4}), \quad \ldots, \quad E = A_2 = g(4^{-1})$$

($A_1$ is not used).

One advantage of interpreting the process in terms of interpolation is that we have a readily available error term:

$$\pi - p(0) = \left(-\frac{1}{4^5}\right)\left(-\frac{1}{4^4}\right)\left(-\frac{1}{4^3}\right)\left(-\frac{1}{4^2}\right)\left(-\frac{1}{4}\right)\frac{g^{(5)}(c)}{5!}$$

for some $c \in (0, 1/4)$. Now

$$g^{(m)}(x) = \sum_{k=m}^{\infty} \frac{(-1)^k \pi^{2k+1} k!\, x^{k-m}}{(2k+1)!(k-m)!\, 2^{2k}}$$

and it is easily seen that for given $x \in [0, 1]$ the terms have alternating signs and their moduli form a strictly decreasing sequence. Thus $g^{(5)}$ is negative and strictly increasing on $[0, 1]$ ($g^{(6)}(x) > 0$ for $x \in [0, 1]$). Consequently,

$$|g^{(5)}(x)| \le |g^{(5)}(0)| = \frac{\pi^{11} 5!}{11!\, 2^{10}} \quad (x \in [0, 1])$$

and it follows that $p(0)$ underestimates $\pi$ by an amount less than

$$\frac{1}{4^5} \times \frac{1}{4^4} \times \frac{1}{4^3} \times \frac{1}{4^2} \times \frac{1}{4} \times \frac{\pi^{11}}{11!\, 2^{10}} = \frac{\pi^{11}}{11!\, 4^{20}},$$

which is less than $7 \times 10^{-15}$. In fact, Stirling gives 14 decimal places, all of which are correct.

It is striking that such an accurate figure has been obtained from a set of approximations, none of which gives more than two decimal places. Huygens [31] and James Gregory [22, 23] had found formulae for improving the estimates of $\pi$ obtained from inscribed and circumscribed polygons and Stirling acknowledges that his approximations are of the type already found by them; in fact, the partial sums of Stirling's series with $r = 4$ reproduce some of their results (see [71, p. 119]).

Note that the choice of $r = 4$, which comes out naturally in the above discussion, was justified by Stirling on the grounds that each successive term in the sequence of differences $A - B$, $B - C$, $C - D$, ... is about one quarter of its predecessor. In fact,

$$\frac{A_{n+1} - A_n}{A_n - A_{n-1}} = \frac{1}{4\cos(\pi/2^{n+1})\cos(\pi/2^{n+2})} \to \frac{1}{4} \quad \text{as} \quad n \to \infty.$$

Stirling had already stated the series of Proposition 30 for the cases $r = 4$ and $r = 2$ in his paper [61] of 1719. There he gave no explanation at all of the series, although he did provide two illustrative examples. For $r = 4$ he gave a calculation of $\pi$ similar to that just discussed but using just four points, viz. $4^{-7}$, $4^{-6}$, $4^{-5}$, $4^{-4}$. In the case $r = 2$ he determined $\ln 2$ by using its representation as $\lim_{x\to 0}(2^x - 1)/x$; his calculation is equivalent to finding $p(0)$, where $p(x)$ is the interpolating polynomial for $g(x) = (2^x - 1)/x$ at the five points $2^{-10}$, $2^{-9}$, $2^{-8}$, $2^{-7}$ and $2^{-6}$. For further details of these calculations see [71, pp. 118–121].

As pointed out by Schoenberg the ideas contained in Proposition 30 were developed further by Karl Schellbach [53][68] and in consequence Schoenberg refers to the numerical process as the "Stirling–Schellbach algorithm"; he also links it to algorithms of Romberg and Runge.

There is no direct evidence in Stirling's version that he understood his Proposition 30 as a result about interpolation. It may be significant that the preceding Proposition 29 gives the general Newton–Lagrange interpolation series, but Stirling makes no reference to this in his discussion. His rather vague remarks concerning the summation of power series, which appear within the *example*, perhaps suggest ideas of interpolation since the partial sums have to be interpreted as particular ordinates of a suitable function for the application of the process as Stirling has explained it.

**Proposition 31 and Its Scholion (pp. 160–164).** The instruction of the proposition is to take the interpolating polynomial determined by the given ordinates and to use its definite integral over the interval cut off by the extreme left and right ordinates to approximate to the area contained by the graph of the function over this interval.

To facilitate this process Stirling provides three tables in the *scholion*. The first of these, the *Table of Areas*, gives the closed Newton–Cotes quadrature formulae for 3 (Simpson's Rule), 5, 7 and 9 ordinates. Stirling had already given these formulae for 3, 5, 7, 9 and 11 ordinates in his 1719 paper [61].[69] He remarked in the paper that he had not provided a table for an even number of ordinates "since other things being equal the area is more accurately defined by an odd number of them". This phenomenon is partly explained by the

[68] In particular, Schellbach introduced a special difference scheme from which Stirling's series can be easily constructed in practice (see [56], also [71, p. 117]).

[69] Cotes developed these formulae at least for 3 up to 11 ordinates in 1711 after seeing the case of four ordinates in the edition of Newton's *Methodus Differentialis* which was published by Jones in that year [47]. Cotes's work was only published posthumously in *Harmonia Mensurarum* (1722) [13]. It is therefore conceivable that Stirling was unaware of Cotes's results when he wrote his paper. See [21] for further information on Cotes and his contributions.

fact that the error term for the formula involving $2n+1$ ordinates can be expressed in terms of the $(2n+2)$-th derivative of the integrand, while that for the formula involving $2n$ ordinates requires the $2n$-th derivative; thus, in particular, the formulae with $2n+1$ ordinates and $2n+2$ ordinates are both exact for polynomials of degree at most $2n+1$ (see [27, pp. 71–74; 58, pp. 154–165]).

The *Table of Corrections*, which does not appear in the paper or in Cotes's work, is of some interest. Suppose that in the interval $[\alpha, \beta]$ we have $2n+1$ equally spaced points

$$\gamma + \frac{k}{2n}(\beta - \alpha) \quad (k = 0, \pm 1, \pm 2, \ldots, \pm n) \quad \text{where} \quad \gamma = \tfrac{1}{2}(\alpha + \beta);$$

now introduce two new points, which lie outside $[\alpha, \beta]$, by giving $k$ the values $\pm(n+1)$. Let $p_n$ and $q_n$ be respectively the interpolating polynomials for a given function $f$ at the original set of $2n+1$ points and at the extended set of $2n+3$ points. The proposed corrections are the quantities $\int_\alpha^\beta p_n(x) - q_n(x)\,dx$, so that the effect of the correction is to replace $\int_\alpha^\beta p_n(x)\,dx$ with $\int_\alpha^\beta q_n(x)\,dx$. Put

$$f_k = f\Big(\gamma + \frac{k}{2n}(\beta - \alpha)\Big) \quad (k = 0, \pm 1, \pm 2, \ldots, \pm(n+1)).$$

With an obvious change of variable we can show that

$$\int_\alpha^\beta p_n(x) - q_n(x)\,dx = \frac{\beta - \alpha}{(2n+2)!\,n}$$

$$\times \left\{ \sum_{k=0}^{n} \binom{2n+2}{k} (-1)^{k+1} (f_{n-k+1} + f_{-n+k-1}) + \binom{2n+2}{n+1} (-1)^n f_0 \right\}$$

$$\times \int_0^n s^2(s^2 - 1^2)(s^2 - 2^2)\ldots(s^2 - n^2)\,ds.$$

The expression given by Stirling for 3 ordinates is correct, but in the cases of 5, 7 and 9 ordinates the factors

$$\frac{1}{470}, \quad \frac{1}{930}, \quad \frac{1}{1600}$$

are convenient approximations to the true values,[70] which are respectively

$$\frac{2}{945}, \quad \frac{3}{2800}, \quad \frac{296}{467775}.$$

In his third table Stirling gives the quadrature formulae in terms of central differences. These versions can be obtained directly from "Stirling's interpolation formula" (Proposition 20, First Case). Consider, for example, the case of 5 ordinates. On putting $x = \gamma + z(\beta - \alpha)/4$ where $\gamma = (\alpha + \beta)/2$, we have

[70] Cf. Stirling's remarks concerning his third table.

$$\int_\alpha^\beta f(x)\,dx = \frac{\beta-\alpha}{4}\int_{-2}^{2} f\Big(\gamma + \frac{z}{4}(\beta-\alpha)\Big)\,dz.$$

Then, using the interpolating polynomial for $g(z) = f(\gamma + z(\beta-\alpha)/4)$ at the points $-2$, $-1$, 0, 1, 2, we have in the notation of Proposition 20 (which is different from that of Proposition 31)

$$\int_\alpha^\beta f(x)\,dx \approx \frac{\beta-\alpha}{4}\int_{-2}^{2} a + \frac{Bz+bz^2}{2} + \frac{2Cz+cz^2}{24}(z^2-1)\,dz$$

$$= \frac{\beta-\alpha}{2}\int_0^2 a + \frac{b}{2}z^2 + \frac{c}{24}z^2(z^2-1)\,dz.$$

A straightforward calculation followed by the appropriate notational change[71] now produces the entry for 5 ordinates in the third table. As Stirling notes the true coefficients of the final terms in the expressions for 9, 11 and 13 ordinates have been replaced by simpler approximations. The changes are as follows:

| # of ordinates | Coeffn. of | True value | Stirling uses |
|---|---|---|---|
| 9 | $E$ | $\frac{989}{28350} = .034885\ldots$ | $\frac{3}{86} = .034883\ldots$ |
| 11 | $F$ | $\frac{16067}{598752} = .026834148\ldots$ | $\frac{94}{3503} = .026834142\ldots$ |
| 13 | $G$ | $\frac{15011}{693000} = .02166\ldots$ | $\frac{66}{3050} = .02163\ldots$ |

The inclusion of a formula for 1 ordinate is a little incongruous – according to Stirling's general description the interval of integration then has zero length. However, we could also interpret this case as the midpoint rule with arbitrary interval length.

Stirling illustrates the use of the three tables in calculations of

$$\int_0^1 \frac{1}{1+x}\,dx = \ln 2.$$

**Proposition 32 (pp. 164–166).** Here we have $n+1$ equidistant ordinates where $n$ of them are known. An approximation to the unknown ordinate is obtained by setting the expression for the $n$-th difference equal to zero. In the case where the unknown ordinate is the middle one of $2n+1$ equidistant

[71] In Propositions 20 and 31 $A, B, C, \ldots, a, b, c, \ldots$ play different roles. However in the final step we just need to note that the differences denoted by $a, b, c, \ldots$ in Proposition 20 (First Case) are denoted by $A, B, C, \ldots$ in Proposition 31.

ordinates we deduce Stirling's second table from the table in the proposition as follows:

$$\text{(3 ordinates – 2 about the middle)} \quad b = \frac{a+c}{2} = \frac{A}{2};$$

$$\text{(5 ordinates – 4 about the middle)} \quad c = \frac{4(b+d)-(a+c)}{6} = \frac{4A-B}{6};$$

etc.

This technique was also given in Stirling's paper [61] (see pp. 1057–1058). In the *example* Stirling uses logarithms to base 10 with $l, x$ standing for $\log_{10} x$ (cf. Proposition 28).

**Proposition 33 (pp. 166–168).** This is a formula for "interpolating to halves" (see [27, pp. 102–103] or [58, p. 32]) which is obtained directly from the second case of Proposition 20. Here $z = 2t/d$, where $t$ is the signed distance to the varying ordinate from $O$, the midpoint of the interval cut off by the two middle ordinates ${}_1A$ and $A_1$, and $d$ is the length of this interval. Hence we put $z = 0$ to determine the ordinate halfway between ${}_1A$ and $A_1$, producing (in the notation of Proposition 20, Case 2)[72]

$$\begin{aligned} &\frac{A}{2} + \frac{3B}{2}\left(-\frac{1}{4\times 6}\right) + \frac{5C}{2}\left(-\frac{1}{4\times 6}\right)\left(-\frac{9}{8\times 10}\right) + \dots \\ &= \frac{A}{2} - \frac{B}{16} + \frac{3C}{256} - \dots . \end{aligned}$$

Now

$$A_1 = a, \quad {}_1A = \alpha, \quad A_3 = b, \quad {}_3A = \beta, \quad A_5 = c, \quad {}_5A = \gamma, \quad \dots,$$

so that the $A$, $B$, $C$, ... of Proposition 20, Case 2 become

$$\begin{aligned} A &= a + \alpha, \\ B &= (\alpha - 2a + b) + (\beta - 2\alpha + a) \\ &= -(a+\alpha) + (b+\beta), \\ C &= (\gamma - 4\beta + 6\alpha - 4a + b) + (\beta - 4\alpha + 6a - 4b + c) \\ &= 2(a+\alpha) - 3(b+\beta) + (c+\gamma), \end{aligned}$$

and so on, which leads to Stirling's expression. The entries in the table are the partial sums of the series of the proposition; the column on the left gives the number of ordinates involved in each partial sum. Again Stirling uses logarithms to base 10 in his illustration. The material of Proposition 33 is also given in Stirling's 1719 paper [61] (see pp. 1061–1062).

---

[72] Note that the following expression has already appeared in Stirling's treatment of Example 2 of Proposition 21.

# Appendix

## Stirling's Letter to De Moivre Dated 19 June 1729

In this letter Stirling communicated to De Moivre some of his results on the middle-ratio problem. I have translated the letter below using the text which De Moivre reproduced on pp. 170–172 of [43] (see also pp. 46–49 of [74]). It would appear from the first paragraph that Stirling had already told De Moivre about his solutions and that the letter was written in response to a request for more detailed information which De Moivre wished to include in his *Miscellanea Analytica* [43]. The main text is very similar to Stirling's statement and discussion in Proposition 23. The numbers in the second calculation below are not as accurate as those in the corresponding calculation in Proposition 23, although the final result is the same in both places.[73]

### *Translation of the Letter*

About four years ago, when I informed Mr *Alex. Cuming* that problems concerning the Interpolation and Summation of series and others of this type which are not susceptible to the commonly accepted analysis, can be solved by *Newton's* Method of Differences, the most illustrious man replied that he doubted if the problem solved by you some years before about finding the middle coefficient in an arbitrary power of the binomial could be solved by differences. Then, led by curiosity and confident that I would be doing a favour to a most deserving man of Mathematics, I took it up willingly: and I admit that difficulties arose which prevented me from arriving at the desired conclusion rapidly, but I do not regret the labour, if I have in fact finally achieved a solution which is so acceptable to you that you consider it worthy of inclusion in your own writings. Indeed it is as follows.

If the index of the power is an even number, let it be called $n$; but if it is odd, let it be called $n-1$; and as the middle coefficient is to the sum of all the coefficients of the same power, so one will be to the mean proportional between the semicircumference of the circle and the following series:

[73] See the note on Proposition 23 concerning this letter and the series and approximations which it contains.

$$n + \frac{A}{2(n+2)} + \frac{9B}{4(n+4)} + \frac{25C}{6(n+6)} + \frac{49D}{8(n+8)} + \frac{81E}{10(n+10)} \text{ \&c.}$$

For example, if the ratio of the middle coefficient to the sum of all the coefficients in the hundredth or ninety-ninth power is required, there will be $n = 100$, which, when multiplied by the semicircumference of the circle, 1.5707963279, produces the first term $A$ of the series: then there will be

$$B = \frac{1}{204}A, \quad C = \frac{9}{416}B, \quad D = \frac{25}{636}C, \quad \text{\&c.}$$

and by carrying out the calculation as at the side, the sum of the terms will be found to be 157.866984459, whose square root 12.5645129018 is to one as the sum of all the coefficients is to the middle coefficient in the hundredth power, or as the sum of all the coefficients is to either of the middle coefficients in the ninety-ninth power.

| |
|---:|
| 157.079632679 |
| 769998199 |
| 16658615 |
| 654820 |
| 37137 |
| 2734 |
| 246 |
| 26 |
| 3 |
| 157.866984459 |

The problem is also solved by means of the reciprocal of that series, for the sum of all the coefficients is to the middle coefficient as the square root of the ratio of the semicircumference of the circle to the series

$$\frac{1}{n+1} + \frac{A}{2(n+3)} + \frac{9B}{4(n+5)} + \frac{25C}{6(n+7)} + \frac{49D}{8(n+9)} + \frac{81E}{10(n+11)} \text{ \&c.}$$

or which comes back to the same, put $a = .6366197723676$, namely the quantity which results on dividing one by the semicircumference of the circle; and the mean proportional between the number $a$ and this series will be to one as the middle coefficient is to the sum of them all.

Thus if $n = 100$ as before, the calculation will be as you see at the side, where the sum of the terms comes out as .00633444670787, whose square root .0795892373872 is to one as the middle coefficient is to the sum of all the coefficients in the hundredth or ninety-ninth power.

There are also other series for the solution of this problem which are just as simple as those presented so far, but a little less convergent, when the index of the binomial is a small number.

| |
|---:|
| .00630316606304 |
| 3059789351 |
| 65566915 |
| 2553229 |
| 143473 |
| 10470 |
| 934 |
| 98 |
| 12 |
| 1 |
| .00633444670787 |

But in practice there is no need to revert to series; for it suffices to take the mean proportional between the semicircumference of the circle and $n + \frac{1}{2}$; for this will always approximate more closely than the first two terms of the series, of which even the first alone suffices for the most part.

Now the same approximation may be expressed otherwise and in a manner more suited to application as follows. Put $2a = c = 1.2732395447352$; and as the sum of the coefficients is to the middle coefficient, so one will be to $\sqrt{\dfrac{c}{2n+1}}$ approximately, the error being an excess of about $\dfrac{1}{16n^2}\sqrt{\dfrac{c}{2n+1}}$.

If $n = 100$, there will be $\dfrac{c}{2n+1} = .006334525$, and its square root .07958973 is accurate in the sixth decimal place; if this is divided by $16n^2$, that is by 160000, it will give the correction .00000050, and when this has been subtracted from the approximation, it leaves the number sought .07958923, which is exact in the last figure.

Likewise if $n = 900$, there will be $\dfrac{c}{2n+1} = .000706962545$, whose square root .026588767 exceeds the true value by two in the ninth decimal place, but if the correction is computed and subtracted from the approximation, the required number will be obtained accurate in the thirteenth decimal place.

But here is an equally easy and more accurate approximation: let the difference between the logarithms of the numbers $n+2$ and $n-2$ be divided by 16, and let the quotient be added to half the logarithm of the index $n$; then let the constant logarithm .0980599385151, that is half the logarithm of the semicircumference of the circle, be added to this sum, and the final sum is the logarithm of the number which is to one as the sum of all the coefficients is to the middle one. If $n = 900$, the calculation will be

| | |
|---|---|
| $\frac{1}{2}$ log. 900. | 1.4771212547 |
| 16 ) Dif. of log. 902 & log. 898 ( | .0001206376 |
| Constant log. | .0980599385 |
| Sum | 1.5753018308 |

And this sum exceeds the true value by two in the last figure; and it is the logarithm of the number 37.6098698 which is to one as the sum of the coefficients is to the middle coefficient in the power 900 or 899.

And if you wish the reciprocal of that number, take the complement of the logarithm, namely −2.4246981692, and the number corresponding to this will be found to be .0265887652.

And these are the solutions which have come out by means of *Newton's* Method of Differences; I do not touch upon their demonstrations at this time, since I intend shortly to communicate to the public a Treatise which I have composed concerning Interpolation and Summation of series.

# References


1. Anon.: Review of Stirling's Methodus Differentialis (French). Memoires pour l'Histoire des Sciences et des Beaux Arts, 1936–1954, 2038–2050 (1732).
2. Anon.: Review of Stirling's Methodus Differentialis (Latin). Nova Acta Eruditorum, 515–526 (1734).
3. W.N. Bailey: Generalised Hypergeometric Series. Cambridge, 1935.
4. D. Beaver: Bernard Walther: innovator in astronomical observation. J. Hist. Astron. **1**, 39–43 (1970).
5. Jakob Bernoulli: Curvatura Laminae Elasticae. Acta Eruditorum (June), 262–276 (1694) (also in Jacobi Bernoulli Basileensis, Opera I (Geneva, 1744), 576–600).
6. Jakob Bernoulli: Positionum de Seriebus Infinitis earumque usu in quadraturis Spatiorum et rectificationibus Curvarum, Pars Quinta (1704). Jacobi Bernoulli Basileensis, Opera II (Geneva, 1744), 957–975.
7. Johann Bernoulli: Additamentum effectionis omnium quadraturarum et rectificationum curvarum per seriem quandam generalissimam. Acta Eruditorum, 437–441 (1694) (also in Johannis Bernoulli Opera I, (Lausanne and Geneva, 1742), 125–128).
8. Tycho Brahe: Tychonis Brahe Dani Opera Omnia (editor: J.L.E. Dreyer). Copenhagen, 1913–1929.
9. Tycho Brahe: Historia Coelestis [edited by Lucius Barrettus (pseudonym of Albert Curtz)]. Augsburg, 1666.
10. A. von Braunmühl: Historische Untersuchung der ersten Arbeiten über Interpolation. Bibl. Math. (3rd series) **2**, 86–96 (1901).
11. William Brouncker (Viscount): The Squaring of the Hyperbola, by an infinite Series of Rational Numbers, together with its Demonstration. Phil. Trans. **3** (No. 34, April 13), 645–649 (1668).
12. John Collins: Commercium Epistolicum D. Johannis Collins, et aliorum de analysi promota: jussu Societatis Regiae in lucem editum. London, 1712 (second edition 1722).
13. Roger Cotes: Harmonia mensurarum, sive Analysis et Synthesis per Rationum et Angulorum Mensuras promotae (editor: R. Smith). Cambridge, 1722.
14. J. Dutka: The early history of the hypergeometric function. Arch. Hist. Exact Sci. **31**, 15–34 (1984).
15. Leonhard Euler: Methodus generalis summandi progressiones. Comm. Acad. Sci. Petrop. **6** (1732–33), 68–97 (1738) (also in Leonhardi Euleri Opera Omnia (Leipzig, Berlin, 1911– ): Series 1, Opera Mathematica, **14**, 42–72).
16. Leonhard Euler: De summis serierum reciprocarum. Comm. Acad. Sci. Petrop. **7** (1734–35), 123–134 (1740) (also in Leonhardi Euleri Opera Omnia (Leipzig, Berlin, 1911– ): Series 1, Opera Mathematica, **14**, 73–86).
17. G.A. Gibson: Taylor's theorem and Bernoulli's theorem: a historical note. Proc. Edinburgh Math. Soc. **39**, 25–33 (1921).

18. G.A. Gibson: Sketch of the history of mathematics in Scotland to the end of the 18th century. Proc. Edinburgh Math. Soc., Series 2, **1**, 1–18, 71–93 (1927).
19. H.H. Goldstine: A History of Numerical Analysis. Springer-Verlag, New York Heidelberg Berlin, 1977.
20. R.W. Gosper: A calculus of series rearrangements. Algorithms and Complexity (editor: J.F. Traub), 121–151. Academic Press, New York, 1976.
21. R. Gowing: Roger Cotes – natural philosopher. Cambridge, 1983.
22. James Gregory: Vera Circuli et Hyperbolae Quadratura. Padua, 1667.
23. James Gregory: Appendicula ad Veram Circuli et Hyperbolae Quadraturam. Exercitationes Geometriae. London, 1668.
24. Edmund Halley: Methodus nova accurata et facilis inveniendi radices aequationum quarumcumque generaliter, sine praevia reductione. Phil. Trans. **18**, 136–148 (1694) (reproduced in [46, pp. 327–343]).
25. Edmund Halley: An easy quadrature of the circle from $2\sqrt{3}$ or $\sqrt{12}$. Mathematical Tables (Henry Sherwin), 53–54. London, 1706 (also in Scriptores Logarithmici (Francis Maseres) III, 139–141 (London, 1796)).
26. Jakob Hermann: Phoronomia, sive de viribus et motibus corporum solidorum et fluidorum libri duo. Amsterdam, 1716.
27. F.B. Hildebrand: Introduction to Numerical Analysis. McGraw Hill, New York, 1956.
28. J.E. Hofmann: Um Eulers erste Reihenstudien. Sammelband zu Ehren des 250. Geburtstages Leonhard Eulers (editor: K. Schröder), 139–208. Akademie-Verlag, Berlin, 1959.
29. L. Hogben: Mathematics in the Making. Macdonald, London, 1960.
30. Francis Holliday: Syntagma Mathesios. London, 1745.
31. Christiaan Huygens: De Circuli Magnitudine Inventa. Leiden, 1654.
32. John Kersey: The Elements of that Mathematical Art commonly called Algebra. London, 1725.
33. T.A. Krasotkina: The correspondence of L. Euler and J. Stirling (Russian). Istor.-mat. Issled. **10**, 117–158 (1957).
34. H. Krieger: Über Stirlings Bestimmung von $\sum_{n=0}^{\infty} \frac{(-1)^n}{2n+1}$ und $\sum_{n=1}^{\infty} \frac{1}{n^2}$. Arch. Hist. Exact Sci. **5**, 37–46 (1968–69).
35. Gottfried Wilhelm Leibniz: De vera proportione circuli ad quadratum circumscriptum in numeris rationalibus. Acta Eruditorum, 41–46 (1682).
36. Gottfried Wilhelm Leibniz: Leibnizens mathematische Schriften. Berlin (Vols 1–2), Halle (Vols 3–7), 1849–1863 (editor: C.I. Gerhardt).
37. V.V. Lihin: First investigations in the theory of summation of functions (Russian). History Methodology Natur. Sci., No. V, Math. (Russian) Izdat. Moskov. Univ., 219–227 (1966) (Math. Rev. 34#10).
38. Colin MacLaurin: Geometria organica: sive descriptio linearum curvarum universalis. London, 1720.
39. Colin MacLaurin: Treatise of Fluxions. Edinburgh, 1742.
40. L.M. Milne-Thomson: The Calculus of Finite Differences. Macmillan, London, 1933.
41. Abraham De Moivre: The Doctrine of Chances. London, 1718.
42. Abraham De Moivre: De Fractionibus Algebraicis Radicalitate immunibus ad Fractiones Simpliciores reducendis, deque summandis Terminis quarumdam Serierum aequali Intervallo a se distantibus. Phil. Trans. **32** (No. 373, Sept., Oct. 1722), 162–178 (1722–23).
43. Abraham De Moivre: Miscellanea Analytica de Seriebus et Quadraturis. London, 1730.

44. Abraham De Moivre: Miscellaneis Analyticis Supplementum. London, 1730.
45. Pierre Rémond De Montmort: De Seriebus infinitis Tractatus. Pars Prima. Una cum Appendice et Additamento per D. Brook Taylor, R. S. Sec. Phil. Trans. No. 353 (July, August, Sept. 1717) **30**, 633–675 (1717–19) (Appendix, 676–683; Additamentum, 683–689).
46. Isaac Newton: Arithmetica Universalis; sive De Compositione et Resolutione Arithmetica Liber. Cambridge, 1707.
47. Isaac Newton: Analysis per Quantitatum Series, Fluxiones, ac Differentias: cum Enumeratione Linearum Tertii Ordinis (editor: W. Jones). London, 1711.
48. Isaac Newton: Philosophiae Naturalis Principia Mathematica. Third edition. London, 1726.
49. François Nicole: Traité du Calcul des Differences Finies. Memoires de Mathematique et de Physique, 7–21 (1717), 20–37, 181–198 (1723) (in Histoire de l'Academie Royale des Sciences (Paris) but separately paginated).
50. François Nicole: Additions aux deux Memoires sur le Calcul des Differences Finies, imprimés l'année dernière. Memoires de Mathematique et de Physique, 138–158 (1724) (in Histoire de l'Academie Royale des Sciences (Paris) but separately paginated).
51. N. Nielson: Handbuch der Theorie der Gammafunktion. Leipzig, 1906.
52. E. Pairman: On a difference equation due to Stirling. Proc. Edinburgh Math. Soc. **36**, 40–60 (1917–18).
53. K. Schellbach: Die Lehre von der Elliptischen Integralen und den Theta-Funktionen. Reimer, Berlin, 1864.
54. R. Schlapp: The contribution of the Scots to mathematics. Math. Gazette **57**, No. 399, 1–16 (1973).
55. I. Schneider: Der Mathematiker Abraham de Moivre (1667–1754). Arch. Hist. Exact Sci. **5**, 177–317 (1968).
56. I.J. Schoenberg: On polynomial interpolation at the points of a geometric progression. Proc. Roy. Soc. Edinburgh **90A**, 195–207 (1981).
57. Abraham Sharp: The quadrature of the circle, computed from $\sqrt{12}$, or $2\sqrt{2}$. Mathematical Tables (Henry Sherwin), 56–59. London, 1706 (also in Scriptores Logarithmici (Francis Maseres) III, 142–145 (London, 1796)).
58. J.F. Steffensen: Interpolation. London and Baltimore, 1927.
59. L. Stephens and S. Lee (editors): Dictionary of National Biography. Oxford, 1908–.
60. James Stirling: Lineae Tertii Ordinis Neutonianae, sive Illustratio Tractatus D. Neutoni De Enumeratione Linearum Tertii Ordinis. Oxford, 1717.
61. James Stirling: Methodus Differentialis Newtoniana Illustrata. Phil. Trans. **30** (No. 362 Sept.–Oct. 1719), 1050–1070 (1717–1719).
62. James Stirling: Methodus Differentialis: sive Tractatus de Summatione et Interpolatione Serierum Infinitarum. London, 1730.
63. James Stirling: Sir Isaac Newton's Differentials, or Method of Fluxions, Illustrated; A Method to find the values of Arithmetical Series, how slow soever they converge (translation in two parts of [61] by Francis Holliday in [30, pp. 208–224, 225–228]).
64. James Stirling: The Differential Method: or, a Treatise concerning Summation and Interpolation of Infinite Series (translation of [62] by Francis Holliday). London, 1749.
65. K.R. Stromberg: An Introduction to Classical Real Analysis. Wadsworth, Belmont, California, 1981.
66. Brook Taylor: Methodus Incrementorum Directa et Inversa. London, 1715.
67. J. Todd: The lemniscate constants. Comm. ACM **18**, 14–19 (1975).
68. H.W. Turnbull: The Correspondence of Isaac Newton. Cambridge, 1959–77.

69. I. Tweddle: Approximating $n!$. Historical origins and error analysis. Amer. J. Phys. **52**, 487–488 (1984).
70. I. Tweddle: James Stirling: 'This about series and such things'. Scottish Academic Press, Edinburgh, 1988.
71. I. Tweddle: James Stirling's early work on acceleration of convergence. Arch. Hist. Exact Sci. **45**, 105–125 (1992).
72. C. Tweedie: Nicole's contributions to the foundations of the calculus of finite differences. Proc. Edinburgh Math. Soc. **36**, 22–39 (1917–18).
73. C. Tweedie: The Stirling numbers and polynomials. Proc. Edinburgh Math. Soc. **37**, 2–25 (1919).
74. C. Tweedie: James Stirling: a Sketch of his Life and Works along with his Scientific Correspondence. Oxford, 1922.
75. John Wallis: Opera mathematica. Oxford, 1695–99.
76. F.J.W. Whipple: The sum of the coefficients of a hypergeometric series. J. London Math. Soc. **5**, 192 (1930).
77. D.T. Whiteside: The Mathematical Papers of Isaac Newton. Cambridge, 1967–81.
78. E.T. Whittaker and G.N. Watson: A course of Modern Analysis (fourth edition). Cambridge, 1927.

# Index

## Sources in the History of Mathematics and Physical Sciences

*Continued from page ii*

Meyenn K. von, Hermann A., Weisskopf V.F. (Eds)
**Wolfgang Pauli: Scientific Correspondence II: 1930-1939**

Meyenn K. von (Ed.)
**Wolfgang Pauli: Scientific Correspondence III: 1940-1949**

Meyenn K. von (Ed.)
**Wolfgang Pauli: Scientific Correspondence IV, Part I: 1950-1952**

Meyenn K. von (Ed.)
**Wolfgang Pauli: Scientific Correspondence IV, Part II: 1953-1954**

Meyenn K. von (Ed.)
**Wolfgang Pauli: Scientific Correspondence IV, Part III: 1955-1956**

Neugebauer O. (Ed.)
**Astronomical Cuneiform Texts**

Sesiano J.
**Books IV to VII of Diophantus' Arithmetica: In the Arabic Translation Attributed to Qustã ibn Luqã**

Sigler L.
**Fibonacci's Liber Abaci: A Translation into Modern English of Leonardo Pisano's Book of Calculation**

Tweddle I.
**Simson on Porisms: An Annotated Translation of Robert Simson's Posthumous Treatise on Porisms and Other Items on this Subject**

MIX
Papier aus verantwortungsvollen Quellen
Paper from responsible sources
FSC® C105338

If you have any concerns about our products,
you can contact us on
**ProductSafety@springernature.com**

In case Publisher is established outside the EU,
the EU authorized representative is:
**Springer Nature Customer Service Center GmbH**
**Europaplatz 3, 69115 Heidelberg, Germany**

Printed by Libri Plureos GmbH
in Hamburg, Germany